Lecture Notes in Computer Sc

Edited by G. Goos, J. Hartmanis, and J. ·

Springer

Berlin
Heidelberg
New York
Hong Kong
London
Milan
Paris
Tokyo

Edmund Burke
Patrick De Causmaecker (Eds.)

Practice and Theory of Automated Timetabling IV

4th International Conference, PATAT 2002
Gent, Belgium, August 21-23, 2002
Selected Revised Papers

 Springer

Series Editors

Gerhard Goos, Karlsruhe University, Germany
Juris Hartmanis, Cornell University, NY, USA
Jan van Leeuwen, Utrecht University, The Netherlands

Volume Editors

Edmund Burke
The University of Nottingham
School of Computer Science and Information Technology
Jubilee Campus, Wollaton Road
Nottingham, NG8 1BB, UK
E-mail: ekb@cs.nott.ac.uk

Patrick De Causmaecker
Information Technology, KaHo St.-Lieven
Gent, Belgium
E-mail: Patrick.DeCausmaecker@kahosl.be

Cataloging-in-Publication Data applied for

A catalog record for this book is available from the Library of Congress.

Bibliographic information published by Die Deutsche Bibliothek
Die Deutsche Bibliothek lists this publication in the Deutsche Nationalbibliografie;
detailed bibliographic data is available in the Internet at <http://dnb.ddb.de>.

CR Subject Classification (1998): F.2.2, G.1.6, G.2, I.2.8

ISSN 0302-9743
ISBN 3-540-40699-9 Springer-Verlag Berlin Heidelberg New York

Springer-Verlag Berlin Heidelberg New York
a member of BertelsmannSpringer Science+Business Media GmbH

http://www.springer.de

© Springer-Verlag Berlin Heidelberg 2003
Printed in Germany

Typesetting: Camera-ready by author, data conversion by PTP Berlin GmbH
Printed on acid-free paper   SPIN: 10929186   06/3142   5 4 3 2 1 0

Preface

This volume contains a selection of papers from the 4th International Conference on the Practice and Theory of Automated Timetabling (PATAT 2002) held in Gent, August 21–23, 2002.

Since the first conference in Edinburgh in 1995, the range of timetabling applications at the conferences has become broader and more diverse. In the selected papers volume from the 1995 conference, there were just two contributions (out of 22) which did not specifically address school and university timetabling. In the selected papers volume from the 1997 conference in Toronto, the number of papers which tackled non-educational problems increased. Two of the papers addressed more than one timetabling application. In both of these papers, educational applications were considered in addition to other applications. A further three papers were concerned with non-educational applications. The conference steering and programme committees have worked hard to attract a wide range of timetabling applications. In the conference held in Konstanz in 2000, the diversification of timetabling problems increased significantly. Of the 21 selected papers in the postconference volume, just 13 were specifically concerned with educational timetabling. In the previous volumes, the papers had been sectioned according to solution technique. In the Konstanz volume the papers were classified according to application domains. One section of the volume was entitled "Employee Timetabling," while sports timetabling, airfleet scheduling, and general software architectures for timetabling were also represented. In the present volume, more than one-third of the 21 papers discuss problems in application areas other than academic and educational ones. Sports timetabling and hospital timetabling are particularly well represented. Indeed, they have their own sections. This shift into more diverse timetabling application domains reflects the growing maturity of the conference series and the goals set by the steering committee. Educational timetabling is a crucially important application area and it will always play a central role in the PATAT conferences. We would like to see the conference series attract even more interest in educational timetabling but we would also like to see the series continue to attract high-quality submissions from sports timetabling, employee timetabling, transport timetabling, and from across the timetabling application spectrum. Another key aim of the conference series is to foster multidisciplinary research which draws on the strengths of Operational Research, Artificial Intelligence and other disciplines. The timetabling research field has always attracted researchers from across disciplinary divides and one of the main goals of the conference series is to support and extend this multidisciplinary collaboration.

Another important aspect of modern timetabling research is the goal of integrating the human aspect of timetabling with the automation of the problem. In a contribution to the previous PATAT conference, Michael Carter[1] said:

---

[1] Michael W. Carter, A Comprehensive Course Timetabling and Student Scheduling System at the University of Waterloo, in Practice and Theory of Automated Timetabling III (edited by Burke and Erben), pages 64–82.

"Practical course timetabling is 10% graph theory, and 90% politics! When we first began designing the system, we were warned: 'You cannot dictate to professors when they will teach courses!' Consequently, we were told that course timetabling could not work."

And he went on to show that it *did* work, if the human factors alluded to in the warning were taken into account. In his conclusions he said:

"...simply giving timetable reps the facility to make real time on-line changes was the single most important contribution."

These comments have relevance for the design of timetabling decision support systems across the application spectrum. The goal of developing interactive and adaptive systems that build on human expertise and at the same time provide the computational power to reach high-quality solutions continues to be one of the key challenges that currently faces the timetabling research community. While human/machine interaction in timetabling has an important role to play, it is clear that there are exciting research opportunities opening up in the underpinning automation methodologies for timetabling across the application range. The success of the series of international conferences on the Practice and Theory of Automated Timetabling (PATAT) has reflected the interest and activity of the scientists who are working in the area and addressing the above (and many other) significant research issues.

As mentioned above, for this fourth volume, we continued with the practice of organizing the papers around application themes – which was established in the last volume. The papers represent a broad range of practical and theoretical research issues and they cover a variety of techniques and applications.

Conference Series

The meeting in Gent was the fourth in the PATAT series of international conferences. The first three conferences were held in Edinburgh (August/September 1995), Toronto (August 1997) and Konstanz (August 2000). Selected papers from these three conferences appeared in the Springer Lecture Notes in Computer Science series. The full references are:

Edmund Burke and Peter Ross (Eds.): *Practice and Theory of Automated Timetabling*, 1st International Conference, Edinburgh, UK, August/September 1995, Selected Papers. Lecture Notes in Computer Science, Vol. 1153. Springer 1996.

Edmund Burke and Michael Carter (Eds.): *Practice and Theory of Automated Timetabling II*, 2nd International Conference, PATAT 1997, Toronto, Canada, August 1997, Selected Papers. Lecture Notes in Computer Science, Vol. 1408. Springer 1998.

Edmund Burke and Wilhelm Erben (Eds.): *Practice and Theory of Automated Timetabling III*, 3rd International Conference, PATAT 2000, Konstanz, Germany, August 2000, Selected Papers. Lecture Notes in Computer Science, Vol. 2079. Springer 2001.

The fifth conference in the series will be held in Pittsburgh, USA, in August 2004. Future conferences will be held every two years. For further information about the conference series, contact the steering committee (whose members are listed below) or see http://www.asap.cs.nott.ac.uk/ASAP/ttg/patat-index.html.

The PATAT conference series is affiliated with the Association of European Operational Research Societies Working Group on Automated Timetabling. See http://www.asap.cs.nott.ac.uk/ASAP/watt/ for further details about this working group.

Acknowledgements

The Gent conference was a great success and we are indebted to a large number of people for their hard work and commitment. In particular, we would like to thank all the members of the organizing committee (listed below). Their effective administration of the event played a significant contribution in its success. Very special thanks go to Greet Vanden Berghe, whose attention to organization and detail was invaluable to the smooth running of the organization.

The papers that appear in this volume were carefully and thoroughly refereed. Many thanks go to the members of the programme committee (listed below) who spent a significant amount of their valuable time rigorously reviewing the submitted papers.

We are also very grateful to the staff of Springer-Verlag for their support and encouragement. As series editor of the Lecture Notes in Computer Science series, Jan van Leeuwen was (as he has always been since the first volume) particularly helpful throughout the duration of this project. We would also like to particularly thank Piers Maddox for the excellent job he did (as he did with the previous volume) in copy editing the book. His hard work is very much appreciated. Special thanks also go to Alison Payne for all the secretarial support she provided during the preparation of this volume.

Of course, it is the authors, presenters and delegates who ultimately determine the success of a conference. Our thanks go to them for the enthusiasm and support they have given this and previous PATAT conferences. Finally, we would like to thank the steering committee (listed below) for their continuing work in bringing us this and future PATAT conferences. We apologize for any omissions that have been inadvertently made. So many people have helped with this conference and with the series of conferences that is difficult to remember them all.

May 2003   Edmund Burke
                                                          Patrick De Causmaecker

4th International Conference on the Practice and Theory of Automated Timetabling Programme Committee

4th International Conference on the Practice and Theory of Automated Timetabling Organizing Committee

KaHo St. Lieven, Gent, Belgium

Patrick De Causmaecker	Chair
Greet Vanden Berghe	General Administration
Peter Demeester	Website and Technical Support
Filip Thomaes	Technical Support and Proceedings
Geert De Maere	Technical Support
Philippe Aelvoet	Technical Support
Anneleen De Causmaecker	Administrative Support

Noveon Inc., Belgium

Viktor Bardadym	Administrative Support

International Series of Conferences on the Practice and Theory of Automated Timetabling (PATAT) Steering Committee

Edmund Burke (Chair)	University of Nottingham, UK
Ben Paechter (Treasurer)	Napier University, UK
Victor Bardadym	Noveon Inc., Belgium
Michael Carter	University of Toronto, Canada
Patrick De Causmaecker	KaHo St. Lieven, Gent, Belgium
David Corne	University of Reading, UK
Wilhelm Erben	FH Konstanz, University of Applied Sciences, Germany
Jeffrey Kingston	University of Sydney, Australia
Gilbert Laporte	École des Hautes Études Commerciales, Montreal, Canada
Amnon Meisels	Ben-Gurion University, Beer Sheva, Israel
Peter Ross	Napier University, UK
Dominique de Werra	EPF-Lausanne, Switzerland and George White University of Ottawa, Canada

Table of Contents

Examination Timetabling

University Course and School Timetabling

Other Timetable Presentations

Author Index

General Issues

Constraints of Availability in Timetabling and Scheduling

Dominique de Werra

École Polytechnique Fédérale de Lausanne,
Lausanne 1015, Switzerland
dewerra.ima@epfl.ch

Abstract. The basic class–teacher timetabling problem is examined with the additional constraints due to the (un-)availability of source teachers and/or classes at some periods. We mention a generalization of this problem which occurs in image reconstruction problems in tomography. Complexity issues are discussed for both types of problems and some solvable cases are presented which can be derived from the image reconstruction formulation. Reductions to canonical forms are also described. Some other types of unavailability constraints (for classrooms or for lectures) are also reviewed.

1   Introduction and Motivation

Let us imagine for a moment that some classes $c_1, \ldots, c_m$ have to follow a continuing education programme offered by a collection of teachers $t_1, \ldots, t_n$. We observe that each teacher t_j has to meet each class c_i for r_{ij} lectures. If we do not distinguish the various topics on which the teachers are lecturing for the classes, the data can be summarized in an $(m \times n)$ array $R = (r_{ij})$ called a requirement matrix.

Assuming that no teacher can teach two classes at the same time and no class is able to get a lecture from more than one teacher at a time, we may ask ourselves how can one construct a schedule which will minimize the number of periods needed.

This basic model in theoretical timetabling has been almost useless in the real cases since most classes are obviously not available every period for lectures.

Considering the set H of periods of a week where lectures can take place, we can associate with each teacher t_j a subset $T_j \subseteq H$ of periods where (s)he is available for giving lectures.

Having introduced this additional feature to make our model look more realistic, the question now is to determine whether a schedule can be found with the following:

1. all lectures in R are given within the set H of periods of the week,
2. no teacher (resp. no class) is involved in more than one lecture during a period,
3. a teacher t_j can give a lecture only during periods k in T_j.

E. Burke and P. De Causmaecker (Eds.): PATAT 2002, LNCS 2740, pp. 3–23, 2003.

Such a problem can be formulated in terms of arrays in the following way. Given the set H of periods, let $h = |H|$; we define an $(m \times h)$ array A where row i is associated to the class c_i and column k to night k.

For each c_i we consider integers $a(i, 1), \ldots, a(i, n)$ with $a(i, j) = r_{ij}$, i.e. the number of lectures to be given by teacher t_j to c_i.

Similarly, for each period k we consider integers $\alpha(k, 1), \ldots, \alpha(k, n)$ defined as follows:

$$\alpha(k, j) = \begin{cases} 1 & \text{if period } k \in T_j, \\ 0 & \text{otherwise}. \end{cases}$$

This means that $\alpha(k, j)$ is 1 if teacher t_j is available during period k or 0 else.

A timetable is then given by array A where entry a_{ik} contains the name of the teacher t_j giving during a period k a lecture to class c_i (a_{ik} is empty if c_i has no lecture during period k).

Then $a(i, j)$ is the number of occurrences of t_j in row i of A while $\alpha(k, j)$ is the number of occurrences of t_j in column k (0 or 1 depending upon the availability of t_j).

An example is given in Figure 1.

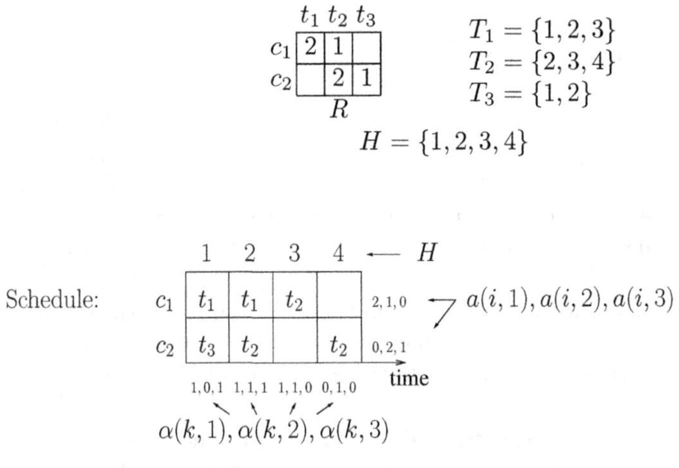

Fig. 1. An example of a schedule

It turns out that this formulation is closely related to a problem of image reconstruction in tomography; we intend to explore here the basic timetabling model where unavailability constraints are present and exploit the analogy with image reconstruction in tomography.

We intend in this paper to explore some complexity issues and some reductions to simple symmetric forms. The purpose is not to describe algorithms for real-life cases of large size, but rather to formalize some basic problems and to explore the boundary between easy and difficult problems.

The reader is referred to [3] for all graph-theoretical notions not defined here and to [4] for a short introduction to complexity and for basic algorithms in scheduling.

2   Some Reductions for the Timetabling Problem

For our purposes the basic timetabling problem with unavailability constraints will be defined by

- a set $C = \{c_1, \ldots, c_m\}$ of classes,
- a set $T = \{t_1, \ldots, t_m\}$ of teachers,
- a set $H = \{1, 2, \ldots, h\}$ of periods,
- an $(m \times n)$ requirement matrix $R = (r_{ij})$,
- a collection $\mathcal{T}$ (resp. $\mathcal{C}$) of subsets $T_j \subseteq H$ (resp. $C_i \subseteq H$) of periods of availability for each teacher t_j (resp. each class c_i).

A graph-theoretical model is often used to represent this problem $TT(C, T, H, R, \mathcal{T}, \mathcal{C})$ of timetabling: it consists of a bipartite multigraph $G = (C, T, R)$ where each class c_i (resp. each teacher t_j) is associated to a node of G; furthermore, nodes c_i and t_j are linked by r_{ij} parallel edges.

There is a one-to-one correspondence between solutions of $TT(C, T, H, R, \mathcal{T}, \mathcal{C})$ and edge h-colourings of G where no two adjacent edges have the same colour and the colour c of each edge $[c_i, t_j]$ is in $C_i \cap T_j$.

We shall now show that the problem TT of timetabling can be reduced to some canonical form.

Figure 2 shows an example of TT with the associated bipartite multigraph G.

Observe that $a(k, j) = 0$ or 1 is the *maximum* number of occurrences of t_j in column k while $a(i, j)$ is the *exact* number of occurrences of t_j in row i. This is due to the fact that for some t_j we may have $|T_j| > \sum_i r_{ij}$, i.e. teacher t_j is not *tight*.

Property 1. $TT(C, T, H, R, \mathcal{T}, \mathcal{C})$ *can be transformed into a restricted problem* $RTT(C', T', H, R', \mathcal{T})$ *such that*

(a) *RTT has solution if an only if TT has one,*
(b) $C_i' = H$ *(i.e. no unavailability) for each class c_i',*
(c) $|T_j| = \sum_i r_{ij}'$ *(teacher t_j' is tight) for each teacher t_j',*
(d) *the sets C', T', R' of classes, teachers, lectures resp. satisfy*

$$|\dot{C}| \leq |C| + |T|,$$
$$|\dot{T}| \leq 2|T| + |C|,$$
$$|\dot{R}| \leq |H|(|T| + |C|) - |R|.$$

Proof. Let us construct problem RTT from TT as follows:

(i) For each class c_i with $C_i \neq H$ introduce a new teacher t_i^+ with $|H| - |C_i|$ new lectures to be given by t_i^+ to c_i and set $T_i^+ = H - C_i$.

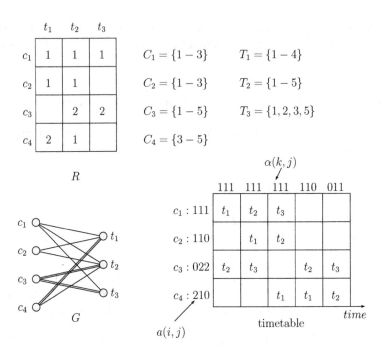

Fig. 2. An example of a TT

(*ii*) Now for every teacher t_j which is not tight, i.e. with $d_j = |T_j| - \sum_i r_{ij} > 0$, introduce a new class c_j^+ and d_j new lectures to be given by t_j to c_j^+.
These lectures should ideally occur in periods of T_j; to force this we introduce a new teacher t_j^* with $|H| - |T_j|$ new lectures to give to class c_j^+ and we set $T_j^* = H - T_j$.

(*iii*) We rename t_j' and c_i' the teachers and classes of the new problem RTT; we set $C_i' = H$ for each c_i' and R' is the new requirement matrix.

(*iv*) One can verify that the resulting RTT has a solution if and only if TT has one. Furthermore, all classes c_i' of RTT have $C_i' = H$; in addition every teacher t_j' is tight.

(*v*) We have introduced in the construction at most $|C|$ new teachers and at most $|C| \cdot |H| - |R|$ new lectures in (*i*). Also in (*ii*) we have introduced at most $|T|$ new classes, at most $|T|$ new teachers and at most $|T| \cdot |H| - |R|$ new lectures. Here $|R| = \sum_i r_{ij} =$ number of lectures. Hence (*iv*) holds.

$\square$

This construction is illustrated in Figure 3 for the example given in Figure 2.

Let us now transform the problem RTT into an equivalent problem RTT* where each teacher t_j^* will have to meet at most once each class c_i^*, i.e. the requirement matrix $R^* = (r_{ij})$ will be such that r_{ij} is 0 or 1. For this purpose, we will use the graph-theoretical model described above.

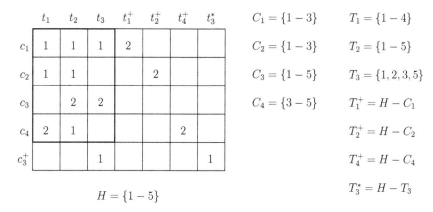

	t_1	t_2	t_3	t_1^+	t_2^+	t_4^+	t_3^*
c_1	1	1	1	2			
c_2	1	1		2			
c_3		2	2				
c_4	2	1				2	
c_3^+			1				1

$C_1 = \{1 - 3\}$ $T_1 = \{1 - 4\}$

$C_2 = \{1 - 3\}$ $T_2 = \{1 - 5\}$

$C_3 = \{1 - 5\}$ $T_3 = \{1, 2, 3, 5\}$

$C_4 = \{3 - 5\}$ $T_1^+ = H - C_1$

$T_2^+ = H - C_2$

$T_4^+ = H - C_4$

$T_3^* = H - T_3$

$H = \{1 - 5\}$

Fig. 3. Construction of RTT for the example of Figure 2

So we are given a timetabling problem $RTT(C, T, H, R, \mathcal{T})$ where as before all teachers are tight and all classes c_i satisfy $C_i = H$. G is the associated bipartite multigraph. When $r_{ij} > 1$, nodes c_i and t_j are linked by r_{ij} parallel edges $[c_i, t_j]_1, [c_i, t_j]_2, \ldots$.

Property 2. $RTT(C, T, H, R, \mathcal{T})$ can be transformed into a restricted problem $RTT^*(C^*, T^*, H, R^*, \mathcal{T})$ such that

1. RTT^* has a solution if and only if RTT has one;
2. The requirement matrix R^* has $r_{ij} = 0$ or 1 for all i, j;
3. The sets C^*, T^*, R^* of classes, teachers, lectures resp. satisfy

$$|C^*| \leq |C| + h^2 \min\{|C|, |T|\},$$
$$|T^*| \leq |T| + h^2 \min\{|C|, |T|\},$$
$$|R^*| \leq |R| + h^3 \min\{|C|, |T|\}.$$

Proof. We shall describe the transformation directly in the associated graph G. To keep notation simple, let us consider an edge $[c_i, t_j]$ belonging to a family of parallel edges between nodes c_i and t_j. We replace $[c_i, t_j]$ by the following graph: as before let $h = |H|$ be the number of periods; we introduce h new teachers $t_i^*, t_{ij,1}^*, \ldots, t_{ij,h-1}^*$ and h new classes $c_j^*, c_{ij,1}^*, \ldots, c_{ij,h-1}^*$. For all these teachers t_x^* and all classes c_y^* we set $T_x^* = C_y^* = H$.

Then each one of these new teachers has to give one lecture to each one of these new classes (except t_i^* which gives a lecture to c_i instead of c_j^*).

Finally, we replace the lecture of t_j to c_i by a lecture of t_j to c_j^*; so we have removed one lecture and introduced $h^2 + 1$ new lectures. The collection R of lectures has been increased by an amount of h^2.

The construction is illustrated in Figure 4. If we repeat this for all but one lecture in every family of parallel edges of G, we will get a simple bipartite graph G^* associated with a new problem RTT^* with $r_{ij}^* \in \{0, 1\}$ for all i, j.

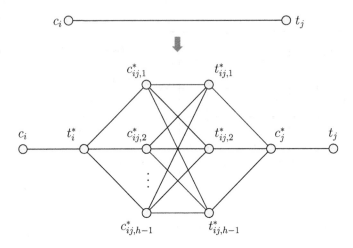

Fig. 4. Transformation of RTT into RTT*

It is easy to see that in any edge h-colouring of G^*, edges $[c_i, t_i^*]$ and $[c_j^*, t_j]$ will have the same colour; so from any h-colouring of G^* we can derive an h-colouring of G and the converse is also true as can be verified easily.

Hence, we will have a problem RTT* having a solution if and only if RTT has one.

The number of edges $[c_i, t_j]$ to transform is at most $h \min\{|C|, |T|\}$, so we have introduced

at most $h^2 \min\{|C|, |T|\}$ new teachers,
at most $h^2 \min\{|C|, |T|\}$ new classes,

and at most $h^3 \min\{|C|, |T|\}$ new lectures. Hence statement 3 holds; since G^* is now simple, we have statement 2. $\qquad\square$

With the reduction described above, our timetabling problem RTT*$(C, T, H, R, \mathcal{T})$ can be stated as follows. We are given sets C, T, H of classes, teachers and periods respectively; $|C| = m$, $|T| = n$, $|H| = h$. In addition, we have an $(m \times n)$ requirement matrix R with $r_{ij} \in \{0, 1\}$ for all i, j and a family $\mathcal{T} = (T_1, \ldots, T_n)$ of subsets $T_j \subseteq H$ giving the availabilities of the teachers t_j.

From these data, we define numbers

$$a(i, j) = r_{ij} \text{ for } i = 1, \ldots, m \text{ and } j = 1, \ldots, n,$$

$$\alpha(k, j) = \begin{cases} 1 & \text{if period } k \in T_j, \\ 0 & \text{else}, \end{cases} \text{ for } k = 1, \ldots, h \text{ and } j = 1, \ldots, n.$$

We have to construct an $(m \times h)$ array A by introducing one of the symbols $t_1, \ldots, t_n$ (teachers) in the entries a_{ik} of A in such a way that for all $j = 1, \ldots, n$ the following hold:

1. t_j occurs exactly $a(i,j)$ times in row i of A (for $i = 1, \ldots, m$);
2. t_j occurs exactly $\alpha(k,j)$ times in column k of A (for $k = 1, \ldots, h$).

The array A will define a schedule in a unique way: a_{ik} contains t_j if and only if teacher t_j gives a lecture to class c_i at period k.

Notice that due to the fact that all teachers are tight, we have exactly $\alpha(k,j)$ occurrences of t_j in column k. In addition, we now have $a(i,j) \in \{0,1\}$ for all i,j (and $\alpha(k,j) \in \{0,1\}$ as before).

3 The Basic Image Reconstruction Problem

In this form the timetabling problem RTT* is very close to the image reconstruction problem which arises in tomography. We will now give its formulation. An image is usually decomposed into a collection of pixels of different colours. Basically it consists of an $(m \times n)$ array A where each entry a_{ij} contains a pixel of some colour s chosen in a set of p colours.

In order to compress the data representing an image, one may for instance give simply the numbers $a(i,s)$ (resp. $\alpha(j,s)$) of pixels of colour s occurring in row i (resp. column j) for $i = 1, \ldots, m, j = 1, \ldots, n$ and $s = 1, \ldots, p$.

The main question is to determine whether to given values $a(i,s), \alpha(j,s)$ for $1 \leq i \leq m, 1 \leq j \leq n, 1 \leq s \leq p$ corresponds an $(m \times n)$ array A having exactly $a(i,s)$ (resp. $\alpha(j,s)$) pixels of colour s in row i (resp. column j). Moreover, the question of uniqueness of the reconstruction image is important. We shall concentrate here on the question of existence of an image. This problem is denoted by RP(m,n,p) (reconstruction of an $(m \times n)$ array from the values $a(i,s), \alpha(j,s)$ with p colours).

The similarity with the restricted timetabling problem is now obvious: the p colours correspond to the n teachers of RTT*, the rows i are associated with the classes c_i and the n columns with the h periods of the timetable to be constructed.

While $a(i,s)$ and $\alpha(j,s)$ are usually non-negative integers in problem RP(m, n, p), they take values 0 or 1 in the timetabling problem RTT*.

As in [6], we call *unary* the colours s for which $a(i,s)$ and $\alpha(j,s)$ are 0 or 1 for all i,j. Problem RP(m,n,p) is denoted by RPU(m,n,p) when all p colours are unary; so there is equivalence between RPU(m,n,p) and RTT*. This fact can be exploited for deriving complexity properties of these problems.

At this stage, we should mention a slight difference between RP(m,n,p) and RTT*: while entry a_{ik} may be empty in the array A associated to a timetabling problem (at period k class c_i has no lecture), in the image reconstruction problem an entry a_{ij} which does not have a pixel of some colour $s \leq p$ is said to have the "ground colour" $p+1$. But we can easily forget it and work with the first p colours.

The analogy between RPU(m,n,p) and RTT* where $r_{ij} \in \{0,1\}$ for all i,j suggests the following: in RPU(m,n,p) the rows and the columns play a symmetric role; it is also the case for RTT* (but only when $r_{ij} \in \{0,1\}$ for all i, j). We can interchange the roles of classes and of periods. More precisely, consider a teacher t_j characterized by

  - a subset $T_j \subseteq H$ of periods k where (s)he has to give lectures,
  - a subset T_j^* of classes c_i to which t_j has to give a lecture,

$$T_j^* = \{c_i | r_{ij} = 1\}.$$

We may consider that H is the set of all classes; each teacher t_j is characterized by

  - a subset T_j^* of periods c_i where (s)he has to give lectures,
  - a subset $T_j \subseteq H$ of classes k to which t_j has to give a lecture.

If RTT** is this last problem, then clearly RTT** has a solution if and only if RTT* has one. This simple observation may be useful for deriving complexity results for RTT in general. One should also remark that when $r_{ij} \notin \{0,1\}$ for some pairs c_i, t_j, then one cannot interchange the role of classes and of periods in RTT.

Let $RTT^*(C, T, H, R, \mathcal{T})$ denote the timetabling problem where all teachers are tight and $r_{ij} \in \{0,1\}$ for all i, j. We recall some complexity results.

Proposition 1 ([7]). $RTT^*(C, T, H, R, \mathcal{T})$ is NP-complete even when $|H| = 3$.

From this we derive immediately the following proposition.

Proposition 2 ([6]). $RTT^*(C, T, H, R, \mathcal{T})$ is NP-complete even when $|C| = 3$.

Notice that when $|T_j| \leq 2$ for each teacher t_j, then $RTT^*(C, T, H, R, \mathcal{T})$ can be solved in polynomial time [7] by transforming it to a 2-SAT problem. It follows that $RTT^*(C, T, H, R, \mathcal{T})$ can be solved in polynomial time if $|H| = 2$.

As above, we may also deduce the following proposition.

Proposition 3 ([6]). $RTT^*(C, T, H, R, \mathcal{T})$ can be solved in polynomial time if $|C| = 2$.

The following result has been obtained for $RPU(m, n, p)$.

Proposition 4 ([6]). $RPU(m, n, p = 3)$ can be solved in polynomial time.

Interpreting this in terms of timetabling gives the following proposition.

Proposition 5 ([6]). $RTT^*(C, T, H, R, \mathcal{T})$ can be solved in polynomial time if $|T| \leq 3$.

One should recall here that while $RPU(m, n, p = 3)$ can be solved in polynomial time, the complexity of $RP(m, n, p = 2)$ is unknown.

Also for $RTT^*(C, T, H, R, \mathcal{T})$ with $|H| \geq 4$, the complexity is unknown.

4   More on the Basic Timetabling Problem

We have shown how a basic timetabling problem $TT(C, T, H, R, \mathcal{T}, \mathcal{C})$ could be transformed into a problem RTT where $C_i = H$ for each class c_i and where all teachers were "tight" ($\sum_i r_{ij} = |T_j|$ for each teacher t_j). We can also transform the basic problem RTT into other regular forms. We shall just mention the following property.

Property 3. *Problem $TT(C, T, H, R, \mathcal{T}, \mathcal{C})$ can be transformed into a problem $RTT'(C', T', H, R', \mathcal{P})$ satisfying the following:*

1. *RTT' has a solution if and only if TT has one,*
2. $|T'| = |T| + |C|$,
3. $|C'| = |C| + |T|$,
4. $\sum_i r'_{ij} = \sum_j r'_{ij} = |H|$ *for all i, j,*
5. $C'_i = T'_j = H$ *for all i, j,*
6. *some families of lectures $c'_i - t'_j$ have to be scheduled at some fixed periods.*

Proof. Let $m = |C|, n = |T|$; we construct an $(m + n) \times (m + n)$ requirement matrix R' by inserting the initial matrix R in the upper left corner and the transposed matrix R^{T} in the lower right corner of R'. Then we may introduce in entries $(m + 1, 1), \ldots, (m + n, n)$ values $r'_{m+j,j} = |H| - \sum_{i=1}^{m} r_{ij} \geq 0$ ($j = 1, \ldots, n$) and similarly in entries $(1, n + 1), \ldots, (m, n + m)$ we introduce values $r'_{i,n+i} = |H| - \sum_{j=1}^{n} r_{ij} \geq 0$.

We now have a requirement matrix R' where all row sums and all column sums are equal to $|H|$.

This corresponds to introducing a set of new classes $\bar{c}_1, \ldots, \bar{c}_n$ (corresponding to the initial teachers $t_1, \ldots, t_n$) and a set of new teachers $\bar{t}_1, \ldots, \bar{t}_m$ (corresponding to the initial classes $c_1, \ldots, c_m$).

We set $\bar{T}_i = C_i$ for $i = 1, \ldots, m$ and $\bar{C}_j = T_j$ for $j = 1, \ldots, n$.

Consider now a pair $\bar{c}_j, t_j$: by construction t_j has to give $r'_{m+j,j}$ lectures to $\bar{c}_j$. Let $d_j = |\bar{C}_j| - \sum_{i=1}^{m} r_{ij} \leq |H| - \sum_{i=1}^{m} r_{ij} = r'_{m+j,j}$.

We assign $r'_{m+j,j} - d_j$ lectures of t_j to $\bar{c}_j$ to the $r'_{m+j,j} - d_j$ periods of $H - \bar{C}_j = H - T_j$; this is the preassignment constraint $\mathcal{P}_j$. The remaining d_j lectures of t_j to $\bar{c}_j$ are not preassigned. We repeat this for all pairs $t_j, \bar{c}_j$, thus obtaining preassignments $\mathcal{P}_1, \ldots, \mathcal{P}_n$.

Similarly, we consider each pair $c_i, \bar{t}_i$: we define $e_i = |\bar{T}_i| - \sum_{j=1}^{n} r_{ij} \leq |H| - \sum_{j=1}^{n} r_{ij} = r'_{i,n+i}$ and preassign $r'_{i,n+i} - e_i$ lectures of $\bar{t}_i$ to c_i to the $r'_{i,n+i} - e_i$ periods of $H - \bar{T}_i = H - C_i$; the remaining e_i lectures of $\bar{t}_i$ to c_i are not preassigned. This gives the preassignment constraint $\mathcal{P}_{n+i}$. Repeating this for all pairs $c_i, \bar{t}_i$ we define preassignment constraints $\mathcal{P}_{n+1}, \ldots, \mathcal{P}_{n+m}$. It is then easy to check that the new problem RTT' satisfies all statements 1–6. □

The construction is illustrated in Figure 5 and timetables for RTT and RTT' are shown in Figure 6.

In terms of the image reconstruction problem, we can interpret RTT' as follows. We have an $(m \times n)$ array A and p colours (corresponding to the teachers):

	t_1	t_2	t_3
c_1	2		1
c_2		1	2

$$m = 2, \quad n = 3$$

$$H = \{1-4\} \qquad T_1 = \{1-3\} = \bar{C}_1$$
$$C_1 = \{1-4\} = \bar{T}_1 \quad T_2 = \{1-2\} = \bar{C}_2$$
$$C_2 = \{1-3\} = \bar{T}_2 \quad T_3 = \{1-4\} = \bar{C}_3$$

Regularization

	t_1	t_2	t_3	$\bar{t}_1$	$\bar{t}_2$	
c_1	2		1	1		$e_1 = 4 - 3 = 1$
c_2		1	2		1	$e_2 = 3 - 3 = 0$
$\bar{c}_1$	2		2			$d_1 = 3 - 2 = 1$
$\bar{c}_2$		3			1	$d_2 = 2 - 1 = 1$
$\bar{c}_3$			1	1	2	$d_3 = 4 - 3 = 1$

$$R'$$

$\mathcal{P}_1$: one lecture $\bar{c}_1 - t_1$ preassigned in $H \setminus \bar{C}_1 = \{4\}$

$\mathcal{P}_2$: two lectures $\bar{c}_2 - t_2$ preassigned in $H \setminus \bar{C}_2 = \{3, 4\}$

$\mathcal{P}_3$: -

$\mathcal{P}_4$: -

$\mathcal{P}_5$: one lecture $c_2 - \bar{t}_2$ preassigned in $H \setminus \bar{T}_2 = \{4\}$

Fig. 5. Construction of problem RTT'

the rows correspond to classes and the columns to periods. All colours s satisfy $\alpha(j, s) = 1$ for all columns j: each colour occurs exactly once in each column; in each row i, colour s occurs exactly $a(i, s)$ times.

Finding such an array would be an easy problem (solvable by classical edge colouring techniques); but here we have a collection of preassignment requirements to take into account: some entries of A have already been assigned some colour.

This is what makes the problem difficult. Notice that in these requirements each colour is preassigned in only one row of A (see the timetable of RTT' in Figure 6 where for instance colour t_2 is preassigned only in the row associated with $\bar{c}_2$).

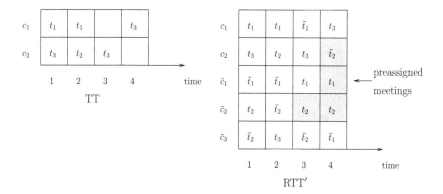

Fig. 6. A timetable for TT and the corresponding one for RTT′

$a(i,s)$						$\alpha(j,s)=1$ $\forall j,s$
1 0 2 0 1	3	3	1	5		
0 1 0 1 2	5	4	5	2		
2 0 2 0 0	1	1	3	3		
0 1 0 3 0	4	2	4	4		
1 2 0 0 1	2	5	2	1		

Fig. 7. The solution of problem $RP(m,n,p)$ associated with the solution of TT in Figure 6

Let us consider in the example of Figure 6 that the classes c_1, c_2, $\bar{c}_1$, $\bar{c}_2$, $\bar{c}_3$ correspond to rows $1,\dots,5$ and the teachers $\bar{t}_1, \bar{t}_2, t_1, t_2, t_3$ to colours $1,\dots,5$. Then the timetable of Figure 6 corresponds to array A of Figure 7 associated with a problem $RP(m,n,p)$. Notice that in this problem every colour occurs exactly once in each column (so that $p=m$); such a problem $RP(p,n,p)$ where $\alpha(j,s)=1$ for all j and all s will be called *regular*. In addition, we notice that the preassigned meetings are such that each corresponding entry (i,k) contains colour i. For instance, entry $(4,3)$ contains colour 4. We shall say that when an entry (i,k) of the array contains colour i, it is a *coincidence*.

So we have established the following theorem.

Theorem 1. *Problem $TT(C,T,H,R,\mathcal{C},\mathcal{T})$ can be polynomially transformed into a regular problem $RP(p,n,p)$ with a subset S of entries which must have coincidences.*

In the example of Figure 6, S consists of entries $(2,4)$, $(3,4)$, $(4,3)$ and $(4,4)$. We next state an additional result related to the complexity of RTT.

Property 4. $RTT(C,T,H,R,\mathcal{T})$ is NP-complete even if every $T_j \subseteq H$ is an interval.

Proof. It has been shown that the following problem is NP-complete [2]. Given a bipartite multigraph $G = (C,T,E)$ with maximum degree $\Delta(G) = 3$, does there exist an edge 3-colouring of G such that for each node $v \in T$ the colours are $1, 2, \ldots, d_G(v)$ (i.e. the first colours)?

It follows immediately that it is also NP-complete to decide whether in such a graph G the edges adjacent to node $v \in T_j$ have colour 2 (if the degree is 1) or colours $2,3$ (if the degree is two), or colours $1,2,3$ (for degree three).

We consider the problem $RTT(C,T,H,R,\mathcal{T})$ with $|H| = 3$ and we may assume as before that all teachers are "tight". We transform the above edge colouring problem into RTT where all T_j are intervals in H: consider a node $v \in T$ of degree 1 and introduce an edge $[v, c_{1v}]$ and an edge $[v, c_{2v}]$ as well as double edges $[c_{1v}, t_{1v}]_1$, $[c_{1v}, t_{1v}]_2$ and $[c_{2v}, t_{2v}]_1$, $[c_{2v}, t_{2v}]_2$ where $c_{1v}, c_{2v}, t_{1v}, t_{2v}$ are new nodes.

Setting $T_{1v} = \{2,3\}, T_{2v} = \{1,2\}, T_v = \{1,2,3\}$, we get a new bipartite graph with maximum degree 3 (the construction is given in Figure 8). We repeat this for all nodes $v \in T$ with $d_G(v) = 1$.

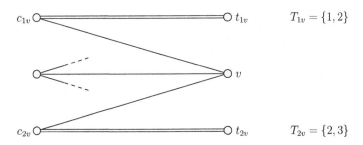

Fig. 8. Transformation of a node v of degree 1

Then we get a new bipartite graph G' with $\Delta(G') = 3$; we set $T_v = \{2,3\}$ for all nodes v of degree 2 in T for which it has not been defined yet. All nodes in T have now degree 2 or 3.

G' corresponds to a timetabling problem $RTT(C,\mathcal{T},H,R,\mathcal{T})$ where all teachers are tight; furthermore, all T_j are intervals $\{1,3\}, \{2,3\}$ or $\{1,2,3\}$. Furthermore, it follows from the construction of G' that RTT has a solution if and only if G has an edge 3-colouring where the colours at each node $v \in T$ are $1, 2, \ldots, d_G(v)$. The result follows.   □

Again in terms of the image reconstruction problem, we notice that $RP(m, n, p)$ remains NP-complete even if each colour s is known to occur only in a subset of consecutive colours.

In some situations, problem $TT(C,T,H,R,\mathcal{T})$ has to be solved while taking into account availability constraints related to the classrooms. We will also assume that $T_j = H$ for each teacher t_j, so that our problem is simply $TT(C,T, H,R)$. The availability constraints are defined by values $\mathcal{H} = (h_1, h_2, \ldots, h_h)$ where h_k represents the number of classrooms (they are all identical) available for lectures at period k.

For $TT(C,T,H,R,\mathcal{H})$ a solution is represented in the associated bipartite multigraph $G = (C,T,R)$ by an edge colouring $(M_1, \ldots, M_h)$ where the matching M_k has at most h_k edges, i.e. $|M_k| \leq h_k$ for $k = 1, \ldots, h$.

The following result is a consequence of properties stated in [8].

Property 5. $TT(C,T,H,R,\mathcal{H})$ is NP-complete even if $|H| = 3$.

The problem can also be reduced to a canonical form which will exhibit its relation with the other timetabling problems discussed earlier.

Consider $TT(C,T,H,R,\mathcal{H})$ and apply the same transformation as in Property 3. Then we call $\mathcal{M}$ the set of all parallel edges of the form $[c_i, \bar{t}_i]$ and $[\bar{c}_j, t_j]$. Such a collection of node disjoint families of parallel edges is called a *multimatching*.

Property 6. $TT(C,T,H,R,\mathcal{H})$ can be transformed into problem $RTT'(C', T', H, R', \mathcal{H}, \mathcal{M})$ satisfying the following:

1. RTT' has a solution if and only if TT has one,
2. $|T'| = |T| + |C|$,
3. $|C'| = |C| + |T|$,
4. $\sum_i r'_{ij} = \sum_j r'_{ij} = |H| = h$ for all i, j,
5. $C'_i = T'_j = H$ for all i, j,
6. In $\mathcal{M}$ at least $|T| + |C| - 2h_k$ edges (not specified in advance) receive colour k (for $k = 1, \ldots, h$).

Proof. We only have to establish statement 1 when statement 6 is given as constraint for RTT'.

Suppose TT has a solution; this means that there exists an edge 4-colouring $(M_1, \ldots, M_h)$ of G with $|M_k| \leq h_k$ for $k = 1, \ldots, h$. We can extend each M_k to a perfect matching in G': if an edge $[c_i, t_j]$ is in M_k, then we also introduce $[\bar{c}_j, t_i]$ into M_k; then one can introduce $|T| + |C| - 2h_k$ edges of the form $[\bar{c}_j, t_j]$ or $[c_i, \bar{t}_i]$ into M_k in order to obtain a perfect matching M'_k of G' (which contains M_k).

Repeating this for $k = 1, \ldots, h$, we get an edge h-colouring $(M'_1, \ldots, M'_h)$ of G' which satisfies the requirement that in $\mathcal{M}$ at least $|T| + |C| - 2h_k$ edges (not specified in advance in the data!) receive colour k ($k = 1, \ldots, h$).

Conversely, assume that for RTT' there is a solution; it is represented by an edge h-colouring $(M'_1, \ldots, M'_h)$ of G' where

$$|M'_k \cap \mathcal{M}| \geq |T| + |C| - 2h_k \quad (k = 1, \ldots, h).$$

Remove the edges of $M'_k \cap \mathcal{M}$ from M'_k; we get a partial matching M^*_k in G'. Furthermore, we observe that $[c_i, \bar{t}_i]$ (resp. $[\bar{c}_j, t_j]$) is in M'_k iff both c_i and $\bar{t}_i$

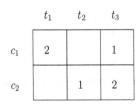

	t_1	t_2	t_3
c_1	2		1
c_2		1	2

$H = \{1, \ldots, 4\}$

$\mathcal{H} = (h_1, h_2, h_3, h_4) = (2, 2, 1, 1)$

Is there a timetable with $\leq h_k$ lectures at period k $(k = 1, \ldots, 4)$?

Regularisation

	t_1	t_2	t_3	$\bar{t}_1$	$\bar{t}_2$
c_1	2		1	1	
c_2		1	2		1
$\bar{c}_1$	2			2	
$\bar{c}_2$		3			1
$\bar{c}_3$			1	1	2

	t_1	t_2	t_3	$\bar{t}_1$	$\bar{t}_2$
c_1				1	
c_2					1
$\bar{c}_1$	2				
$\bar{c}_2$		3			
$\bar{c}_3$			1		

R' $\mathcal{M}$

G' regular $\Delta(G) = 4$

solution: edge coloring (M_1', M_2', M_3', M_4') of G' with

$$|\mathcal{M} \cap M_1'| \geq |T| + |C| - 2h_1 = 1$$
$$|\mathcal{M} \cap M_2'| \geq |T| + |C| - 2h_2 = 1$$
$$|\mathcal{M} \cap M_3'| \geq |T| + |C| - 2h_3 = 3$$
$$|\mathcal{M} \cap M_4'| \geq |T| + |C| - 2h_4 = 3$$

Fig. 9. Construction of RTT for classroom availabilities

(resp. $\bar{c}_i$ and t_j) are not adjacent to any edge of M_k^*. If $[c_i, \bar{t}_j]$ (resp. $[\bar{c}_j, t_j]$) is not in M_k' then both c_i and $\bar{t}_i$ (resp. $\bar{c}_j$ and t_j) are adjacent to some edge of M_k'. Hence M_k^* contains the same number of edges $[c_i, t_j]$ with $i \leq m$, $j \leq n$ as of edges $[\bar{c}_j, \bar{t}_i]$ with $i \leq m$, $j \leq n$. Since $|M_k^*| = |M_k'| - |M_k' \cap \mathcal{M}| \leq 2h_k$, we will have $|M_k^*|/2 \leq h_k$ edges of M_k^* with endpoints in $C = \{c_1, \ldots, c_m\}$ and in $T = \{t_1, \ldots, t_n\}$. Let M_k be the set of these edges.

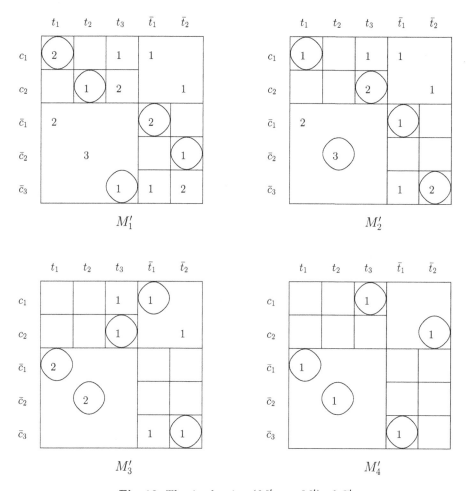

Fig. 10. The 4-colouring $(M'_1, \ldots, M'_4)$ of G'

Repeating this for $k = 1, \ldots, h$, we will get an edge h-colouring $(M_1, \ldots, M_h)$ of G with $|M_k| \le h_k$ for $k = 1, \ldots, h$; this defines a solution of TT. □

This construction is illustrated in Figure 9.

Figure 10 gives a 4-colouring $M'_1, \ldots, M'_4$ corresponding to the example in Figure 9. One sees that by considering the submatrix R of R' (first m rows and first n columns) we get a 4-colouring $(M_1, \ldots, M_4)$ of G with $|M_k| \le h_k$ ($k = 1, \ldots, 4$).

The timetable shown in Figure 11 may be viewed as the matrix $\mathcal{A}$ of the associated $RP(m, m, p)$. In each one of the first two columns, we have at least one row $\bar{c}_j$ containing the associated colour t_j; in each one of the last two columns of $\mathcal{A}$ we have at least two rows $\bar{c}_j$ (or c_i) containing the associated colour t_j (or

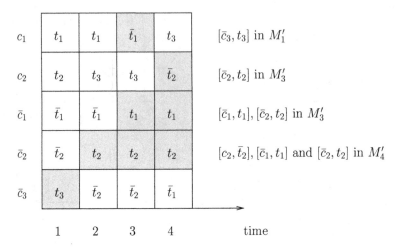

Fig. 11. The timetable associated with the colouring of Figure 10

c_1	t_1	t_1	$\bar{t}_1$	t_3	$[\bar{c}_3, t_3]$ in M_1'
c_2	t_2	t_3	t_3	$\bar{t}_2$	$[\bar{c}_2, t_2]$ in M_3'
$\bar{c}_1$	$\bar{t}_1$	$\bar{t}_1$	t_1	t_1	$[\bar{c}_1, t_1], [\bar{c}_2, t_2]$ in M_3'
$\bar{c}_2$	$\bar{t}_2$	t_2	t_2	t_2	$[c_2, \bar{t}_2], [\bar{c}_1, t_1]$ and $[\bar{c}_2, t_2]$ in M_4'
$\bar{c}_3$	t_3	$\bar{t}_2$	$\bar{t}_2$	$\bar{t}_1$	

Figure 12 (as shown):

$a(i,s)$					$\alpha(j,s)=1 \quad \forall j, s$
1 0 2 0 1	3	3	1	5	
0 1 0 1 2	4	5	5	2	
2 0 2 0 0	1	1	3	3	
0 1 0 3 0	2	4	4	4	
1 2 0 0 1	5	2	2	1	

Fig. 12. The solution of problem RP(m, n, p) associated with the solution in Figure 11

$\bar{t}_i$): in column 4 for instance, row c_2 contains colour $\bar{t}_2$, row $\bar{c}_1$ contains colour t_1 and row $\bar{c}_2$ contains colour t_2. These are precisely coincidences.

Since this timetabling problem is NP-complete, we can observe that it is also difficult to decide the existence of an edge h-colouring in a regular bipartite multigraph with $\Delta(G) \le h$ and with the constraint that in a given multimatching $\mathcal{M}$ there are at least p_k edges (not specified in advance) which must get colour k $(k = 1, \ldots, h)$.

In terms of image reconstruction problem RP(m, n, p), we have the following formulation. A multimatching $\mathcal{M}$ corresponds to a collection of disjoint pairs $[\bar{c}_j, t_j]$ or $[c_i, \bar{t}_i]$ where $\bar{c}_j$ (resp. c_i) is a row and t_j (resp. $\bar{t}_i$) is a colour. By calling $1, 2, \ldots, m$ the rows of the array (corresponding to classes $c_1, \ldots, c_m, \bar{c}_1, \ldots, \bar{c}_m$), and $1, 2, \ldots, p$ the colours (corresponding to teachers $\bar{t}_1, \ldots, \bar{t}_m, t_1, \ldots, t_m$), we have the following formulation: in a problem $RP(m, m, p)$ where each colour occurs once in each column (so that $p = m$), we have to take into account the additional requirement: given integers $p_1, \ldots, p_n$ in each column k there are at least p_k rows i which contain precisely colour i (see Figures 11 and 12 for an illustration), i.e. there are at least p_k entries (i, k) which have a coincidence.

From these observations, we can deduce the following theorem.

Theorem 2. *Problem $TT(C, T, H, R, \mathcal{H})$ can be polynomially transformed into a regular problem $RP(p, n, p)$ with the additional requirement that in each column k, there must be at least $|C| + |T| - 2h_k$ coincidences.*

In fact we have essentially shown that some basic problems of timetabling with unavailability constraints can be reduced to edge colourings in bipartite multigraphs where all nodes have the same degree. In addition, we have to satisfy some requirements on "horizontal" edges: in a regular problem $RP(m, n, p)$, the coincidences (entry (i, k) gets colour i) may indeed be viewed as requirements saying that class c_i^* has to get at period k a lecture from teacher t_i^* (with the same index i). We have shown that unavailabilities of classes, teachers and/or classrooms can all be formulated in terms of requirements on "horizontal" edges.

It follows from this observation that all these types of unavailability constraints may be present simultaneously in the timetabling problem without essentially changing the model $RP(m, n, p)$ which will just have some constraints of coincidences.

5   Double Lectures

We shall finally examine the case where all lectures do not have the same length. More precisely, we shall assume that in addition the two one-period lectures given by a teacher t_j to a class c_i, there are also *double lectures* which are also given by a single teacher to a single class but they have a length of two periods; they consist of two normal lectures which have to be scheduled on two consecutive periods (in general, when we have lunch breaks or even breaks between a day and the next one, we have to avoid scheduling double lectures on the last period before a break; we shall not examine this here).

These double lectures may be viewed as creating some type of dynamic unavailability constraints: if a double lecture of t_j to c_i is scheduled to start at period k, then t_j and c_i become unavailable for other lectures at period $k + 1$.

We shall assume for the moment that there are no other unavailability constraints, i.e. given sets C, T, H of classes, teachers and periods, we have $C_i = T_j = H$ for all i, j.

In addition, we are given a requirement matrix $R = (r_{ij})$; to keep notation simple we shall suppose that $r_{ij} \in \{0, 1, 2\}$ and whenever $r_{ij} = 2$, this means that t_j has to give a double lecture to c_i.

Let $TT(C, T, H, R, M_2)$ be the corresponding timetabling problem where M_2 defines the set of double lectures, i.e. the set of entries (i, j) of R with $r_{ij} = 2$.

We shall first transform the problem into a regular form in a similar way to the reductions described in the previous sections.

Notice that the following reduction is valid also if M_k is a set of k-tuple lectures.

Property 7. $TT(C, T, H, R, M_k)$ *characterized by a requirement matrix R with $r_{ij} \leq k$ for all i, j can be transformed into a problem $TT'(C', T', H, R', M_k')$ satisfying the following:*

1. *TT' has a solution if and only if TT has one,*
2. *$|C'| = |T'| = |T| + |C|$,*
3. *$\sum_j r_{ij}' = |H|$ for all classes c_i',*
   $\sum_i r_{ij}' = |H|$ for all teachers t_j',
4. *$|M_k'| = 2|M_k|$.*

Proof. We transform R into R' exactly as in Property 3. The multiple edges between $\bar{c}_j, t_j$ and between $c_i, \bar{t}_i$ correspond to single lectures; so statements 2, 3 and 4. are satisfied. To show that statement 1 holds, we observe that from any solution to TT' we can trivially deduce a solution of TT by considering only the first m rows and the first n columns of R'.

Conversely, if TT has a solution, we can take the symmetric solution for the submatrix of R' located in the last n rows and the last m columns:

> If t_j gives a lecture to c_i at period k, then $\bar{t}_i$ gives a lecture to $\bar{c}_j$ at the same period; moreover, if c_i (resp. t_j) has no lecture at period k, then c_i meets $\bar{t}_i$ (resp. t_j meets $\bar{c}_j$) at this period.
> This will give us a solution for TT' which will respect the requirements of M_k' if those of M_k are respected. Hence statement 1 holds.

$\square$

We shall restrict the problem to the special case of double lectures in the following complexity result.

Property 8 ([9]). *Consider problem $TT(C, T, H, R, M_2)$ with $h = |H| \geq 4$ periods, where TT is reduced as above. It is NP-complete to decide whether TT has a solution in h periods.*

There is, however, a solvable case that can be stated as follows.

Property 9. *Consider a reduced problem $TT(C, T, H, R, M_{h-1})$ with $h = |H|$ periods. Assume M_{h-1} contains only $(h-1)$-tuple lectures. Then TT has a solution in at most $2h - 2$ periods. Moreover, there is a polynomial algorithm for constructing a solution in h periods if there is such a solution.*

From this we have immediately the following corollary.

Corollary 1 ([9]). *Consider a problem $TT(C, T, H, R, M_2)$ with $h = |H| = 3$ periods. Then there exists a polynomial algorithm to decide whether TT has a solution in 3 periods. If there is no such solution, then one can get one in 4 periods.*

Proof (Property 9). Let G be the multigraph associated with the problem $TT(C, T, H, R, M_{h-1})$. We can assume that TT has been reduced so that G is regular; if a teacher t_j (resp. a class c_i) has a multiple lecture (i.e. a $(h-1)$-tuple lecture), then it has to be scheduled either in the first $h-1$ periods or in the last $h-1$ periods.

In both cases, at periods $2, 3, \ldots, h-1$ the multiple lecture will be running.

There will be a timetable in exactly h periods if and only if G contains a perfect $(h-2)$-matching $\mathcal{M}$ (i.e. a partial subgraph having exactly $h-2$ edges at each node of G) that uses exactly $h-2$ edges of each multiple edge.

This can be seen easily by colouring with colours $2, 3, \ldots, h-1$ the edges of $\mathcal{M}$; the remaining edges of G form a 2-regular graph which can be coloured with colours 1 and h. This gives a timetable in h periods where each multiple lecture is scheduled in h consecutive periods. $\mathcal{M}$ can be constructed in polynomial time by using network flow techniques (see [3] or [4]).

If there is no such 2-matching $\mathcal{M}$, we can construct a perfect $(h-2)$-matching as before; these edges are the lectures scheduled in periods $2, 3, \ldots, h-1$. The remaining graph has degree 2 and corresponds to periods 1 and h. All lectures have been scheduled, but some multiple edges do not have consecutive colours.

In fact, each $(h-1)$-tuple lecture has been scheduled within an interval of h periods. There is exactly one period l where no part of a given multiple lecture $c_i - t_j$ is scheduled. If $l = 1$ or h, we are done. If $2 \le l \le h-1$, we reschedule the $l-1$ first periods of $c_i - t_j$ in periods $h+1, h+2, \ldots, h+l-1$. This can be done for all such multiple lectures independently, because from our assumption no teacher and no class is involved in more than one multiple lecture. Since $l \le h-1$, we will in all cases get a timetable in at most $2h-2$ periods.   □

By using the regularization technique of Property 3 one can obtain the following result; here M_* consists of a collection of multiple lectures (whose multiplicities, i.e. durations, may be different).

Theorem 3. *$TT(C, T, H, R, M_*)$ can be polynomially transformed into a regular problem $RP(p, n, p)$ where in each row i the occurrences of every colour are consecutive.*

Sketch of proof. There are no unavailabilities in the problem TT; we can regularize it as in the proof of Property 3 (this doubles the number $|M_*|$ of multiple lectures). Then we can transform the problem in such a way that each teacher will give at most one lecture (simple or multiple) to any class. This is done by applying the technique used in the proof of Property 2 for each multiple lecture $[c_i, t_j]$; suppose its multiplicity is k: we introduce as before t_i^* and c_j^*. In addition, we have to introduce $c_{ij,1}^*, \ldots, c_{ij,h-k}^*$ and $t_{ij,1}^*, \ldots, t_{ij,h-k}^*$. The $c_{ij,d}^*$ and $t_{ij,d}^*$ are linked by k parallel edges (corresponding to k simple lectures) for $d = 1, \ldots, h-k$. The c_i and t_i^* (resp. c_j^* and t_j) are linked by a multiple edge representing a multiple lecture with duration k. The remaining edges are introduced as in the proof of Property 2.

In order to destroy the multiple edges between nodes $c^*_{ij,d}$ and $t^*_{ij,d}$ we may again apply the technique of Property 2 to the first $k-1$ parallel edges between these nodes. Then we get a bipartite graph where all nodes have degree h.

It is a simple matter to verify that it has an h-colouring satisfying the requirements on the multiple lectures if and only if the initial problem has a solution   □

6   Extensions and Conclusions

We have restricted our attention to some variations of the basic timetabling problem; this has allowed us to exploit the similarity with some elementary image reconstruction problems, like $RP(m, n, p)$.

We have not mentioned in an explicit way the analogies with other types of scheduling problems, such as open shop scheduling (see [4]). In particular, from the timetabling problem $TT(C, T, H, R, M_2)$ one may derive the fact that the open shop problem with processing times of value $0, 1$ or 2 is difficult (when the total processing time is at least 4).

Our purpose was to examine the simple variations of the class teacher model, to reduce them to some "canonical" forms and to derive some complexity properties.

The knowledge of these properties will be useful when real cases have to be solved, for instance by decomposing them into smaller problems which can be solved in polynomial time.

One could as well start from other basic timetabling models like the ones taking the classroom assignment into account, as in [5] or the models where group lectures (involving several classes at a time) are present, as in [1]. For these problems, the image reconstruction $RP(m, n, p)$ may not be the most natural neighbour problem to consider. It will be an interesting research area to explore this field with the objective of exploiting analogies either for the design of heuristics or for complexity studies.

Many other basic timetabling problems could have been discussed here, but we decided to concentrate on some basic models to explore the boundary between easy and difficult problems.

References

[1] Asratian, A.S., de Werra, D.: A Generalized Class–Teacher Model for Some Timetabling Problems. Eur. J. Oper. Res. (2002) (to appear)
[2] Asratian, A.S., Kamalian, R.R.: Interval Edge Coloring of Multigraphs. Appl. Math. Yerevan University (in Russian) 5 (1987) 21–34
[3] Berge, C.: Graphs and Hypergraphs. North-Holland, Amsterdam (1973)
[4] Blazewicz, J., Ecker, K., Pesch, E., Schmidt, G., Weglarz, J.: Scheduling Computer and Manufacturing Processes. Springer-Verlag, Berlin Heidelberg New York (1996)
[5] Carter, M.W., Tovey,C.A.: When Is the Classroom Assignment Problem Hard? Oper. Res. 40 (Suppl. 1) (1996) S28–S39

[6] Costa, M.-C., de Werra, D., Picouleau, C.: On Some Image Reconstruction Problems. (2002) (submitted)
[7] Even, S., Itai, A., Shamir, A.: On the Complexity of Timetable and Multicommodity Flow Problems. SIAM J. Comput. **5** (1976) 691–703
[8] Gabow, T., Nishizeki O., Kariv, D., Leven, O., Terada, O.: Algorithms for Edge-Coloring Graphs. Unpublished manuscript, University of Colorado (Boulder) (1983)
[9] Williamson, D.P., Hall, L.A., Hoogeven, J.A., Hurkens, C.A.J., Lenstra, J.K., Sevastianov, S.V., Shmoys, D.B.: Short Shop Schedules. Oper. Res. **45** (1997) 288–294

A Standard Framework for Timetabling Problems

Matthias Gröbner[1], Peter Wilke[2], and Stefan Büttcher[1]

[1] Lehrstuhl für Informatik II,
Universität Erlangen-Nürnberg,
Martensstrasse 3, 91058 Erlangen, Germany
Groebner@informatik.uni-erlangen.de
[2] Centre for Intelligent Information Processing Systems (CIIPS),
Department of Electrical & Electronic Engineering,
The University of Western Australia,
35 Stirling Highway, Crawley WA 6009, Australia
wilke@ee.uwa.edu.au

Abstract. When timetabling experts are faced with a new timetabling problem, they usually develop a very specialised and optimised solution for this new underlying problem.

One disadvantage of this strategy is that even slight changes of the problem description often cause a complete redesign of data structures and algorithms. Furthermore, other timetabling problems cannot be fit to the data structures provided.

To avoid this, we have developed a standardised framework which can describe arbitrary timetabling problems such as university timetabling, examination timetabling, school timetabling, sports timetabling or employee timetabling. Thus, a general timetabling language has been developed which enables the definition of resources, events and constraints.

Furthermore, we provide a way to apply standard problem solving methods such as branch-and-bound or genetic algorithms to timetabling problems defined by means of the general timetabling language. These algorithms can be improved by problem-specific user-defined hybrid operators.

In this paper we present a generalised view on timetabling problems from which we derive our timetabling framework. The framework implementation and its application possibilities are shown with some concrete examples. The paper concludes with some preliminary results and an outlook.

1 Introduction

There exist many different timetabling problems such as university or examination timetabling, school timetabling, sports timetabling or employee timetabling. Furthermore, there exist many problem solving methods, which usually use the concepts of standard optimisation algorithms such as Backtracking [14] Evolutionary Algorithms [1,4,6,8] or Constraint Logic Programming [10,13].

E. Burke and P. De Causmaecker (Eds.): PATAT 2002, LNCS 2740, pp. 24–38, 2003.

Unfortunately, these standard algorithms often do not yield acceptable timetables or cannot compute solutions within a reasonable amount of time. So the standard algorithms have to be adapted to be able to handle the special concrete timetabling problem. Therefore, the search strategy is changed and special operators and problem-specific data structures are developed.

The disadvantage of these optimised implementations is that slight changes of the problem description often cause radical changes of the data structures and algorithms have to be redesigned to get acceptable solutions again. In addition, these data structures cannot be used to describe other new timetabling problems.

A further problem of missing standard timetabling descriptions is that new proposed optimisation algorithms cannot be reliably compared to existing ones with respect to performance and solution quality.

Moreover, from a theoretical point of view it would be interesting to compare the structure of different timetabling problems to each other or different problem solving strategies. Open questions such as the phase transition niche [16] could be analysed. This could be a step towards a better understanding of the timetabling research field.

In recent years this problem has been tackled, and first attempts have been made to standardise the description of timetabling problems [3,5,11,12]. In this paper we present a new way to generalise the timetabling problem that adopts some ideas from these known approaches and introduces new points of view. From this general view we derive an object-oriented standard timetabling framework which is able to describe arbitrary timetabling problems and can apply standard optimisation algorithms to the problem.

2   The General View

2.1   A Generic Timetable Scheme

As mentioned before, many types of timetabling problems exist. But all these problems have several properties in common.

One of these similarities is that certain entities have to be scheduled. For example, the German high school timetabling problem [1,15] has several entities such as classes or single students, teachers, subjects, lessons and rooms. All these entities have properties. For example classes are linked to the subject the students of this class are taught.

Usually, these entities are differentiated into *resources* and *events* (or sometimes called *meetings*). In addition, *constraints* have to be considered.

In the employee timetabling case, for instance, we find those entities, too. There are employees with different qualifications and monthly target hours or there are shifts the employees have to be assigned to.

As already mentioned, some of these entities are linked with others. There exist links from the shifts to the employees assigned to these shifts or from the students to their teachers. Some of these links are fixed, such as the links from the shifts to the employees with the qualifications required to work on these

Table 1. Different categories of resources in timetabling (TT) problems

High school TT	Employee TT	Arbitrary course TT
Classes/Students	Employees	Participants
Teachers	Shifts	Courses
Rooms	Qualifications	Organiser
Subjects	Employer(s)	(Rooms)
Lessons		

Employer's production process

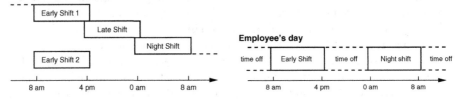

Fig. 1. The same time stream from the employer's point of view and from an employee's point of view, respectively

shifts, and cannot be changed. Others have to be assigned during a planning process, e.g. linking a lesson to a suitable room.

A planning algorithm has to construct a timetable, so we have to define what a timetable consists of. A timetable can be interpreted as an arbitrary sequence of *events*. To every event a certain number of time intervals is assigned, each having a starting and an ending point.

Each timetable can be seen from different points of view: for example, an employer has a different view compared to the view of his employees, as shown in Figure 1.

2.2   The Object-Oriented View

The considerations of Section 2.1 can be used to describe timetabling problems in an object-oriented manner:

There are different *resources* whose instances have references to each other, e.g. an instance of the *subject* class refers to instances of the *teacher* class who are able to teach that subject.

Moreover, there are entities with a certain property, called *events*. This property is a certain time interval (or several time intervals) that is assigned to these events, as shown in Figure 2.

2.3   Planning

An algorithm for constructing a timetable has to assign instances of the different resource classes to the event class instances. Some of these assignments are

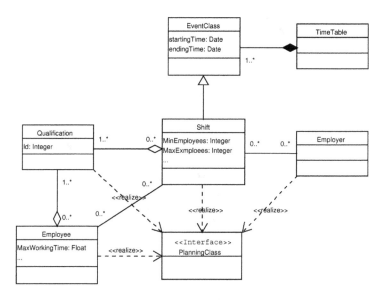

Fig. 2. Object-oriented view of the employee timetabling problem in Unified Modelling Language (UML)

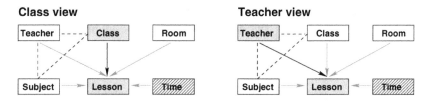

Fig. 3. Class view and teacher view. The viewing instance `class` or `teacher` is fixed (black arrows), whereas instances of the other classes to be planned have to be assigned (light arrows) to the event class `lesson`. The assignment of a time interval to each event is mandatory for all timetabling problems

predetermined and cannot be changed, and some have to be done during the planning phase.

To construct a timetable, one of the *views* mentioned in Section 2.1 is used. In the school timetabling case our algorithm might use the `class` view to assign a subject, teacher and room to a `lesson`. In this case the class is fixed and the other instances have to be assigned to (see Figure 3). Additionally, a time interval has to be assigned to each event class instance.

For each viewing perspective there are as many timetables as instances of this class to be planned exist. If we have t teachers at a high school, for example, t different teacher timetables belong to them. That is, if there exist l lessons in a high school timetabling problem, furthermore t teachers, r rooms and c classes, the number of instances of event classes including all views will be $(t + r + c) \times l$.

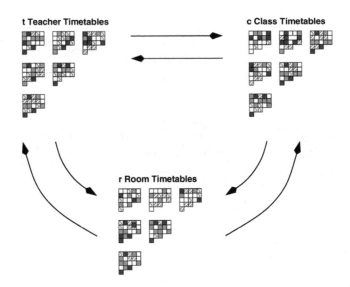

Fig. 4. The different timetables can be mapped to each other

The timetables of the instances of one planning class contain all information necessary to construct the timetables for the instances of the other planning classes: i.e. the timetables of the different views can be mapped to timetables of other views. From the employees' timetables the employer's timetable can be constructed or in the school timetabling case the t teachers' timetables can be mapped to the r rooms' timetables (see Figure 4).

That is why it is usually sufficient for a timetabling program to save the timetables of one resource type only. This avoids data redundancy caused by storing the same event information in different places, i.e. from different views (Figure 5). Nevertheless, to be able to check constraint violations (see the next section), translations to other views have to be done, for example to compute the number of assigned lessons of a teacher when working with the class *view*. Otherwise expensive computing time has to be accepted in order to compute the necessary information.

2.4   Constraints

Assignments usually cannot be done arbitrarily, but many constraints have to be considered. We distinguish two different types, namely hard and soft constraints. A solution is feasible if no hard constraints are violated. A feasible solution is better than another if fewer soft constraints are violated.

A timetabling algorithm can use different strategies to get a solution without violations of hard constraints. Violations can either be avoided from the outset [13,14] or penalised to lead the algorithm towards better solutions and introduce repair mechanisms [7].

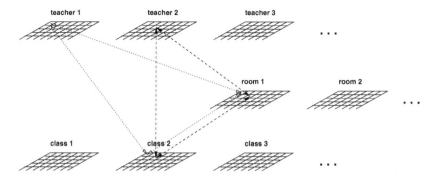

Fig. 5. Information redundancy is necessary to compute constraint violations of all resources

3   A General Timetabling Language

To describe the concepts introduced in Section 2, we implemented a general timetabling language (GTL). A GTL file starts with the declaration of the name of the timetabling problem, e.g. **TimetablingProblem** School { ... }.

3.1   Resources and Events

The language describes resources as instances of the `Resource` class and events as instances of the `Event` class. The super-class of both `Resource` and `Event` is the so-called `PlanningClass` class.

To declare members of these derived classes, the standard data types `boolean`, `float`, `double`, `int`, `short`, `long` and `string` are provided. The `reference` keyword declares a reference to objects of a certain type, where multidimensional arrays are indicated by the dimension count included in brackets.

To give an impression, the declaration of an *employee* resource in an employee timetabling example is shown:

```
class Staff extends Resource {
    string firstName;
    double workLoad;   // target working hours

    // list of shifts the employee cannot be assigned to
    reference(Shift)[1] absent;
} // end of class Staff
```

Additional standard member variables, such as an array containing references to the assigned events, are automatically provided by the timetabling framework as described in Section 4.

Events are declared in the same way. The concept of links from the events to the resources is covered by the timetabling framework, thus arrays containing references to the assigned resources need not be declared, as shown in the shift example:

```
class Shift extends Event {
    int shiftType;
} // end of class Shift
```

3.2 Constraints

In GTL constraints are introduced by a `Constraint` class. Instances of this class are assigned at a later stage in the timetabling framework to each planning class. A constraint can be derived either from the `HardConstraint` or from the `SoftConstraint` class. The declaration of a constraint class itself does not imply any direct consequences yet, but the different handling of the `Constraint` instances as hard or weak has to be managed by the algorithms applied. The `Constraint` class has a `compute` method to compute the constraint violations. This method must be implemented in the GTL file by the user. It returns a `ConstraintViolation` object which contains information about the penalty points and information for a `repair` method of the corresponding constraint class that can be implemented, too. The syntax of GTL is Java-like.

The `repair` method can be called by timetabling algorithms and uses problem-specific knowledge, which is usually necessary to get feasible solutions with Genetic Algorithms [2,8,15,17]. The example below shows the GTL definition of `ClashConstraint` and its `compute` method. Methods like `getAssignments` and `clashes` are provided by the standard library of the framework (Section 4):

```
class ClashConstraint extends HardConstraint {
    public ConstraintViolation compute(PlanningClass owner) {
        double penalty = 0.0;
        List assignments = ((Resource) o).getAssignments();
        List clashList = new ArrayList();
        for (int j = 0; j < assignments.size() - 1; j++) {
            Assignment ass1 = (Assignment)assignments.get(j);
                for (int k = j + 1; k < assignments.size(); k++) {
                    Assignment ass2 = (Assignment)assignments.get(k);
                    if (ass1.event.clashes(ass2.event)) {
                        penalty += 10.0; clashList.add(ass2.resList);
        } } }
        if (penalty > 0) {
            return new ConstraintViolation(this, owner, "Clash at "
                    + owner + ": " + clashList, penalty, clashList);
        } else return null;
    } // end of compute(PlanningClass)
    public boolean repair(ConstraintViolation violation) {
    ...
    } // end of repair(ConstraintViolation)
} // end of class ClashConstraint
```

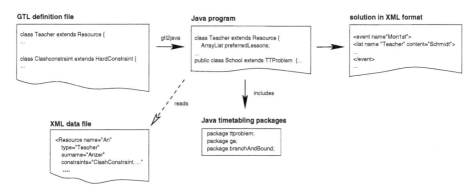

Fig. 6. The road-map to get a solution for a specific timetabling problem from a GTL description and an XML data file

4   The Timetabling Framework

All necessary classes of a timetabling problem can be declared with the time-tabling language GTL. But to be able to store all necessary information about a specific instance of a timetabling problem, a framework is needed that instanti-ates the actual resources and events, applies algorithms to get solutions and for that purpose computes the constraint violations.

In order to ensure this, we developed a framework [9] that is able to read information about the actual instances of the corresponding timetabling problem using the GTL class declarations of the underlying GTL file. This information has to be defined in an XML data file. Thus, in contrast to STTL [12] the information about the specific instances of the timetabling problem is separated from problem definition itself.

The framework is then able to apply standardised algorithms to compute a solution of the timetabling problem, which can be exported to XML, HTML or text format. The solution contains information about the timetables of the different views and all remaining constraint violations.

Currently the framework is available in a beta version providing the complete functionality described in this section. The GTL description file (declaration of resources and the definition of the constraints) and the XML data file has to be created by the user. A friendly user interface does not exist at the current stage of development. Especially for the design of the `compute` method of the constraints some understanding of the data structures is required.

4.1   XML Data File

The syntax of the XML input data file is defined as follows. The file starts with the definition of the timetabling problem, where `layout` describes the structure of the timetable. That information is only used for graphical output, e.g. inside an HTML file. The meaning of the three parameters is *Days per Period, Time Intervals per Day* and *Periods*.

```
<TTProblem layout="5,11,1" description="Name of the
school">
```

Definition of the resources can be done by a `Resource` tag which contains arbitrary elements and their values according to the member variables defined in the GTL file. `type` describes the special type a resource is of and `name` is a kind of ID. The `constraints` list contains the constraints that have to be considered for this resource instance. The example shows the definition of a specific employee of a rostering problem in a hospital:

```
<Resource type="Staff" name="1126"
description="Schlueter, Anna"
    workLoad="15.0" features="Nurse,Chief"
    absent="" constraints="ClashConstraint,DiffHourMonthConstraint,
            OnlyOneDayFreeConstraint,OnlyOneWorkingDayConstraint,..."/>
```

There exists a pre-defined resource type `TimeSlot` with the variables *name, from, until* and *layout* which are defined in the same way. *layout* defines a position in the timetable grid as explained above.

Events also have a name, type (i.e. the event class) and constraints. In addition, events have a number of resource lists which contain all possible assignable resources, indicated by the `takefrom` keyword. In the following example the assigned time slot is fixed and cannot be changed, so the variable `fixed` is set to true. During the planning process, to all resource lists which are not fixed arbitrary resources are assigned taking members from the `takefrom` list. The minimum and maximum number of resources to be selected for assignment from the resource list by the timetabling algorithm are described by the `min` and `max` keywords:

```
<Event type="Shift" name="February 26th (early
shift)"
        constraints="OutOfMinMaxConstraint" shiftType="0">
  <ResourceList name="TimeSlot" min="1" max="1"
                takeFrom="February 26th 06:30" fixed="true"/>
  <ResourceList name="Nurse" min="2" max="4" target="3"
                takeFrom="2176,2277,2282,2570,2770,2793,2807,2829,..."/>
  <ResourceList name="Student" min="0" max="2" target="1"
                takeFrom="3065,3069,3173,3178,3295,3296,3297"/>
  <ResourceList name="Chief" min="0" max="1" target="1" takeFrom="2277"/>
</Event>
```

4.2   Timetabling Packages

In order to generate an executable program from the GTL description as indicated in Figure 6 we need some standard timetabling packages which provide the basic timetabling class structures. These packages are included into the main program.

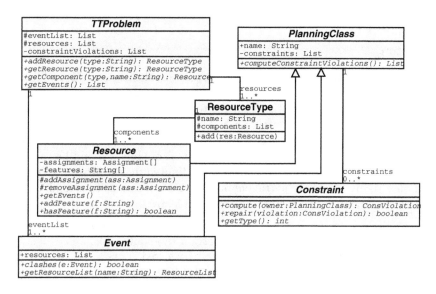

Fig. 7. Data structure of timetabling problem as provided by the `ttclasses` package

The structure of the `ttclasses` package is shown in Figure 7. A timetabling problem `TTProblem` holds an event list with all events and lists with resources of different types. Both `Resource` and `Event` are derived from the basic class `PlanningClass`. Each instance of `PlanningClass` refers to a number of constraints which should not be violated for this instance. The actual resource and event instances are read from the XML data file, are instantiated and then added to the corresponding `eventList` or `resourceList`, respectively.

Our implementation has been done in Java. One disadvantage of Java is its low run time performance. Nevertheless, Java allows straight object-oriented design of the data structures and provides a lot of useful standard packages such as lists and containers which facilitate handling complex data as found in timetabling problems. Furthermore, Java byte code can be executed on any operating system which allows an easy application of the framework on any computer. A comparison of the run time performance of the framework and an optimised implementation for a single problem is made in the next section.

To get an executable Java class file, we implemented a GTL-to-Java compiler which translates the GTL description into Java source code and adds some additional code such as import statements, constructors of the classes and a main method for starting the program.

The framework currently provides the possibility to apply a standard hybrid Genetic Algorithm and a simple branch-and-bound algorithm to the timetabling problem. The framework utility packages are included in the main Java file. Both algorithms use problem-specific knowledge defined by means of the repair operators of the constraint violations of the corresponding GTL files. Implementation details of the algorithms are omitted here for lack of space.

5   Example

5.1   GTL

To demonstrate the ability of our framework to solve real-world problems we will have a closer look at the time scheduling for a special course week at our university. The intention of the course is to attract more students to technical-oriented studies. So a whole week packed with lab classes and social function is organised. Each participant selects four courses with first priority and another four with second priority. The participant can also name a friend he would like to join him in the lab classes. As lab space is limited most courses are run more than once to give as many participants as possible a chance to attend.

The constraints when setting up a timetable are:

- participants can be assigned only once at the same time (clash constraint),
- participants may join a course of a certain type only once,
- participants should be assigned to their preferred courses,
- participants should be assigned to exactly four courses,
- some participants have to be assigned to courses taking place at the same time due to car pools,
- the number of participants assigned to the courses must be at least *min* and at most *max*.

In 2001, 251 participants were registered and 187 courses of 37 different types were offered. Courses take place either in the morning or in the afternoon, but there exist two-part courses which comprise two sessions and thus can take place in the morning and afternoon or even on two different days.

The GTL description of the resource `Participant` and the event `Course` is given as follows:

```
class Participant extends Resource {
    string firstName;

    reference(Project)[1] firstChoice;
    reference(Project)[1] secondChoice;
    reference(Participant)[1] friends;
}

class Project extends Event {
    string signature;
}
```

`firstChoice` and `secondChoice` are containers holding the preferred and alternative projects of the participant. `friends` are the friends that should be assigned to courses taking place at the same time as the courses the participant has been assigned to. Each event has a signature which defines the course type of the project.

In addition, the constraints presented above have been defined with corresponding `compute` and `repair` methods.

5.2 XML Data File

To get an XML input database of the practical week instances, we wrote a converter program that translates the available ASCII data into the required XML input format. This XML file describes a concrete participant my means of a `Resource` tag of the type `Participant` and elements defining the values of the firstname, name, the selected projects, the friends and the constraints to be considered.

For each day timeslots are defined for morning and afternoon courses, so 10 timeslots have been defined.

Each `Project` event has the variables *signature*, which is used to identify the type a certain project is of, and *name*, which specifies the course ID. An event owns two resource lists. One resource list holds the corresponding time slot and is fixed, because each course is assigned to one of two (two-part courses) fixed time slots. The other resource list contains all participants that can be assigned to that event because the course type is on their selection list.

5.3 Results

We applied the branch-and-bound algorithm as well as the standard hybrid Genetic Algorithm to the problem. Figure 8 shows several runs of the Genetic Algorithm with different parameter values. Computing time was 75 min on a 1.2 GHz computer. As expected, the steady-state Genetic Algorithm without usage of repair operators converges much more slowly and does not reach such good results as the hybrid runs.

All results computed by the hybrid Genetic Algorithm yielded feasible timetables with less than 10 courses having few more than *max* participants. Due to high request for some courses, on average about 30 participants were assigned to only three instead of four courses. Finally, there remained about 200 participants who did not get their first-choice courses but had to join at least one of the alternative courses selected. These constraint violations explain the remaining penalty points in Figure 8.

In addition, we applied our simple branch and bound algorithm to the problem. The final results were only marginally different with respect to penalty points and number of constraint violations. But the branch and bound algorithm outperformed the Genetic Algorithm with respect to time and computed the results in only about 25 s.

To get further information about the appropriateness of the Genetic Algorithm we introduced room resources with allocation clashes to be avoided. In this case the Genetic Algorithm yielded feasible solutions where the branch and bound algorithm failed to find solutions of the same quality in reasonable time.

Furthermore, we compared the timetables computed by our framework to those created by a special Genetic Algorithm-based application. The algorithm has been developed for this specific course timetabling problem in C++ programming language. The special application stopped after 10 min computing time and on average created timetables with about 60 participants assigned to

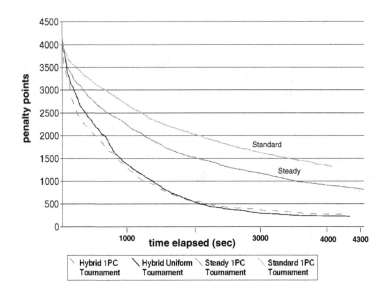

Fig. 8. Application of the Genetic Algorithm to the practical week timetabling problem. Hybrid: hybrid Genetic Algorithm with application of the repair operators; Steady: steady-state Genetic Algorithm; Standard: without any elitism or hybrid components. Crossover method is either one-point-crossover (1PC) or uniform. Selection method is always tournament selection

too few courses and more than 200 participants assigned to some alternative courses. Thus computing timetables using our framework did not result in a loss of quality, but computing time clearly increased compared to specially designed programs.

6   Conclusion and Outlook

In this paper we have presented a general object-oriented view on timetabling problems and how this description has been realized by means of a timetabling framework. The framework is able to describe arbitrary timetabling problems using the general timetabling language GTL. Actual instances are defined by an XML input file. Furthermore, standardised timetabling algorithms are provided by the timetabling framework.

One advantage of the framework is that it ensures that the application of timetabling algorithms to new timetabling problems neither enforces the redesign of the data structure nor the redesign of the algorithms. Furthermore, the widespread use of the framework could ensure a better comparability of new timetabling algorithms because it introduces standardised input and output formats.

However, one disadvantage of the presented timetabling framework is that the existing timetabling data has to be transformed into the defined XML data format which requires some amount of preliminary work. But it is usually much less costly to convert data to a standard format than developing special data structures in a programming language.

Our next steps will be the implementation of further algorithms such as Simulated Annealing or Tabu Search so as to compare the applicability of these algorithms to different problems. To date, we have defined a school timetabling problem, a practical week problem and some employee timetabling problems using GTL and XML input format and successfully applied the Genetic Algorithm.

First results show that it is possible to compare the solubility of timetabling problems with respect to different timetabling algorithms. As a next project we want to analyse the structure of arbitrary timetabling problems with the help of the timetabling framework to get a better understanding of which algorithms should be preferably used to solve which type of problems.

Acknowledgement. This work has been partly supported by a grant from the Lehrstuhl für Programmiersprachen-und-methodik, Institut für Informatik, Friedrich-Alexander-Universitaet Erlangen-Nuernberg, Martensstrasse 3, 91058 Erlangen, Germany. We would like to thank them for their support.

References

1. Bufe, M., Fischer, T., Gubbels, H., Häcker, C., Haspirch, O., Scheibel, C.,Weicker, K., Wenig, M.: Automated Solution of a Highly Constrained School Timetabling Problem – Preliminary Results. In: Boers, E.J.W. et al. (Eds.): Proc. EvoWorkshops 2001. Springer-Verlag, Berlin Heidelberg New York (2001) 431–440
2. Burke E., Elliman, D., Weare, R.: Specialised Recombinative Operators for Timetabling Problems. In: Proc. AISB (AI and Simulated Behaviour) Workshop Evolut. Comput. Springer-Verlag, Berlin Heidelberg New York (1995) 75–85
3. Burke E.K., Kingston J.H.: A Standard Data Format for Timetabling instances. In: Burke, E, Carter, M. (Eds.): Practice and Theory of Automated Timetabling II (PATAT 1997, Toronto, Canada, August, selected papers). Lecture Notes in Computer Science, Vol. 1408. Springer-Verlag, Berlin Heidelberg New York (1998) 213–222
4. Caldeira, J.P., Rosa, A.C.: School Timetabling Using Genetic Search. In: Burke, E, Carter, M. (Eds.): Practice and Theory of Automated Timetabling II (PATAT 1997, Toronto, Canada, August, selected papers). Lecture Notes in Computer Science, Vol. 1408. Springer-Verlag, Berlin Heidelberg New York (1998) 115–122
5. Colorni, A., Dorigo, M., Maniezzo, V.: Genetic Algorithms and Highly Constrained Problems: The Time-Table Case. In: Proc. 1st Int. Workshop on Parallel Problem Solving from Nature. Springer-Verlag, Berlin Heidelberg New York (1990) 55–59
6. Corne, D., Ross, P., Fang, H.-L.: Evolutionary Timetabling: Practice, Prospects and Work in Progress. In: Prosser, P. (Ed.): Proc. UK Planning and Scheduling SIG Workshop. University of Strathclyde (1994)
7. Fernandes, C., Caldeira, J.P., Melicio, F., Rosa, A.: High School Weekly Timetabling by Evolutionary Algorithms. In: Proc. 14th Annual ACM Symp. on Applied Computing (San Antonio, TX, 1999)

8. Gröbner, M., Wilke, P.: Optimizing Employee Schedules by a Hybrid Genetic Algorithm. In: Boers, E.J.W. et al. (Eds.): Proc. EvoWorkshops 2001. Springer-Verlag, Berlin Heidelberg New York (2001) 463–472

9. Gröbner, M.: GTL, A General Timetabling Language, Beta Version, and a Bibliography of Timetabling Publications. Available at http://www2.cs.fau.de/Research/Activities/Soft-Computing/Timetabling (2002)

10. Gueret, C., Jussien, N., Boizumault, P., Prins, C.: Building University Timetables Using Constraint Logic Programming. In: Burke, E, Ross, P. (Eds.): Practice and Theory of Automated Timetabling I (PATAT 1995, Edinburgh, Aug/Sept, selected papers). Lecture Notes in Computer Science, Vol. 1153. Springer-Verlag, Berlin Heidelberg New York (1996) 393–408

11. Kingston, J.H.: A User's Guide to the STTL Timetabling Language, Version 1.0. Basser Department of Computer Science, The University of Sydney (1999)

12. Kingston, J.H.: Modelling Timetabling Problems with STTL. In: Burke, E, Erben W. (Eds.): Practice and Theory of Automated Timetabling III (PATAT 2000, Konstanz, Germany, August, selected papers). Lecture Notes in Computer Science, Vol. 2079. Springer-Verlag, Berlin Heidelberg New York (2001) 433–445

13. Lever, J., Wallace, M., Richards, B.: Constraint Logic Programming for Scheduling and Planning. British Telecom Technol. J. **13** (1995) No 1, January

14. Meisels, A., Lusternik, N.: Experiments on Networks of Employee Timetabling Problems. In: Burke, E, Carter, M. (Eds.): Practice and Theory of Automated Timetabling II (PATAT 1997, Toronto, Canada, August, selected papers). Lecture Notes in Computer Science, Vol. 1408. Springer-Verlag, Berlin Heidelberg New York (1998) 215–228

15. Oster, N.: Stundenplanerstellung für Schulen mit Evolutionären Verfahren. Thesis, Universität Erlangen-Nürnberg (2001)

16. Ross, P., Corne, D., Terashima, H.: The Phase Transition Niche for Evolutionary Algorithms in Timetabling. In: Burke, E, Ross, P. (Eds.): Practice and Theory of Automated Timetabling I (PATAT 1995, Edinburgh, Aug/Sept, selected papers). Lecture Notes in Computer Science, Vol. 1153. Springer-Verlag, Berlin Heidelberg New York (1996) 269–282

17. Weare, R., Burke, E., Elliman, D.: A Hybrid Genetic Algorithm for Highly Constrained Timetabling Problems. In: Eshelman, L.J. (Ed.): Proc. 6th Int. Conf. Genetic Algorithms (Pittsburg, PA). Morgan Kaufmann, San Mateo, CA (1995) 605–610

Solving Dynamic Resource Constraint Project Scheduling Problems Using New Constraint Programming Tools

Abdallah Elkhyari[1], Christelle Guéret[1,2], and Narendra Jussien[1]

[1] École des Mines de Nantes, BP 20722
F-44307 Nantes Cedex 3, France
{aelkhyar,gueret,jussien}@emn.fr
[2] IRCCyN,
Institut de Recherche en Communications et
Cybernétique de Nantes, France

Abstract. Timetabling problems have been much studied over the last decade. Due to the complexity and the variety of such problems, most work concerns static problems in which activities to schedule and resources are known in advance, and constraints are fixed. However, every timetabling problem is subject to unexpected events (for example, for university timetabling problems, a missing teacher, or a slide projector breakdown). In such a situation, one has to quickly build a new solution which takes these events into account and which is preferably not too different from the current one. We introduce in this paper constraint-programming-based tools for solving dynamic timetabling problems modelled as Resource-Constrained Project Scheduling Problems. This approach uses explanation-based constraint programming and operational research techniques.

1 Introduction

Timetabling problems have been much studied over the last decade. Due to the complexity and the variety of such problems, most work concern static problems in which activities to schedule and resources are known in advance, and constraints are fixed. However, every timetabling problem is subject to unexpected events (for example, for university timetabling problems, a missing teacher, or a slide projector breakdown). In such a situation, one has to quickly build a new solution which takes these events into account and which is preferably not too different from the current one.

In this paper, we present an exact approach for solving dynamic timetabling problems which uses explanation-based constraint programming and operational research techniques. In this approach, timetabling problems are modelled as Resource-Constrained Project Scheduling Problems (RCPSPs).

The paper is organized as follows. Section 2 introduces timetabling problems and RCPSPs, and explains how timetabling problems can be modelled as

E. Burke and P. De Causmaecker (Eds.): PATAT 2002, LNCS 2740, pp. 39–59, 2003.

RCPSPs. Section 3 presents the basics of explanation-based constraint programming. Our approach is presented in Section 4. Dynamic events that can be taken into account in our system are listed in Section 5 and computational results are reported in Section 6.

2   Timetabling Problems and RCPSPs

2.1   Timetabling Problems

Timetabling problems can be defined as the scheduling of a certain number of activities (lectures, tasks, etc.) which involve specific groups of people (students, teacher, employees, etc.) over a finite period of time, requiring certain resources (rooms, materials, etc.) in conformity with the availability of resources and fulfilling certain other requirements. A huge variety of timetabling problems exist: school timetabling, examination timetabling, employee timetabling, university timetabling, etc.

Due to the complexity and the variety of such problems, most work concerns static problems in which both activities to schedule and resources are known in advance, and constraints are fixed. The different techniques used are graph colouring [8], integer programming [32], genetic algorithms [28], tabu search [17], etc. Several different authors have presented *constraint programming* techniques for solving timetabling problems [4,14,15,23,26]. To our knowledge, no work concern the resolution of dynamic timetabling problems.

2.2   The RCPSP

The RCPSP can be defined as follows: let $A = \{1, 2, \ldots, n\}$ be a set of activities, and $R = \{1, \ldots, r\}$ a set of renewable resources. Each resource k is available in a constant amount R_k. A resource k is called disjunctive if $R_k = 1$. Otherwise it is called cumulative. Each activity i has a duration p_i and requires a constant amount a_{ik} of the resource k during its execution. Pre-emption is not allowed. Activities are related by precedence constraints, and resource constraints require that for each period of time and for each resource, the total demand of resource does not exceed the resource capacity. The objectives considered here are to find a feasible solution or a solution for which the end of the schedule is minimized (see for example Figure 1). The problem is *NP-hard* [3].

Let S_i be the starting time of activity i. Several extensions of the RCPSP can be considered:

- Generalized precedence constraints: such a constraint imposes that an activity j must be executed after another activity i and that there are exactly d units of time between the end of i and the starting of j, or that there are at least d_{min} and at most d_{max} units of time between them. These constraints will be denoted $i \rightarrow^d j$ and $i \rightarrow^{d_{max}}_{d_{min}} j$ respectively.

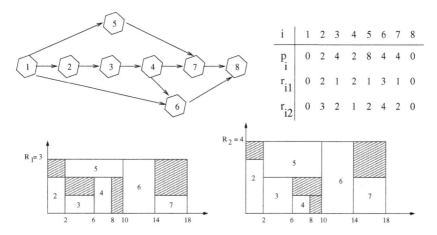

Fig. 1. One project with eight activities and two resources, and a feasible solution

- Generalized disjunctive constraints: such a constraint imposes that i and j are in disjunction and that there are exactly d, or at least d_{min} and at most d_{max} units of time between them. These constraints will be noted $i \leftrightarrow^d j$ and $i \leftrightarrow^{d_{max}}_{d_{min}} j$ respectively.
- Generalized overlapping constraints: two activities may have to overlap (be executed during at least a common unit of time) during exactly d, or at least d_{min} and at most d_{max} units of time. These constraints will be noted $i \|^d j$ and $i \|^{d_{max}}_{d_{min}} j$ respectively.
- Generalized resource constraints: the capacity of resources may change during certain periods of time.

The static RCPSP has been extensively studied [5,25]. Currently, the most competitive exact algorithms for the RCPSP are the ones of Brucker et al. [6], Demeulemeester and Herroelen [12], Mingozzi et al. [27] and Sprecher [31]. One of the main difficulty in the RCPSP is to maintain the resource limitation during the planning period. The classical deduction rules used for that purpose are *core times* [24] and *task interval* [9,10].

The dynamic RCPSP is seldom studied. Two classical methods are used to solve it:

- Recomputing a new schedule each time an event occurs: this is quite time consuming and may lead to a solution very different from the previous one.
- Constructing a partial schedule and completing it progressively as time goes by (like in online scheduling problems) [30]: this is unacceptable in the context of timetabling problems since the complete timetable must be known in advance.

Recently, Artigues et al. [1] introduced a formulation of the RCPSP based on a flow network model. The authors developed a polynomial algorithm based on this model to insert an unexpected activity.

2.3   Timetabling Problems as RCPSPs

Timetabling problems can be thought of as special cases of RCPSPs. In [7], Brucker and Knust show that the high school timetabling can easily be formulated as a RCPSP. Dignum et al. [13] translated the particular timetabling problem of an educational establishment of Maastricht into a Time- and Resource-Constrained Scheduling Problem, i.e. a RCPSP in which the objective is to determine a schedule which is completed on time and such that the total additional costs are minimized.

Let us consider the following example of a French university timetable [15]: m classes have to follow a set of lectures given by n teachers over an horizon of T time periods. All lectures are of the same length. Each lecture may require specific material (slide-projector, specific room, etc.).

The constraints are the following:

- C1: a class cannot follow more than one lecture at a time,
- C2: a teacher cannot give more than one lecture at a time,
- C3: some teachers or classes are not available in some periods,
- C4: material availability must be respected,
- C5: some lectures have to be scheduled at the same time (shared gymnastic rooms, class divided in several groups for foreign language lectures, etc.),
- C6: two lectures concerning the same subject should not be scheduled on too close periods,
- C7: some lectures are linked by precedence constraints (prerequisite, etc.).

Clearly, this example can be modelled as a RCPSP in which lectures are the activities to schedule and resources are teachers, classes and materials. Some of these resources are disjunctive (teachers, for example), others are cumulative (if several slide-projectors are available, for example). These lectures are subject to (generalized) precedence constraints (C7), (generalized) overlapping constraints (C5) and (generalized) disjunctive constraints (C1, C2 and C6). Resource constraints must be respected (C4), and some resource capacities are time dependent (C3).

3   Explanation-Based Constraint Programming

Constraint programming techniques have been widely used to solve scheduling problems. A constraint satisfaction problem (CSP) consists in a set V of variables defined by a corresponding set of possible values (the domains D) and a set C of constraints. A solution for the network is an assignment of a value to each variable such that all the constraints are satisfied. An extension of classical constraint programming has been recently introduced. It is called explanation-based constraint programming (*e-constraints*) and it has already proved to be of interest in many applications [18] including dynamic constraint solving. We recall in this section what it is and how it can be used.

In the following, we consider a CSP (V, D, C). Decision making during the enumeration phase (variable assignments) amounts to add (e.g. upon decision making) or remove (e.g. upon backtracking) constraints from the current constraint system. Enumeration is therefore considered as a dynamic process.

3.1 Explanations

A *contradiction explanation* (also known as *nogood* [29]) is a subset of the current constraints system of the problem that, left alone, leads to a contradiction. Thus, no feasible solution contains a *nogood*. A contradiction explanation is composed of two parts: a subset of the original set of constraints ($C' \subset C$) and a subset of decision constraints introduced so far in the search:

$$C \vdash \neg (C' \wedge v_1 = a_1 \wedge \cdots \wedge v_k = a_k) . \tag{1}$$

In a contradiction explanation composed of at least one decision constraint, a variable v_j is selected and the previous formula is rewritten as[1]

$$C \vdash C' \wedge \bigwedge_{i \in [1...k] \setminus j} (v_i = a_i) \rightarrow v_j \neq a_j . \tag{2}$$

The left-hand side of the implication constitutes an *eliminating explanation* for the removal of value a_j from the domain of variable v_j and is denoted $\mathtt{expl}(v_j \neq a_j)$.

Classical CSP solvers use domain-reduction techniques (removal of values). Recording eliminating explanations is sufficient to compute contradiction explanations. Indeed, a contradiction is identified when the domain of a variable v_j is emptied. A contradiction explanation can easily be computed with the eliminating explanations associated with each removed value:

$$C \vdash \neg \left(\bigwedge_{a \in d(v_j)} \mathtt{expl}(v_j \neq a) \right) . \tag{3}$$

There exist generally several eliminating explanations for the removal of a given value. Recording all of them leads to an exponential space complexity. Another technique relies on *forgetting* (erasing) eliminating explanations that are no longer relevant[2] to the current variable assignment. By doing so, the space complexity remains polynomial. We keep only *one* explanation at a time for a value removal.

3.2 Computing Explanations

During propagation, constraints are awoken (like agents or daemons) each time a variable domain is reduced (this is an event) possibly generating new events

---

[1] A contradiction explanation that does not contain such a constraint denotes an over-constrained problem.

[2] A *nogood* is said to be relevant if all the decision constraints in it are still valid in the current search state.

(value removals). A constraint is fully characterized by its behaviour regarding the basic events such as value removal from the domain of the variables and domain bound updates. Explanations for events are computed when the events are generated.

Explanations for basic constraints. It is easy to provide explanations for basic constraints. The following example shows how to compute them.

Example 1. Let us consider a two-variable toy problem: x and y with the same set of possible values $[1, 2, 3]$. Let us state the constraint $x > y$. The resulting sets of possible values are $[2, 3]$ for x and $[1, 2]$ for y. An explanation for this situation is the constraint $x > y$. Now, let us suppose that we choose to add the constraint $x = 2$. The only resulting possible value for x is 2. The explanation of the modification is the constraint $x = 2$. The other consequence is that the remaining value for y is 1. The explanation for this situation is twofold: a direct consequence of the constraint $x > y$ and also an indirect consequence of constraint $x = 2$.

Explanations for global constraints. Computing a precise explanation for global constraints may not be easy because it is necessary to study the algorithms used for propagation. However, there always exists a generic explanation: the current state of the domains of each variable of the constraints. In Section 4, we will describe how to provide more precise explanations for timetabling-related constraints.

3.3   Using Explanations

Explanations are useful in many situations [18]. The following sections detail some of them: providing user information, improving search strategies and handling dynamic problems.

Providing User Information

- Explanations can be scanned to determine past effects of selected constraints: these are the events for which the associated explanation contains one of the selected constraints;
- Considering the union of the explanations of the currently removed values in the current solution is a justification of that situation;
- When encountering a contradiction, a contradiction explanation will provide a subset of the constraint system that justifies the contradiction and that can be provided to the user;
- etc.

Improving search strategies. Explanations can also be used to efficiently guide searches. Indeed, classical backtracking-based searches only proceed by backtracking to the last choice point when encountering failures. Explanations can be used to improve standard backtracking and to exploit information gathered to improve the search: to provide intelligent backtracking [16], to replace standard backtracking with a jump-based approach à la *dynamic backtracking* [20], or even to develop new local searches on partial instantiations [21].

The common idea of these techniques is, upon encountering a contradiction, to determine an explanation from which a constraint will be selected either to determine a relevant backtracking point (for intelligent backtracking) or to only be dynamically removed and replaced with its negation (as in dynamic backtracking).

Dynamically adding/removing constraints. We can use the explanations for adding or removing constraints because they help in pointing out past effects of constraints that can be effortlessly undone without a complete re-computation from scratch. Notice that adding a constraint to a problem is a well known issue but removing it is not so easy. To dynamically remove constraints [2,11] one needs to disconnect the constraint from the constraint network, set back values by undoing the past events (which are easily accessible thanks to the recorded explanations, see above) and re-propagate to get back to a consistent state.

4   Solving Dynamic Timetabling Problems

Using explanations provides efficient solving techniques for dynamic problems [19]. In this section, we describe a branch-and-bound algorithm used to solve RCPSPs and the addition of explanation capabilities into it in order to provide a dynamic timetabling problem solver.

4.1   The Principle

We have developed an environment for solving dynamic RCPSP and timetabling problems which is based upon

- A branch-and-bound algorithm for RCPSPs (inspired by [6]) within a constraint programming solver: in each node, deduction rules are applied in order to determine redundant information (constraint[3] propagation).
- An extensive use of *explanations*. Explanations are recorded during search and propagation (i.e. using the propagation rules – namely *core times* and *task interval*) which has been upgraded in order to provide a precise explanation for every deduction made (see Section 4.3). They are used to handle

---

[3] By *constraint* we mean each initial constraint of the problem (precedence and resource constraints), but also each decision taken by the branching scheme, and each deduction made (thanks to propagation rules) during the search as mentioned in Section 3.

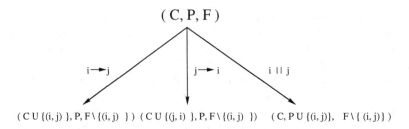

Fig. 2. Branching scheme

dynamic events. Indeed, an unexpected event leads to addition, modification or removal of a constraint in the system. In the first two cases, if the current solution is no longer valid, then the explanations tell us which are the constraints responsible for the contradiction. Repairing is done by removing at least one constraint (preferably a search decision) from the explanation of the contradiction, and by adding its negation. The resulting solution is generally quite similar to the previous one, and is found faster than if we have had solved the problem from scratch.

4.2   Branch-And-Bound Algorithm

The branch-and-bound algorithm used in our approach is inspired by Brucker et al. [6].

The branching scheme. Each node of the tree search is defined by three disjoint sets: a *conjunction* set, a *parallel* set and a *flexible* set. The conjunction set C contains all pairs of activities (i, j) in conjunction, i.e. that satisfy one of the relations $i \to j$ or $j \to i$). The parallel set P consists in all pairs of activities (i, j) that overlap (i.e. that satisfy the relations $i \parallel j$), and finally the flexible set F contains all the remaining pairs of activities (i, j).

Branching is done by transferring one pair of activities (i, j) from F either to set C by imposing the relation $i \to j$ or $j \to i$, or to set P (see Figure 2). This branching is repeated until set F becomes empty.

In each node, constraint propagation updates sets C and P by removing pairs from set F when constraints are added. Repairs (or backtracks) put back pairs from C and P to F.

Simple expression of relations: notion of distance. In order to implement our approach, it is necessary to be able to easily express both a decision taken in the search tree and its negation. For example, if we consider a possible decision $i \to j$, its negation is the disjunction $j \to i \vee i \parallel j$. Disjunctions are not easily posted or handled when performing a search in constraint programming. To overcome that problem, we introduced the notion of *distance*[4].

---

[4] This notion is a generalization of the one introduced by Brucker et al. [6].

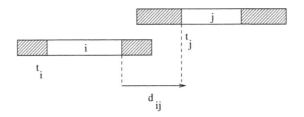

Fig. 3. Distance between two activities

Fig. 4. Positions of two activities i and j according to the value of d_{ij}

Definition 1. *Let i and j be two distinct activities. The distance d_{ij} between i and j is defined as the time between the ending date of i and the starting date of j. We have $d_{ij} = t_j - t_i - p_i$ (see Figure 3).*

The relative positions of activities i and j can easily be deduced according to the value of the distance d_{ij} (see Figure 4):

- $d_{ij} \geq 0$ iff $i \to j$,
- $d_{ij} \leq -p_i - p_j$ iff $j \to i$,
- $-p_i - p_j < d_{ij} < 0$ iff i and j overlap.

We can deduce that

- $d_{ij} \times d_{ji} \leq 0$ iff i and j are in disjunction,
- $d_{ij} \times d_{ji} > 0$ iff i and j overlap,
- d_{ij} and d_{ji} cannot both take a strictly positive value at the same time.

Using this notion of distance, the decisions taken in the search tree and their negations can easily be translated in terms of mathematical constraints (see Table 1). Furthermore, as we will see in Section 5.1, generalized temporal constraints can also easily be translated using distances.

The next two sections describe how explanations are added in the constraints that are used to enforce decisions made during search and also in the initial constraints of the problem (both temporal and resource-related).

4.3   Adding Explanations to Temporal Constraints

Providing explanations for temporal binary constraints is straightforward. Therefore, here are the explanations for basic temporal constraints (notice that explanations for generalized temporal constraints can easily be deduced):

Table 1. Decisions as constraints on distances

Decision	Constraint	Opposite decision	Opposite constraint
$i \to j$	$d_{ij} \geq 0$	$i \nrightarrow j$	$d_{ij} < 0$
$j \to i$	$d_{ij} \leq -p_i - p_j$	$j \nrightarrow i$	$d_{ij} > -p_i - p_j$
$i \parallel j$	$d_{ij} \times d_{ji} > 0$	$i \leftrightarrow j$	$d_{ij} \times d_{ji} \leq 0$

- $d_{ij} \geq 0$ (resp. $d_{ij} > -p_i - p_j$). The lower bound of the variable d_{ij} is updated to 0 (resp. $-p_i - p_j + 1$). This modification is only due to the constraint itself, and hence the explanation of the modification is the constraint itself.
- $d_{ij} < 0$ (resp. $d_{ij} \leq -p_i - p_j$). The upper bound of the variable d_{ij} is updated to -1 (or $-p_i - p_j - 1$). This modification is only due to the constraint itself, and hence the explanation of this modification is the constraint itself.
- $d_{ij} \times d_{ji} > 0$. If the upper bound of the variable d_{ij} (resp. d_{ji}) becomes strictly negative then the upper bound of the variable d_{ji} (resp. d_{ij}) is updated to -1 (we manipulate here integer variables). This modification is due first to the use of the constraint itself and second to the previous modification of the upper bound of variable d_{ij} (resp. d_{ji}), and hence the explanation of this modification is twofold: the explanation of the previous modification for the upper bound of the variable d_{ij} (resp. d_{ji}) and the constraint itself.
- $d_{ij} \times d_{ji} \leq 0$. If the lower bound of the variable d_{ij} (resp. d_{ji}) becomes positive then the upper bound of the variable d_{ji} (resp. d_{ij}) is updated to 0. This modification is due to the use of the constraint itself and the previous modification of the lower bound of the variable d_{ij} (resp. d_{ji}). The explanation of this modification is twofold: the explanation of the previous modification for the lower bound of the variable d_{ij} (resp. d_{ji}) and the constraint itself.

4.4   Adding Explanations to Resource Constraints

Explanations for resource management constraints are not that easy. It is necessary to study the algorithms used for propagation. Classical techniques for maintaining resource limitations for scheduling problems are *core times* [24], *task interval* and *resource histogram* [9,10].

Resource histogram constraints. The principle of the resource histogram technique is to associate to each resource k an array `level(k)` in order to keep a timetable of the resource requirements. This histogram is used for detecting a contradiction and reducing the time windows of activities.

The core time technique [24] is used for detecting contradictions. A core time $\mathtt{CT}(i)$ is associated with each activity i. It is defined as the interval of time during which a portion of an activity is always executed whether it starts at its earliest or latest starting time. A lower bound of the schedule is obtained when considering only the core time of each activity.

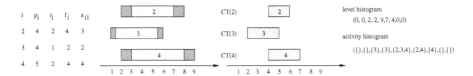

i	p_i	r_i	f_i	a_{i1}
2	4	2	4	3
3	4	1	2	2
4	5	2	4	4

Fig. 5. Example of timetable

Combining these two techniques provides an efficient resource-conflict detecting constraint. A timetable for each resource is computed as follows: for each activity i, its core time $\texttt{CT}(i) = [f_i, r_i + p_i)$ (f_i is the latest starting time of i, r_i its earliest starting time) is computed and the amount a_{ik} of resource k for each time interval $[f_i, f_i + 1), \ldots, [r_i + p_i - 1, r_i + p_i)$ is reserved. We associate with each timetable constraint two histograms (see Figure 5):

- a *level histogram*, which contains the amount of resource required at each time interval $[t - 1, t)$;
- an *activity histogram*, which contains the sets S_t of activities which require any amount of resource for each time period $[t - 1, t)$. It will essentially be used to provide explanations.

These histograms are used in the following ways:

- Detecting resource conflicts when the required level of a resource k at one time period t exceeds the resource capacity at that time.[5] The conflict set associated with this contradictory situation is constituted from the set of activities (S_t) stored in slot t of the activity histogram. Let t_v be the variable representing the starting time of activity v. The explanation of this conflict is given by the following equation (c being the histogram constraint itself):

$$\left(\bigwedge_{v \in S_t} \left(\bigwedge_{a \in d(t_v)} \texttt{expl}(t_v \neq a) \right) \right) \bigwedge c . \qquad (4)$$

- Tightening the time window of an activity. It may occur that the current bounds of the time window of an activity are not compatible with the other activities. In that situation, the tightening of the time window will be explained by the set S_t of activities requiring the resources during the incompatible time period $[t - 1, t)$. The following equation is used to provide an explanation for the modification of a time window (c being the histogram constraint itself):

$$\left(\bigwedge_{v \in S_t} \left(\bigwedge_{a \in d(t_v)} \texttt{expl}(t_v \neq a) \right) \right) \bigwedge c . \qquad (5)$$

---

[5] It may not be a constant value when considering resources whose capacity is variable during time.

Task interval constraints. The technique of *task intervals* [9,10] used for managing cumulative resources can detect conflicts, deduce precedences and tighten time windows.

A task interval $T = [i, j]$ is associated with each pair of activities (i, j) which require the same resource k. It is defined as the set of activities ℓ which share the same resource and such that $r_i \leq r_\ell$ and $d_\ell \leq d_j$ (r_i being the earliest starting time of the activity i and d_i its due date), i.e. that need to be scheduled between tasks i and j. The set $inside(TI)$ represents the set of activities constituting the task interval, while $outside(TI)$ contains the remaining activities. Let $energy(TI) = \sum_{\ell \in inside(TI)} (d_\ell - r_\ell) \times a_{\ell k}$ be the energy required by a task interval.

Several propagations rules can be defined upon *task interval*. We consider here the following two.

Integrity rule. If $energy(TI)$ is greater than the total energy $(d_j - r_i) \times R_k$ available in the interval then a conflict is detected. The conflict explanation set is built from the set of activities *inside* the task interval, i.e. (c being the constraint associated with this rule)

$$\left(\bigwedge_{v \in inside(TI)} \left(\bigwedge_{a \in d(t_v)} \texttt{expl}(t_v \neq a) \right) \right) \bigwedge c. \qquad (6)$$

Throw rule. This rule consists in tightening the time window of an activity intersecting a given task interval TI. This is done by comparing the necessary energy shared by the activities in $inside(TI)$ and another activity u intersecting TI, and the energy available during the *task interval*. The explanation of this update is built from the set of activities *inside* the task interval (c being the constraint associated with this rule):

$$\left(\bigwedge_{v \in inside(TI)} \left(\bigwedge_{a \in d(t_v)} \texttt{expl}(t_v \neq a) \right) \right) \bigwedge c \bigwedge \left(\bigwedge_{a \in d(t_u)} \texttt{expl}(t_u \neq a) \right). \qquad (7)$$

5   Dynamic Events Taken into Account

This section lists the various dynamic events taken into account in our system.[6]

5.1   Temporal Events

Adding constraints. Adding a new (generalized) precedence, disjunctive or overlapping constraint between two lectures can be done thanks to the different following procedures adding temporal events in our system:

---

[6] Notice that events that can be modelled as simple arithmetic constraints (such as Teacher A cannot give his/her lecture on March 24th) are automatically handled by the underlying constraint solver we use. Only specific complex constraints introduced for solving RCPSPs and timetabling problems are presented here.

Table 2. Constraints added by the procedure add(i, j, Relation) for each possible relation between i and j

Relation	Constraint
i → j	$d_{ij} \geq 0$
j → i	$d_{ij} \leq -p_i - p_j$
i ↔ j	$d_{ij} \times d_{ji} \leq 0$
i ∥ j	$d_{ij} \times d_{ji} > 0$

Table 3. Constraints added by the procedure add(i, j, Relation, d) for each possible relation between i and j

Relation	Constraint
i → j	$d_{ij} = d$
j → i	$d_{ij} = -d - p_i - p_j$
i ↔ j	$d_{ij} = d \vee d_{ji} = d$
i ∥ j	$d_{ij} = -d$

Table 4. Constraints added by the procedure add(i, j, Relation, Dmin, Dmax) for each possible relation between i and j

Relation	Constraint
i → j	$D_{min} \leq d_{ij} \leq D_{max}$
j → i	$-D_{max} - p_i - p_j \leq d_{ij} \leq -D_{min} - p_i - p_j$
i ↔ j	$D_{min} \leq d_{ij} \leq D_{max} \vee -D_{max} - p_i - p_j \leq d_{ij} \leq -D_{min} - p_i - p_j$
i ∥ j	$-D_{max} \leq d_{ij} \leq -D_{min}$

- add(i, j, Relation): imposes the relationship "Relation" between activities i and j. Table 2 gives the constraint added for each possible relationship.
- add(i, j, Relation, d): imposes the relationship "Relation" between activities i and j and that there are exactly d units of time between i and j. Table 3 gives the constraint added for each possible relationship.
- add(i, j, Relation, Dmin, Dmax): imposes the relationship "Relation" between activities i and j and that there are at least D_{min} and at most D_{max} units of time between i and j. Table 4 gives the constraint added for each relationship.
- add(i, Dmin, Dmax): imposes a time window $[D_{min}, D_{max}]$ for activity i. The constraint added is $D_{min} \leq i \leq D_{max}$.

Removing constraints. Our system is able to dynamically remove all these temporal events thanks to the following procedures: remove(i, j, Relation), remove(i, j, Relation, d), remove(i, j, Relation, Dmin, Dmax) and remove(i, Dmin, Dmax).

Modifying constraints. Each of the temporal events that can be added or removed can also be replaced by another one:

- modify(i, j, Old_Relation, New_Relation) replaces relation "Old_Relation" by "New_Relation" between i and j. Actually, this procedure removes the constraint associated to the relationship "Old_Relation" and adds the constraint associated to the relationship "New_Relation".

5.2   Activity-Related Events

The procedures concerning the activity related events are

- add(i, Duration, Res_requirements, Predecessors, Successors): this procedure adds an activity i with duration "Duration", resources requirements "Res_requirements". Its predecessors are "Predecessors" and its successors "Successors". This procedure creates a new variable i, connects it to the constraints network, adds the temporal constraints associated and the disjunction constraints resulting from the capacity limitations, and inserts variable i in the *timetable* and in the *task interval* constraints.
- remove(i): this procedure removes activity i from the problem. It disconnects variable i from the constraints network, removes all the constraints related to either i, any d_{ij} or any d_{ji}. It also removes variable i from the *timetable* and the *task interval* constraints, and removes all *task interval* constraints having i as starting or ending activity.

Activity modification is handled as an activity removal followed by an activity addition (the modified one).

5.3   Resource-Related Events

It is possible to add or remove a resource:

- add(Resource, Capacity, Capacities requirements) adds a new resource to the problem and the disjunctive constraints which result from its capacity limitation. It also adds the *timetable* and *task interval* constraints associated to it.
- remove(Resource) removes a resource from the problem, as well as the disjunctive constraints which results from the capacity limitation of this resource, the *timetable* and all the *task interval* constraints associated.

As for activities, modifying a resource is handled as a resource removal followed by a resource addition.

5.4   Translating Timetabling-Related Events into Our Event System

Obviously, the events described so far are not meant to be directly used by a final user of our system. A translation between the real-life events and our events needs to be done. For example, when considering timetabling problems, the following events could be handled:

- The addition of *new lectures for an existing course*, for example because a teacher needs more hours than expected, can be handled as the addition of new activities which may be related to other activities by (generalized) precedence, disjunctive or overlapping constraints.
- The suppression of one or several lectures of a teacher who finishes his/her course earlier than expected implies the removal of one or several activities.
- The installation of a new classroom equipped with computers, or the purchase of video-conference materials corresponds to the addition of a new resource or to the modification of the capacity of a resource.
- The fact that *a teacher is no longer available on the afternoons* can be handled as a modification (namely its capacity level) of a resource.
- etc.

5.5 The Special Case of Over-Constrained Problems

Uncontrolled addition, removal or modification of constraints can lead to an over-constrained problem. Our explanation-based system is then able to provide a contradiction explanation which will help the user either to select a constraint/relation to be removed or modified, or to allow an increase of the resulting makespan of the project.

6 Computational Experiments

We present here our first experimental results. Our experiment consists in comparing our dynamic scheduler with a static scheduler: a scheduling problem P is first optimally solved. Then, an event e is added to this problem. We then compare a re-execution from scratch (as a static scheduler would do) and the dynamic addition of the related constraints in terms of CPU time.

All experiments[7] were conducted on RCPSPs as our system was primarily designed for such problems. Experiments on real-life timetabling problems are planned in the near future.

6.1 First Benchmark: Patterson RCPSP Instances

We use here some RCPSP test problems introduced by Patterson[8].

For this set of problems, we tried to evaluate the impact of a single modification when comparing static and dynamic scheduling approaches. The following events were evaluated:

- Adding a precedence constraint: add(i, j, →). Table 5 presents the results obtained on original problems from Patterson from which a single precedence

---

[7] Experimental data and results are available at http://www.emn.fr/jussien/RCPSP
[8] www.bwl.uni-kiel.de/Prod/psplib/dataob.html

Table 5. Adding a precedence constraint

Name	#Act	#Res	t_p	t_e	t_{pe}	$\frac{t_{pe}-t_e}{t_{pe}}\%$
T1P1a	14	3	7.26	1.16	6.96	83.3
T1P1b	22	3	6.75	0.23	6.96	96.7
T1P2	7	3	0.23	0.02	0.19	89.5
T1P3a	13	3	25.82	4.16	12.58	67.0
T1P3b	13	3	15.64	1.21	12.58	90.4
T1P4a	22	3	8.01	1.46	3.63	59.8
T1P4b	22	3	4.50	0.50	3.63	86.2
T1P5a	22	3	20.37	8.70	13.80	**37.0**
T1P5b	22	3	11.34	1.76	13.80	87.3
T1P6a	22	3	135.11	19.60	78.61	75.1
T1P6b	22	3	36.70	7.81	78.61	90.1
T1P6c	22	3	87.80	0.96	78.61	**98.8**
T1P8	9	1	2.11	0.32	2.30	86.1
T1P10	8	2	0.34	0.06	0.25	76.0

constraint has been removed (which gives the starting problem) then added (which gives the final problem – the original Patterson one[9]).

- Adding a generalized precedence constraint: add(i, j, →, d). Table 6 presents the results obtained on original problems from Patterson to which a generalized precedence constraint is added (this constraint is an existing precedence in the optimal solution for which the existing time lag is increased).
- Adding a generalized precedence constraint: add(i, j, →, Dmin, Dmax). Table 7 presents the results obtained on original problems from Patterson to which a generalized precedence constraint is added (this constraint is an existing precedence in the optimal solution for which the existing time lag is increased).
- Adding an overlapping constraint: add(i, j, ‖). Table 8 presents the results obtained on original problems from Patterson to which a randomly chosen overlapping constraint is added.

In all the tests, activities i and j are randomly selected. In all the tables, we designate as follows:

- #Act: the number of activities of the problem;
- #Res: the number of resources of the problem;
- t_p: the CPU time in seconds needed to optimally solve problem P;
- t_e: the CPU time in seconds spent to solve this problem after dynamically adding the event e;

---

[9] This explains that some results obtained for the static solver are the same: the final problem is the same but not the original one as a different constraint has been removed.

Table 6. Adding a generalized precedence constraint

Name	#Act	#Res	t_p	t_e	t_{pe}	$\frac{t_{pe}-t_e}{t_{pe}}\%$
T2P1a	14	3	7.31	0.87	8.84	90.2
T2P1b	14	3	6.58	0.82	9.55	91.4
T2P5a	22	3	14.90	3.45	8.33	58.6
T2P5b	22	3	16.90	3.44	8.19	58.0
T2P5c	22	3	13.82	3.50	8.22	57.4
T2P6a	22	3	62.45	5.14	24.43	79.0
T2P6b	22	3	64.42	1.87	19.56	90.5
T2P6c	22	3	62.50	1.24	23.40	**94.7**
T2P8	9	1	2.19	0.22	2.58	91.5
T2P10	8	2	0.18	0.02	0.26	92.3
T2P11	8	2	0.39	0.20	0.42	**52.4**

Table 7. Adding a generalized precedence constraint

Name	#Act	#Res	t_p	t_e	t_{pe}	$\frac{t_{pe}-t_e}{t_{pe}}\%$
T3P1a	14	3	7.12	0.81	8.50	90.5
T3P1b	14	3	9.69	0.61	9.81	93.8
T3P3	13	3	17.49	0.45	14.34	**96.9**
T3P5a	22	3	14.46	4.82	7.59	36.5
T3P5b	22	3	15.25	3.10	7.81	60.3
T3P6a	22	3	68.38	9.63	19.30	50.1
T3P6b	22	3	75.88	2.52	21.12	88.1
T3P6c	22	3	66.00	17.74	27.34	**35.1**
T3P6d	22	3	65.45	16.25	30.65	47.0
T3P8	9	1	2.90	0.12	2.84	95.8

Table 8. Adding an overlapping constraint

Name	#Act	#Res	t_p	t_e	t_{pe}	$\frac{t_{pe}-t_e}{t_{pe}}\%$
T4P1	14	3	7.43	0.38	10.62	**96.4**
T4P5	22	3	15.16	16.39	24.84	34.0
T4P6a	22	3	77.73	8.44	14.60	42.2
T4P6b	22	3	70.45	6.63	18.59	64.3
T4P6c	22	3	64.87	9.85	12.86	**23.4**
T4P8a	9	1	2.37	0.38	2.35	83.9
T4P8b	9	1	2.27	0.12	2.64	95.5

Table 9. Some Kolish, Sprecher and Drexl instances (four consecutive dynamic events): relative speed-up (in %)

12Act/4Res	Mod. 1	Mod. 2	Mod. 3	Mod. 4
T1KSD1	12.76	26.09	35.84	35.58
T1KSD2	46.24	54.92	62.68	63.13
T1KSD3	35.88	44.12	45.82	45.87
T1KSD5	32.01	67.90	75.15	75.65
T1KSD6	3.81	26.38	46.62	55.92
T1KSD8	22.13	10.64	21.08	22.21
T1KSD9	46.72	52.58	47.75	49.27
T1KSD10	−29.21	−0.35	15.74	30.27

- t_{pe}: the CPU time in seconds needed to obtain an optimal solution of the problem $P \cup \{e\}$ from scratch;
- $\frac{t_{pe} - t_e}{t_{pe}}$: the overall interest of using a dynamic scheduler compared to a static scheduler expressed as the percentage of improvement.

These results clearly show that using a dynamic scheduling solver is of great use compared to solving a series of static problems. In our first experiments, the improvement is never less than 23% and can even get to 98.8%!

6.2   Second Benchmark: Kolish, Sprecher and Drexl RCPSP Instances

Our second set of experiments performed on Kolish, Sprecher and Drexl RCPSP instances[10] considers four consecutive modifications (any kind of event)[11] to an original problem. As can be seen in Table 9, our results are quite promising. Even bad results (instance T1KSD10) get better in the long run. However, notice that some results (see Table 10 – results for five consecutive modifications) show that dynamic handling is not always a panacea and rescheduling from scratch can be very quick.

7   Conclusion

In this paper, we have presented the integration of explanations within scheduling-related global constraints and their interest for solving dynamic timetabling problems. We presented first experimental results of a system that we developed using those techniques. These results demonstrate that incremental constraint solving for scheduling problem is useful.

---

[10] www.wior.uni-karlsruhe.de/RCPSP/ProGen.html
[11] See http://www.emn.fr/jussien/RCPSP for the details of each instance.

Table 10. Other Kolish, Sprecher and Drexl instances (five consecutive dynamic events): relative speed-up (in %)

22Act/4Res	Mod. 1	Mod. 2	Mod. 3	Mod. 4	Mod. 5
T2KSD11	47.15	12.49	5.87	6.89	−1.98
T2KSD13a	34.81	43.55	43.04	49.89	50.98
T2KSD13b	39.19	−126.79	−174.22	−216.67	−146.42
T2KSD15	−47.62	−42.50	−41.00	−65.06	−88.20
T2KSD17a	−2.40	1.65	−5.90	−16.23	−43.06
T2KSD17b	3.58	13.11	−12.76	−10.51	−74.91
T2KSD18a	30.61	26.76	−52.76	−19.07	0.85
T2KSD18b	40.39	56.09	40.54	42.56	45.15

We are currently improving our system with user-interaction capabilities still using explanations following [22]. Moreover, our main goal now is to give our system to the timetabling service in our institution in order to evaluate it in a real-world situation.

References

1. Artigues, C., Roubellat, F.: A Polynomial Activity Insertion Algorithm in a Multi-resource Schedule with Cumulative Constraints and Multiple Modes. Eur. J. Oper. Res. **127** (2000) 179–198
2. Bessière, C.: Arc Consistency in Dynamic Constraint Satisfaction Problems. In: Proc. AAAI'91 (1991)
3. Blazewicz, J., Lenstra, J.K., Rinnoy Kan, A.H.G.: Scheduling Projects Subject to Resource Constraints: Classification and Complexity. Discr. Appl. Math. **5** (1983) 11–24
4. Boizumault, P., Delon, Y., Péridy, L.: Constraint Logic Programming for Examination Timetabling. J. Logic Programming: Appl. Logic Programming (Special Issue) **26** (1996) 217–233
5. Brucker, P., Drexl, A., Möring, R., Neumann, K., Pesch, E.: Resource-Constrained Project Scheduling: Notation, Classification, Models and Methods. Eur. J. Oper. Res. **112** (1999) 3–41
6. Brucker, P., Knust, S., Schoo, A., Thiele, O.: A Branch and Bound Algorithm for the Resource-Constrained Project Scheduling Problem. Eur. J. Oper. Res. **107** (1998) 272–288
7. Brucker, P., Knust, S.: Resource-Constrained Project Scheduling and Timetabling. In: Burke, E, Erben W. (Eds.): Practice and Theory of Automated Timetabling III (PATAT 2000, Konstanz, Germany, August, selected papers). Lecture Notes in Computer Science, Vol. 2079. Springer-Verlag, Berlin Heidelberg New York (2001) 277–293
8. Burke, E.K., Elliman, D.G., Weare, R.F.: A University Timetabling System Based on Graph Colouring and Constraint Manipulation. J. Res. Comput. in Ed. **27** (1994) 1–18

9. Caseau, Y., Laburthe, F.: Cumulative Scheduling with Task-Intervals. In: JIC-SLP'96: Joint Int. Conf. Symp. on Logic Programming (1996)
10. Caseau, Y., Laburthe, F.: Improving CLP Scheduling with Task Intervals. In: Van Hentenryck, P. (Ed.): Proc. 11th Int. Conf. Logic Program. (ICLP'94). MIT Press, Cambridge, MA (1994) 369–383
11. Debruyne, R., Ferrand, G., Jussien, N., Lesaint, W., Ouis, S., Tessier, A.: Correctness of Constraint Retraction Algorithms. In: FLAIRS'03: 16th Int. Florida Artif. Intell. Res. Soc. Conf. (St Augustine, FL, May) AAAI Press (2003)
12. Demeulemeester, E., Herroelen, W.: A Branch and Bound Procedure for the Multiple Resource-Constrained Project Scheduling Problem. Management Sci. **38** (1992) 1803–1818
13. Dignum, F.P.M., Nuijten, W.P.M., Janssen, L.M.A.: Solving a Time Tabling Problem by Constraint Satisfaction. Technical Report, Eindhoven University of Technology (1995)
14. Goltz, H.J.: Combined Automatic and Interactive Timetabling using Constraint Logic Programming. In: Proc. PATAT 2000 (Konstanz, Germany, August)
15. Guéret, C., Jussien, N., Boizumault, P., Prins, C.: Building University Timetables Using Constraint Logic Programming. In: Burke, E, Ross, P. (Eds.): Practice and Theory of Automated Timetabling I (PATAT 1995, Edinburgh, Aug/Sept, selected papers). Lecture Notes in Computer Science, Vol. 1153. Springer-Verlag, Berlin Heidelberg New York (1996) 130–145
16. Guéret, C., Jussien, N., Prins, C.: Using Intelligent Backtracking to Improve Branch and Bound Methods: an Application to Open-Shop Problems. Eur. J. Oper. Res. **127** (2000) 344–354
17. Hertz, A.: Tabu Search for Large Scale Timetabling Problems. Eur. J. Oper. Res. **54** (1991) 39–47
18. Jussien, N.: E-constraints: Explanation-Based Constraint Programming. In: Workshop on User-Interaction in Constraint Satisfaction (CP'01, Paphos, Cyprus, December) (2001)
19. Jussien, N., Boizumault, P.: Dynamic Backtracking with Constraint Propagation – Application to Static and Dynamic CSPs. In: Workshop on The Theory and Practice of Dynamic Constraint Satisfaction (CP'97, Schloss Hagenberg, Austria, November) (1997)
20. Jussien, N., Debruyne, R., Boizumault, P.: Maintaining Arc-Consistency within Dynamic Backtracking. In: Principles and Practice of Constraint Programming (CP 2000, Singapore, September). Lecture Notes in Computer Science, Vol. 1894. Springer-Verlag, Berlin Heidelberg New York (2000) 249–261
21. Jussien, N., Lhomme, O.: Local Search with Constraint Propagation and Conflict-Based Heuristics. Artif. Intell. **139** (2002) 21–45
22. Jussien, N., Ouis, S.: User-Friendly Explanations for Constraint Programming. In: ICLP'01 11th Workshop on Logic Programming Environments (WLPE'01, Paphos, Cyprus, December) (2001)
23. Kang, L., White, G.M.: A Logic Approach to the Resolution of Constraint in Timetabling. Eur. J. Oper. Res. **61** (1992) 306–317
24. Klein R., Scholl, A.: Computing Lower Bounds by Destructive Improvement: an Application to Resource-Constrained Project Scheduling Problem. Eur. J. Oper. Res. **112** (1999) 322–345
25. Kolisch, R., Hartmann, S.: Heuristic Algorithms for Solving the Resource-Constrained Project Scheduling Problem: Classification and Computational Analysis. In: Weglarz, J. (Ed.): Handbook on Recent Advances in Project Scheduling. Kluwer, Dordrecht (1998) 147–178

26. Lajos, G.: Complete University Modular Timetabling Using Constraint Logic Programming. In: Burke, E, Ross, P. (Eds.): Practice and Theory of Automated Timetabling I (PATAT 1995, Edinburgh, Aug/Sept, selected papers). Lecture Notes in Computer Science, Vol. 1153. Springer-Verlag, Berlin Heidelberg New York (1996) 148–161

27. Mingozzi, A., Maniezzo, V., Ricciardelli, S., Bianco, L.: An Exact Algorithm for Project Scheduling with Resource Constraints Based on a New Mathematical Formulation. Management Sci. **44** (1998) 714–729

28. Ross, P., Corne, D., Fang., H-L.: Improving Evolutionary Timetabling with Delta Evaluation and Directed Mutation. In: Parallel Problem Solving in Nature III. Lecture Notes in Computer Science, Vol. 866. Springer-Verlag, Berlin Heidelberg New York (1994) 565–566

29. Schiex, T., Verfaillie, G.: Nogood Recording for Static and Dynamic CSP. In: Proc. 5th IEEE Int. Conf. on Tools with Artif. Intell. IEEE, Boston, MA (1993) 48–55

30. Sgall, J.: On-line scheduling – a Survey. In: Fiat, A., Woeginger, G.J. (Eds.): On-Line Algorithms: The State of the Art. Lecture Notes in Computer Science, Vol. 1442. Springer-Verlag, Berlin Heidelberg New York (1998) 196–231

31. Stinson, J.P., David, E.W., Khamawala, B.M.: Multiple Resource-Constrained Scheduling using Branch and Bound. IIE Trans. **1** (1978) 252–259

32. Tripathy, A.: School Timetabling – a Case in Large Binary Integer Linear Programming. Management Sci. **30** (1984) 1473–1489

Sports Timetabling

Integer and Constraint Programming Approaches for Round-Robin Tournament Scheduling

Michael A. Trick

Graduate School of Industrial Administration,
Carnegie Mellon, Pittsburgh, PA 15213, USA
trick@cmu.edu

Abstract. Real sports scheduling problems are difficult to solve due to the variety of different constraints that might be imposed. Over the last decade, through the work of a number of researchers, it has become easier to solve round-robin tournament problems. These tournaments can then become building blocks for more complicated schedules. For example, we have worked extensively with Major League Baseball on creating "what-if" schedules for various league formats. Success in providing those schedules has depended on breaking the schedule into easily solvable pieces. Integer programming and constraint programming methods each have their places in this approach, depending on the constraints and objective function.

1 Introduction

There has been a lot of work on combinatorial optimization methods for creating sports schedules (see, for instance, [2,3,4,8,11,15,19,21,22,23]. Much of this work has revolved around creating single round-robin schedules, where every team plays every other team once, and double round-robin schedules, where every team plays every other team twice (generally once at its home venue and once at each opposing venue).

Round-robin scheduling is interesting in its own right. Some leagues have a schedule that is a single or double round-robin schedule. Examples of this include many US college basketball leagues and many European football (soccer) leagues. For such leagues, the scheduling problem is exactly a constrained round-robin scheduling problem, where the constraints are generated by team requirements, league rules, media needs, and so on.

For other leagues, the schedule is not a round-robin schedule but it can be divided into sections that are round-robins among subsets of teams. There are good reasons, both algorithmically and operationally, to make these divisions. By creating these sections, schedulers are able to use results about round-robin schedules to optimize the pieces. Operationally, such schedules are often appealing to the leagues since they offer an understandable structure and often are

E. Burke and P. De Causmaecker (Eds.): PATAT 2002, LNCS 2740, pp. 63–77, 2003.

perceived as being fairer than unstructured schedules. For instance, by scheduling four consecutive round-robins, each team will see each other in each quarter of the schedule; an unstructured schedule would not necessarily require that.

This work was originally motivated by looking at a large practical scheduling problem, that of Major League Baseball (MLB). Fully defining the MLB schedule is a daunting task, requiring the collation of more than 100 pages of team requirements and requests, along with an extensive set of league practices. The key insight into effectively scheduling MLB, however, was the recognition that the complicated schedule could generally be broken into various phases, where each phase consists of a round-robin schedule, sometimes among subsets of teams. These round-robin schedules often have additional constraints reflecting team requirements or effects from other phases of the schedule. By better understanding constrained round-robin scheduling, we are better able to schedule MLB.

2   Round-Robin Scheduling

We begin with the most fundamental problem in sports scheduling: designing an unconstrained round-robin schedule. In a round-robin schedule, there are an even number n teams each of whom plays each other team once over the course of the competition. We will work only with *compact* schedules: the number of slots for games equals $n - 1$, so every team plays one game in every slot. We refer to this as a Single Round-Robin (SRR) Tournament.

A related problem is the Bipartite SRR (BSRR) Tournament problem. Here, the teams are divided into two groups X and Y, each with $n/2$ teams. There are $n/2$ slots during which all teams in X need to play all teams in Y, but teams within X (and within Y) do not play each other.

Henz, Müller, and Thiel [12] (who we will refer to as HMT) have recently examined SRR carefully in the constraint programming context. We expand on their work by examining integer programming formulations and further exploring the strengths and weaknesses of the different approaches.

HMT point out that the SRR can be formulated with two major types of constraints. First, the games in every slot correspond to a one-factor (or matching) of the teams. Second, for any team i, its opponents across all of the slots must be exactly the set of teams except for i. We will call the first the *one-factor constraint* and the second the *all-different constraint*. Different formulations have different ways of encoding and enforcing these constraints.

3   Integer Programming Formulation

Our basic integer program for SRR begins with binary variables x_{ijt} which is 1 if teams i and j play each other in slot t, and is 0 otherwise. Since the order of i and j does not matter, we could either define this only for $i > j$ or set $x_{ijt} = x_{jit}$ for all i, j, t.

This leads to the formulation (in the OPL language [25])

```
int n=...;
range Teams [0..n-1];
range Slots [1..n-1];
range Binary 0..1;
var Binary plays[Teams,Teams,Slots];

solve {
   //No team plays itself
   forall (i in Teams, t in Slots) plays[i,i,t] = 0;

   //Every team plays one game per slot
   //One-factor constraints
   forall (ordered i,j in Teams, t in Slots)
      plays[i,j,t] = plays[j,i,t];
   forall (i in Teams, t in Slots)
      sum (j in Teams) plays[i,j,t] = 1;

   //Every team plays every other team
   //All-different constraint
   forall (i,j in Teams: i<>j) sum(t in Slots) plays[i,j,t] = 1;

};
```

We call the above the Base-IP Formulation. We can strengthen this formulation by a stronger modelling of the one-factor constraint. The polyhedral structure of the one-factor polytope is perhaps the most well-studied polytope in combinatorial optimization. Edmonds [9] showed that the polytope is defined by adding *odd-set* constraints. In this context, the odd set constraints are as follows. For a particular slot t, let S be a set of teams, $|S|$ odd. Then

$$\sum_{i \in S, j \notin S} x_{ijt} \geq 1$$

is valid for the one-factor constraint. If we add all of these constraints, then the one-factor constraint is precisely defined in the polyhedral sense (all extreme points of the polyhedron are integer).

We call the formulation with all of the odd-set constraints the Strong-IP formulation. Of course, there are too many odd-set constraints to simply add them all to the integer program. We can solve Strong-IP by using a constraint generation method. In this method, we begin with a limited set of odd-set constraints and solve the linear relaxation of the instance. We then identify violated odd-set constraints (this can be done with a method by Padberg and Rao [16] using cut-trees) and add them to the formulation. We repeat until either we have added "enough" constraints or until all odd-set constraints are satisfied. At that point, we then continue our normal branch and bound approach to integer programs.

Strong-IP is not needed for the BSRR problem. In this case, the odd set constraints are redundant, so need not be added.

4   Constraint Programming Formulation

HMT extensively analyse constraint programming formulations for SRR. Their basic variables are `opponent[i,t]` which gives the opponent i plays in slot t. Their formulation can be represented very simply:

```
int n=...;
range Teams [0..n-1];
range Slots [1..n-1];
var Teams opponent[Teams,Slots];

solve {
   //No team plays itself
   forall (i in Teams, t in Slots) opponent[i,t] <>i;

   //Every team plays one game per slot
   //One-factor constraint
   forall (t in Slots)
      one-factor(all (i in Teams) opponent[i,t]);

   //Every team plays every other team
   //All-different constraint
   forall (i in Teams)
      all-different(all (t in Slots) opponent[i,t]);

};
```

The key issue is to define how the one-factor and all-different constraints are implemented. There are a variety of propagation algorithms available for each. HMT argues convincingly that the all-different constraint should use arc-consistent propagation by the method of Régin [17]. Briefly, arc-consistency means that the domains of the variables are such that for any value in a domain, setting the variable to that value allows settings for all the other variables so that the constraint is satisfied. For more details on the fundamentals of constraint programming, see [14].

For the one-factor constraint, the main emphasis of HMT, they examine three different approaches. The first is the simplest. It uses the constraints

```
forall (i in Teams)
   opponent[opponent[i,t]] = i;
```

This set of constraints is sufficient to define the one-factor constraint, but its propagation properties are not particularly strong. In particular, the only domain reduction that is done is when i is in the domain for j for a particular time period t, but j is not in the domain of i for that time period. In that case, i can be removed for the domain of j.

The propagation properties of this constraint can be improved by adding the redundant constraint

```
all-different(all (i in Teams) opponent[i,t]);
```

HMT give an example where adding the all-different constraint leads to improved domain reduction. We call this combination the *all-different* one-factor approach.

The combination of these constraints do not create an arc-consistent propagation for the one-factor constraint. HMT provide an arc-consistent propagation method using results from non-bipartite matchings. In general, this approach is much better than the all-different approach. The exceptions mimic when the Strong-IP does not improve on Basic-IP: if the underlying graph is bipartite, then the all-different approach is arc-consistent.

To prove this, let D_i be the feasible opponents for i. We say the D_i are bipartite if we can divide the teams into X and Y such that

1. $|X| = |Y| = n/2$
2. $i \in X \rightarrow D_i \subseteq Y$
3. $i \in Y \rightarrow D_i \subseteq X$.

Theorem 1. *If the D_i are bipartite, then arc-consistency for the constraints*

```
forall (i in Teams)
    opponent[opponent[i]] = i;
all-different(opponent[i]);
```

implies arc-consistency for `one-factor(opponent)`.

Proof. Suppose the D_i are consistent for the constraints

```
forall (i in Teams)
    opponent[opponent[i]] = i;
all-different(opponent);
```

Let $j \in D_i$. We will show there is a one-factor that has j as the opponent for i. Without loss of generality, we will assume $j \in Y$, so $i \in X$. By consistency of the all-different constraint, there is a setting of opponent values so that `j=opponent[i]`, and `all-different[opponent]`. Create a new setting of the opponent values `opponent'` such that `opponent'[i]` = `opponent[i]` for $i \in X$ and `opponent'[i]` = `opponent[opponent[i]]` for $i \in Y$. By the consistency requirement, `opponent'[i]` $\in D_i$ for all i. `opponent'` is therefore a one-factor that has `j=opponent[i]` as required.   □

Therefore, for either BSRR or for cases where the home/away pattern for a time slot is fixed (creating a bipartition between teams that need to be home versus those that need to be away), one-factor propagation can be replaced by all-different propagation.

5   Computational Tests

If there were no further requirements on the schedules, creating a SRR schedule would be straightforward. Kirkman (1847) gave a method for creating such a schedule (outlined in [1]) which also shows there is an ordering of the decision variables such that a constraint program would not need to backtrack in assigning variables (provided arc-consistent approaches to the all-different constraint is used).

Most league schedules have a number of additional constraints, however. A few of the most common are

- *Fixed games.* A set of games are fixed to occur in certain slots.
- *Prohibited games.* A set of games are fixed not to occur in certain slots.
- *Home/Away restrictions.* Each team has a home venue, and each game must be assigned to a venue. There are additional constraints on such things as the permitted number of consecutive home or away games.

It might seem that adding constraints would make the problem easier, since it reduces the possible search space. In fact, adding fixed games (or, equivalently, prohibiting games) can make the relatively easy problem of finding a schedule become an NP-complete problem. This has been shown by Colbourn [5] for the bipartite SRR case, where a schedule is equivalent to a Latin Square, and by Easton for the general SRR case. So it is clear that there may be very difficult instances of these restricted problems.

There may also be an objective function to be optimized. For instance, there may be an estimate c_{ijt} for the number of people who would attend a game between i and j during time slot t. Can we maximize the total number of people who attend games during the tournament?

All the testing in this paper was done using ILOG's OPL Studio, version 3.5 ([13]) running under Windows XP on a 1.8 GHz Pentium IV processor computer with 512 MB of memory.

5.1   Tightly Constrained Round Robin Tournaments

For our first test, we use a data set from HMT, called "Tightly Constrained Round-Robin Tournaments". For these instances, there are random forbidden opponents, with a sufficient number to lead to instances with very few or no feasible schedules.

For this test, we compare two codes: Basic-IP and the all-different constraint program. Our codes were implemented within the OPL system, version 3.5 and used default branching strategies for the integer program and default search strategies for the constraint program.

In addition, we repeat the computational results of HMT for their arc-consistent one-factor method. To offset different machine capabilities, we divided their computation times by 4.5 to represent a rough approximation of the difference between their 400 MHz machine and our 1.8 GHz machine.

For each instance, we give the number of failures (F) in the search tree (for the constraint programs) or number of nodes (N) in the search tree (for the integer program) as well as the computation time in seconds. In the following the instances that end in "yes" are feasible, though generally with a small number of solutions; those that end in "no" are infeasible. For feasible problems, the

Table 1. Benchmarks on tightly constrained SRR

Problem	n	all-different		Basic-IP		HMT	
		F	Time	N	Time	F	Time
s_6_yes	6	5	0.00	0	0.01	4	0.01
s_8_yes	8	17	0.01	4	0.04	10	0.04
s_10_yes	10	4	0.02	1	0.09	1	0.02
s_12_yes	12	376	0.41	57	0.46	179	1.39
s_14_yes	14	862	1.24	276	5.54	527	4.53
s_6_no	6	3	0.00	0	0.01	4	0.00
s_8_no	8	11	0.01	10	0.04	6	0.01
s_10_no	10	23	0.02	4	0.10	6	0.03
s_12_no	12	24	0.07	0	0.14	25	0.17
s_14_no	14	135	0.23	50	1.02	69	0.56
s_16_no	16	79	0.30	0	0.39	86	1.19
s_18_no	18	43	0.32	0	0.42	30	0.50
s_20_no	20	696.30	5.47	0	0.78	254	5.11

timing for HMT includes work needed to find all solutions, while those for the IP and the all-different CP only find the first solution. This suggests that the "no" instances, where the codes perform the same task, is the fairer comparison.

Basic-IP is competitive with the constraint programming approaches, and can do markedly better in proving infeasibility (as in s_20_no).

This table might lead to the conclusion that these tightly constrained SRR problems are relatively easy to solve in this size range. That is not the case. The instance s_16_no is actually just one of a series of instances, corresponding to varying numbers of prohibited games. The instance begins with 1192 prohibited games. The timing above corresponds to prohibiting all but the final seven of the prohibited games. We can create new instances by varying the number of prohibited games. The instances remain infeasible through prohibiting all but the final 54 games, and which point the instance becomes feasible.

The computational effort for these instances varies tremendously for the Basic-IP and the all-different approaches. Basic-IP does poorly for a broad range of prohibitions (the good behaviour above is an anomaly). For instance, the computation time for the feasible instance that is created by prohibiting all but the final 80 games is more than 9000 s. The all-different constraint program never does that poorly but can still take more than 300 s for various instances. Clearly, many of these instances are difficult for our codes, even for these relatively small-sized instances.

5.2   Divisional Schedules

There is a natural set of restrictions that are extremely difficult for the constraint-based formulations but are solved much more quickly by the integer programs. Many leagues are divided into two or more divisions. In such

Table 2. Divisional schedules (30 min time limit)

Size	all-different		Basic-IP		Strong-IP	
	F	Time	N	Time	N	Time
10	116	0.20	393	0.19	0	0.20
14	–	–	–	–	0	0.34
18	–	–	–	–	0	0.32
22	–	–	–	–	0	0.38

cases, some leagues like to begin by playing games between divisions and finish the schedule with games between divisional opponents. For two equally sized divisions, this approach works fine if n is divisible by 4, but does not work well for cases where $n = 2$ mod 4. In that case, divisions have an odd number of teams, so no compact round-robin schedule is possible with only divisional play at the end.

We can create difficult instances by fixing a large number of games. Suppose there are n teams, with $n = 2$ mod 4, numbered 0 up to $n - 1$. Divide the teams into two groups $X = [0, \ldots, n/2 - 1]$ and $Y = [n/2, \ldots, n-1]$. For the first $n/2 - 1$ slots, play X and Y as a bipartite tournament, leaving one game between X and Y unplayed for each team. Then, in slot $n/2$ fix two of the remaining games between X and Y. For six teams, the schedule might be

```
                    Team
    Slot    0    1    2    3    4    5
    ---    ---  ---  ---  ---  ---  ---
    1       3    4    5    0    1    2
    2       4    5    3    2    0    1
    3       5    3         1         0
    4
    5
```

With these fixtures, the schedule is infeasible. In the above example, teams 2 and 4 need to play in slot 3, and then 0, 1, and 2 need to play a round-robin among themselves in the remaining two slots, which is impossible.

The size 6 example is easy for any approach; things get more interesting for larger problems, as shown in Table 2; the dashes mean that no proof of infeasibility was found in half an hour of computation time (1800 s).

For the Strong-IP, the only needed constraints are for the odd sets associated with each division in each time period. That is sufficient for the linear relaxation to be infeasible for every n, which gives the near-constant computation times. For the one-factor method of HMT, there is no immediate proof of infeasibility for $n \geq 10$ by domain reduction, so at least some branching is needed (and we believe the amount of work will be significant).

Table 3. Maximum value schedules (30 min time limit)

Size	all-different		Basic-IP	
	F	Time	N	Time
8	84 962	5.33	0	0.03
10	–	–	66	0.29
12	–	–	402	3.59
14	–	–	7263	133.03
8b	1 458	0.04	0	0.02
10b	3 832	0.36	0	0.04
12b	4800 172	216.75	0	0.09
14b	–	–	0	0.10

5.3   Maximum Value Schedules

As a final test for SRR, we randomly generated values for each game in each slot and tried to find the maximum value schedule. For an instance of size n, we independently generated a value uniformly among the integers $1, \ldots, n^2$ for each game (i, j, t). There were no other restrictions on the schedule. We also generated bipartite versions of these problems (denoted b in Table 3).

For these sorts of optimizations, the search strategy is critical for constraint programming. The strategy used was to set the variables slot by slot, beginning with the first slot. The opponents for each team are ordered in decreasing order by value, and high-value opponents are tried first.

The results are shown in Table 3. Not surprisingly, the integer programming based approach does much better at this test. In order for constraint programming to be competitive in this test, some sort of cost-based domain reduction would have to be done. Note that even the bipartite problems are difficult for constraint programming despite the arc-consistent propagation we do through the all-different constraint. HMT's **one-factor** constraint is no stronger in the bipartite case.

6   Home/Away Pattern Restrictions

The final set of constraints we would like to consider is extremely important in practice. For some leagues, every team has a home venue and every game is played at the home venue of one of the two teams competing. In this situation, there are often constraints on the home and away patterns of these teams. These constraints might include

- Restrictions that a particular team be home (or away) in a particular slot;
- Limitations on the number of consecutive home (or away) games a team may play;

- Requirements on the number of home games that must appear in some subset of the slots. For instance, a team might want to be at home at least half of the weekend games, or half of the games during the summer;
- Restrictions on pairs of slots. For instance, if a team begins with an away game, the final game of the tournament might be required to be a home game.

There has been much work on scheduling with home/away patterns. Much of this work has concentrated on multiple-phase approaches, where first the home/away pattern is fixed, and then the games are chosen consistent with this pattern (see, for example, [6,7,23,20,15,11]. Alternatively, some work has reversed the process where first the games are chosen and then the home/away pattern chosen [19,24]. While these approaches are often very successful, there are cases where they do not work very well. For instance, depending on the restrictions on home/away patterns, there can be a huge number of feasible patterns, and a correspondingly large number of basic match schedules [23]. Enumerating and searching through all of them can be a computationally prohibitive task.

Is it possible to have one model that contains both game assignment and home/away pattern decisions? Conceptually, the models are straightforward to formulate. We consider a *double round-robin* (DRR) tournament where every team plays every other team twice, once at home and once away.

For the integer program, we reinterpret the x_{ijt} variables to mean that i plays at j during slot t. We also create auxiliary variables h_{it} which is 1 if i is home in slot t and 0 otherwise. Clearly $h_{it} = \sum_j x_{jit}$.

For the constraint program, there are a number of possible formulations. For this test, in order to maximize the effect of the all-different constraint, we used the variables plays[i,j] to be the slot number in which i plays at j. This gives the basic formulation of

```
forall (i in Teams)
    all-different(all (j in Teams) plays[i,j],
                  all (j in Teams) plays[j,i]);
```

(For notational convenience, we actually created an n by $2n$ array of variables with plays[i,j] for j <=n giving the slot where i is at home to j and for j>n giving the slot where i is away to $j - n$, but will continue the exposition with the original variables).

Our home/away variables are given by

```
forall (i, j in Teams)
    home[i,plays[j,i]] = 1;
    home[i,plays[i,j]] = 0;
```

For both the integer and constraint programming approaches, some schedule requirements are easy to formulate with these variables. Fixing teams to be at home (or away) at a particular time is simply a matter of fixing the h (or home) variable to take on the appropriate value. Fixing the number of home games in a subset of slots is simply a linear constraint on the sum of the home variables. Putting an upper bound on the number of consecutive home games can be done as follows: if no more than k consecutive home games are permitted, then for every i and t, add a constraint

```
sum(t1 in Slots: t>=t & t<=t+k) home[i,t1] <= k
```

A similar restriction on consecutive away games can be introduced with the use of `1-home[i,t1]`.

OPL has a stronger way of handling these constraints: the `sequence` constraint allows for the explicit bounding of the number of times a value can appear in a subsequence of an array. We add to the constraint program the redundant constraints that half the teams must be at home in every slot (without them, the constraint program works very poorly).

Our first test is to simply determine whether our programs can find schedules in the absence of additional constraints. It was shown in HMT that constraint programs can find unrestricted schedules quickly for more than 20 teams (and more than 40 teams with their one-factor improvements). How does adding home and away requirements affect that?

In Table 4, we give the time to find one schedule with n teams and an upper bound of k consecutive home or away games. For $k = 1$, there is no feasible schedule, so the time given is the time to prove infeasibility. With just one exception, the constraint programming approach did much better, though the integer program was able to generate solutions in a reasonable amount of time.

The entry for the constraint program for $n = 12$, $k = 3$ is not a misprint: despite the ease at which the constraint program solved the other instances, the search went poorly in this case, and no solution was found within one-half hour. This suggests that either an improved search procedure is needed (we simply instantiated the `play` variable before the `home` variable team-by-team) or a stronger propagation algorithm is needed to ensure consistency in computation time. Still, it is clear that constraint programming is by far superior for this type of instance.

To move closer to the types of schedules needed in practice, we add a constraint that there cannot be any length-one home stands or road trips. This is done by adding constraints of the form

```
home[i,t] <= home[i,t-1]+home[i,t+1];
(1-home[i,t]) <= (1-home[i,t-1])+(1-home[i,t+1]);
```

with the obvious changes for the beginning and end of the schedule. This makes the $k = 2$ instances infeasible.

The results are shown in Figure 5. Clearly this approach to limiting the home/away pattern is not consistent with the rest of the constraint programming model: constraint programming is unable to find any feasible solutions within half an hour. The integer programs are slow, but do find solutions. Given the success the multiple-phase approaches have with instances like this, it is clear

Table 4. Length-constrained H/A schedules (30 min limit)

n	k	Integer program		Constraint program	
		N	Time	F	Time
8	1	4	1.04	40	0.05
8	2	7	1.41	6	0.05
8	3	4	1.04	21	0.04
8	4	0	0.56	4	0.02
10	1	6	8.82	40	0.01
10	2	4	5.92	199	0.24
10	3	1	2.87	462	0.44
10	4	6	3.72	1141	0.98
12	1	6	24.84	220	0.54
12	2	4	17.29	2	0.84
12	3	4	15.11	–	–
12	4	17	32.42	0	0.12
14	1	2	59.71	312	1.21
14	2	9	70.10	11	0.20
14	3	11	82.34	3	0.18
14	4	20	169.42	2	0.19
16	1	2	163.52	420	2.48
16	2	35	604.86	184	0.74
16	3	–	–	197	0.32
16	4	124	1557.02	1	0.03
18	1	2	669.14	544	4.82
18	2	28	892.64	227	1.16
18	3	–	–	9	0.04
18	4	–	–	1	0.05

Table 5. Length-constrained H/A schedules, no singles (30 min limit)

n	k	Integer program		Constraint program	
		N	Time	F	Time
8	2	1427	21.42	2166	0.99
8	3	513	60.12	–	–
8	4	750	83.86	–	–
10	2	115	111.09	42768	26.18
10	3	1354	921.6	–	–
10	4	2214	1290.38	–	–

that a smarter search rule (mimicking the multiple-phase approach) or a better propagation rule should have significant effect on the constraint program.

Finally, we added values for matchups on particular days, generating random values in the range $1, \ldots, n^2$ for each pairing in each slot. Unfortunately, neither code at this stage can solve even the $n = 8$, $k = 3$ instance with a no-singleton

Table 6. Length-constrained H/A schedules, maximum value (30 min limit)

		Integer program		Constraint program	
n	k	N	Time	F	Time
8	3	1 516	22.73	–	–
8	4	77	0.92	–	–
10	3	–	–	–	–
10	4	15 268	594.70	–	–

constraint within 30 min. The results in the table are without the no-singleton constraint.

One final set of requirements we have not yet included has to do with travel requirements on teams. For leagues like MLB where travel is a concern, it is important to minimize the travel distances of each team. Unfortunately, the direct formulation of this does not lead to solvable models. More complicated models involving different variables seems to be needed [10].

We can, within the models given, preclude terrible travel by including constraints that require trips from, say, the East Coast to the West Coast to include at least two West Coast teams before returning. Such constraints are similar to (for both formulation and computation) the constraints that preclude length-one homestands and roadtrips.

7 Conclusions

We have shown that round-robin schedules with constraints of practical interest can be modelled by both constraint and integer programming techniques. The constraint programs were often faster except when there was an objective function, or in certain infeasible cases where the propagation was not strong enough to recognize infeasibility.

Returning to the Major League Baseball example, once the problem has been divided into smaller pieces, it is clear that IP/CP approaches are reasonable methods to solve the sections. Computation times are low enough to allow for multiple iterations of each section. In these iterations, constraints and objectives can be modified to push the process towards a good overall schedule.

In the course of this study, a number of gaps in current knowledge have been identified, and these make interesting future research directions.

- Is it worth adding constraints via the Strong-IP formulation? Henz, Müller, and Thiel [12] show that stronger propagation is generally a good idea for constraint programs. Is it also true that stronger relaxations are good for these integer programs?
- How can costs be better handled for the constraint programs? Handling costs is an active issue in the constraint programming community, and round-robin scheduling makes a good test-bed for these approaches.

- Clearly it would be good to include the strong propagation of HMT to the home/away models. Is there stronger propagation available combining the opponents with the home/away structures? Are there additional constraints that can be added to the integer programs? Is there a better way of handling home/away models which does not require a multi-phase approach?
- Can the integer programming and constraint programming approaches be usefully combined for these problems?

Round-robin scheduling makes an interesting test-bed for exploring algorithmic issues in combinatorial optimization. Success in this scheduling also provides the building blocks for scheduling of real-world sports leagues.

References

1. Anderson, I.: Combinatorial Designs and Tournaments. Oxford University Press Oxford (1997)
2. Ball, B.C., Webster, D.B.: Optimal Scheduling for Even-Numbered Team Athletic Conferences. AIIE Trans. **9** (1977) 161–169
3. Cain, W.O., Jr.: A Computer Assisted Heuristic Approach Used to Schedule the Major League Baseball Clubs. In: Ladany, S.P., Machol, R.E. (Eds.): Optimal Strategies in Sports. North-Holland, Amsterdam (1977) 32–41
4. Campbell, R.T., Chen, D.-S.: A Minimum Distance Basketball Scheduling Problem. In: Machol, R.E., Ladany, S.P., Morrison, D.G. (Eds.): Management Science in Sports. North-Holland, Amsterdam (1976) 15–25
5. Colbourn, C.J.: Embedding Partial Steiner Triple Systems is NP-Complete. J. Combinat. Theory, Series A **35** (1983) 100–105
6. de Werra, D.: Geography, Games, and Graphs. Discr. Appl. Math. **2** (1980) 327–337
7. de Werra, D.: Some Models of Graphs for Scheduling Sports Competitions. Discr. Appl. Math. **21** (1988) 47–65
8. Easton, K.K.: 2002. Using Integer Programming and Constraint Programming to Solve Sports Scheduling Problems, Doctoral Dissertation, Georgia Institute of Technology.
9. Edmonds, J.: Maximum Matching and a Polyhedron with $(0, 1)$ Vertices. J. Res. Natl Bureau Standards **69B** (1965) 125–130
10. Easton, K.K., Nemhauser, G.L., Trick, M.A.: Solving the Travelling Tournament Problem: A Combined Integer Programming and Constraint Programming Approach. In: this volume
11. Henz, M.: Scheduling a Major College Basketball Conference: Revisited. Oper. Res. **49** (2001) 163–168
12. Henz, M., Müller, T., Thiel, S.: Global Constraints for Round Robin Tournament Scheduling. Eur. J. Oper. Res. (to appear)
13. ILOG OPL Studio, User's Manual and Program Guide. ILOG (2000)
14. Marriott, K. Stuckey, P.J.: Programming with Constraints: An Introduction. MIT Press Cambridge, MA (1998)
15. Nemhauser, G.L., Trick, M.A.: Scheduling a Major College Basketball Conference. Oper. Res. **46** (1998) 1–8
16. Padberg, M.W. Rao, M.R.: Odd Minimum Cut-Sets and b-matchings. Math. Oper. Res. **47** (1982) 67–80

17. Régin, J.-C.: A Filtering Algorithm for Constraints of Difference in CSPs. In: Proc. AAA 12th Natl Conf. Artif. Intell. (1994) 362–367
18. Régin, J.-C.: The Symmetric alldiff Constraint. In: Dean, T. (Ed.): Proc. Int. Joint Conf. Artif. Intell., Vol. 1 (1999) 420–425
19. Régin, J.-C.: Minimization of the Number of Breaks in Sports Scheduling Problems using Constraint Programming. DIMACS Workshop on Constraint Programming and Large Scale Discrete Optimization (1999)
20. Russell, R.A., Leung, J.M.: Devising a Cost Effective Schedule for a Baseball League. Oper. Res. **42** (1994) 614–625
21. Schaerf, A.: Scheduling Sport Tournaments using Constraint Logic Programming. Constraints **4** (1999) 43–65
22. Schreuder, J.A.M.: Constructing Timetables for Sport Competitions. Math. Program. Study **13** (1980) 58–67
23. Schreuder, J.A.M.: Combinatorial Aspects of Construction of Competition Dutch Professional Football Leagues. Discr. Appl. Math. **35** (1992) 301–312
24. Trick, M.A.: A Schedule-Then-Break Approach to Sports Timetabling. In: Burke, E, Erben W. (Eds.): Practice and Theory of Automated Timetabling III (PATAT 2000, Konstanz, Germany, August, selected papers). Lecture Notes in Computer Science, Vol. 2079. Springer-Verlag, Berlin Heidelberg New York (2001) 242–253
25. Van Hentenryck, P.: The OPL Optimization Programming Language. MIT Press, Cambridge, MA (1999)

Characterizing Feasible Pattern Sets with a Minimum Number of Breaks

Ryuhei Miyashiro[1], Hideya Iwasaki[2], and Tomomi Matsui[1]

[1] Department of Mathematical Informatics,
Graduate School of Information Science and Technology,
The University of Tokyo, Tokyo 113-8656, Japan
{miya,tomomi}@misojiro.t.u-tokyo.ac.jp
[2] Department of Computer Science,
The University of Electro-Communications, Tokyo 182-8585, Japan
iwasaki@cs.uec.ac.jp

Abstract. In sports timetabling, creating an appropriate timetable for a round-robin tournament with home–away assignment is a significant problem. To solve this problem, we need to construct home–away assignment that can be completed into a timetable; such assignment is called a feasible pattern set. Although finding feasible pattern sets is at the heart of many timetabling algorithms, good characterization of feasible pattern sets is not known yet. In this paper, we consider the feasibility of pattern sets, and propose a new necessary condition for feasible pattern sets. In the case of a pattern set with a minimum number of breaks, we prove a theorem leading a polynomial-time algorithm to check whether a given pattern set satisfies the necessary condition. Computational experiment shows that, when the number of teams is less than or equal to 26, the proposed condition characterizes feasible pattern sets with a minimum number of breaks.

1   Introduction

Constructing a timetable for a sports competition is an important task for the organizers of the competition because the timetable affects the results of games. Since creating an appropriate timetable by hand is difficult, demand for automated timetabling has been increasing.

Sports timetabling is the research region that concerns creating an optimal timetable for a sports competition, constructing timetabling algorithms, and investigating mathematical structure of timetabling problems in sports. Recently, a number of papers on sports timetabling have been published [1,2,3,4,5,6,7, 8,10,11,12,13,14,15,16,17,18,19], and most of them concerned timetabling of a round-robin tournament.

In this paper, we consider a round-robin tournament with home–away assignment. There are mainly two approaches to construct a timetable for such a tournament. One approach [16] is to fix an opponent of each game first and then set the place of the game. The other [8,12,15] is to decide a place of each

E. Burke and P. De Causmaecker (Eds.): PATAT 2002, LNCS 2740, pp. 78–99, 2003.

	1	2	3	4	5	slot
1 :	3	@4	5	@6	2	
2 :	@5	6	4	3	@1	
3 :	@1	5	@6	@2	4	
4 :	@6	1	@2	5	@3	
5 :	2	@3	@1	@4	6	
6 :	4	@2	3	1	@5	
team						

Fig. 1. Timetable of six teams

game first, i.e. fix a pattern set, then assign an opponent. Each approach has its advantage and disadvantage compared to the other. When there are many requirements about places of games, the latter is adequate for timetabling. Constructing a timetable with the latter approach, we often encounter the problem to determine whether it is possible to create a timetable from a fixed pattern set. We call this problem the *pattern set feasibility problem*. It is reported that some enumerative methods, such as integer programming, are fairly effective for this problem. However, few theoretical results of research on this problem are known. In this paper, we investigate the pattern set feasibility problem in detail, and propose a highly efficient algorithm to solve this problem in a particular case.

This paper consists of five sections, including this introductory one. Section 2 formally defines the pattern set feasibility problem. Section 3 proposes a new necessary condition for feasible pattern sets. Section 4 contains three subsections, which concern pattern sets belonging to a particular class and these feasibility. Section 5 describes future work and conclusions.

2   Pattern Set Feasibility Problem

A round-robin tournament is a tournament in which each team (or player, etc.) matches in turn against every other team. In this paper, we consider a round-robin tournament consisting of $2n$ teams with $2n - 1$ slots; each team plays one game in each slot. Each team has its home, and each game is held at the home of one of the teams playing. A game between teams t_1 and t_2 played at the home of t_1 is called a *home game* for t_1 and an *away game* for t_2. In a timetable, each game with "@" means that the game is an away game and each game without "@" is a home game for the team corresponding to the row. For example, in the timetable of Figure 1, team 4 plays against team 3 at the home of team 3 in slot 5.

A *pattern set* is a table showing whether each game is a home game or an away game for each team. Figure 2 is the pattern set corresponding to the timetable of Figure 1. In a pattern set, each "A" means an away game and each "H" means a home game for the team corresponding to the row.

	1	2	3	4	5
1 :	H	A	H	A	H
2 :	A	H	H	H	A
3 :	A	H	A	A	H
4 :	A	H	A	H	A
5 :	H	A	A	A	H
6 :	H	A	H	H	A

Fig. 2. Pattern set corresponding to the timetable of Figure 1

	1	2	3
1 :	A	A	H
2 :	A	H	H
3 :	H	H	A
4 :	H	A	A

Fig. 3. Pattern set of four teams

	1	2	3		1	2	3
1 :	@4	@2	3	1 :	@3	@2	4
2 :	@3	1	4	2 :	@4	1	3
3 :	2	4	@1	3 :	1	4	@2
4 :	1	@3	@2	4 :	2	@3	@1

Fig. 4. Timetables corresponding to the pattern set of Figure 3

Assume that we need to construct a timetable from a given pattern set. From the pattern set of Figure 3, we can construct the two corresponding timetables of Figure 4. If a pattern set has at least one corresponding timetable, we say that the pattern set is *feasible*. Usually, one feasible pattern set produces numerous timetables [12].

Unfortunately, there is a pattern set that cannot generate a timetable; such a pattern set is called an *infeasible* pattern set. Figure 5 is an example of an infeasible pattern set. In the pattern set of Figure 5, teams 4, 5 and 6 cannot play the games among them in slots 3, 4 and 5, because in each of those slots these teams have the same characters ("A" or "H"). Thus, they have to play the three games in slots 1 and 2, which is an impossible situation.

The *pattern set feasibility problem* is to determine the feasibility of a given pattern set. For this problem, polynomial-size characterization of feasible pattern sets has not been found yet, and whether this problem is NP-complete or not is still open. Although a polynomial-time algorithm to solve this problem is not yet known, this problem is solvable with integer programming [12], and constraint programming [8]. Theoretical results from graph theory were described in [3,4, 5,6,15,17].

In the remainder of this paper, we address the pattern set feasibility problem. First, we give three remarks about this problem.

	1	2	3	4	5
1 :	H	A	H	A	H
2 :	A	A	H	A	H
3 :	A	H	H	A	H
4 :	A	H	A	H	A
5 :	H	H	A	H	A
6 :	H	A	A	H	A

Fig. 5. Infeasible pattern set of six teams

Remark 1. Permutation of the rows, i.e. teams, of a pattern set does not change its feasibility.

Remark 2. Permutation of the columns, i.e. slots, of a pattern set does not change its feasibility.

Remark 3. Replacing each "A" with "H" and each "H" with "A" of a pattern set does not change its feasibility.

3   Necessary Condition for Feasible Pattern Sets

In this section, we propose a new necessary condition for feasible pattern sets.
    Every feasible pattern set must satisfy the following two conditions [12]:

(i) in each slot, the number of "A"s and "H"s are equal;
(ii) each team has a different row from those of the other teams.

If a given pattern set does not satisfy the above, the pattern set is infeasible.

Assumption. In the remainder of this paper, all pattern sets satisfy conditions (i) and (ii).

    Of course, there is an infeasible pattern set that meets conditions (i) and (ii). Figure 5 is an example of such a pattern set. Seeing teams 4, 5 and 6, we find that the pattern set is infeasible.
    We show another infeasible pattern set. Figure 6 is a pattern set of 10 teams, showing only teams 1–5. We are able to judge that this pattern set is infeasible without seeing teams 6–10. In each of slots 1, 8 and 9, teams 1, 2, 3, 4 and 5 cannot play against each other. In slot 2, we can hold at most one game among these five teams because there are one "A" and four "H"s; at most two games in slot 3, and so on. In total, we can assign at most nine games among the five teams. However, we need to hold 10 games ($= \binom{5}{2}$). Hence, this pattern set is infeasible.
    We describe the above procedure for a general case. In a given pattern set, let T be an arbitrary subset of teams. In each slot, count the number of "A"s and that of "H"s in T, then take the minimum of the two. If the sum total of the minima is strictly less than $\binom{|T|}{2}$, the pattern set is infeasible.

		1	2	3	4	5	6	7	8	9
1	:	A	A	H	A	H	A	H	A	H
2	:	A	H	H	A	H	A	H	A	H
3	:	A	H	A	H	H	A	H	A	H
4	:	A	H	A	H	A	A	H	A	H
5	:	A	H	A	H	A	H	A	A	H
6	:					⋯				
⋮						⋮				
10	:					⋯				
#A	:	5	1	3	2	2	4	1	5	0
#H	:	0	4	2	3	3	1	4	0	5
min	:	0	1	2	2	2	1	1	0	0

Fig. 6. Infeasible pattern set of 10 teams, showing only teams 1–5

In the rest of this paper, we denote the set of all teams by $U = \{1, 2, \ldots, 2n\}$ and the set of all slots by $S = \{1, 2, \ldots, 2n - 1\}$. For any $T \subseteq U$ and $s \in S$, let functions $A(T, s)$ and $H(T, s)$ return the number of "A"s and that of "H"s in s among T, respectively. We define function α as follows:

$$\alpha(T) \stackrel{\text{def.}}{=} \sum_{s \in S} \min\{A(T, s), H(T, s)\} - \frac{|T|(|T| - 1)}{2}.$$

We propose a new necessary condition for feasible pattern sets in the following theorem.

Theorem 1. *For any feasible pattern set, $\forall T \subseteq U$, $\alpha(T) \geq 0$.*

Proof. For any pattern set, if there exists $T \subseteq U$ such that $\alpha(T) < 0$, the games among T cannot be completed. □

Example 1. For the pattern set of Figure 6, let $T \subseteq U$ be $\{1, 2, 3, 4, 5\}$. Then, $\alpha(T) = \alpha(\{1, 2, 3, 4, 5\}) = (0 + 1 + 2 + 2 + 2 + 1 + 1 + 0 + 0) - 5(5 - 1)/2 = -1$. Since $\alpha(T) < 0$, the pattern set is infeasible.

Remark 4. We have already assumed that any pattern set satisfies the following:

(i) in each slot, the number of "A"s and that of "H"s are equal;
(ii) each team has a different row from those of the other teams.

These conditions can be restated in terms of α as follows:

(i) $\alpha(U) = 0$;
(ii) $\forall T \subseteq U$ such that $|T| = 2$, $\alpha(T) \geq 0$.

We have stated the necessary condition for feasible pattern sets of Theorem 1. In the rest of this section, we show the theorems and a lemma concerning the condition.

We denote the complement of $T \subseteq U$ by $\bar{T}$: $\bar{T} = U \setminus T$. The following theorem shows the relationship between the values of $\alpha(T)$ and $\alpha(\bar{T})$.

Theorem 2. *For any pattern set,* $\forall T \subseteq U$, $\alpha(T) = \alpha(\bar{T})$.

Proof. For each $s \in S$, $\min\{A(T, s), H(T, s)\} + \max\{A(\bar{T}, s), H(\bar{T}, s)\} = n$
because $A(T, s) + A(\bar{T}, s) = H(T, s) + H(\bar{T}, s) = n$.
Thus, $\min\{A(\bar{T}, s), H(\bar{T}, s)\} = |\bar{T}| - \max\{A(\bar{T}, s), H(\bar{T}, s)\}$
$= (2n - |T|) - \left(n - \min\{A(T, s), H(T, s)\}\right) = n - |T| + \min\{A(T, s), H(T, s)\}$.
Hence, $\alpha(\bar{T}) = \sum_{s \in S} \min\{A(\bar{T}, s), H(\bar{T}, s)\} - |\bar{T}|(|\bar{T}| - 1)/2$
$= \sum_{s \in S}\left(n - |T| + \min\{A(T, s), H(T, s)\}\right) - (2n - |T|)(2n - |T| - 1)/2$
$= \sum_{s \in S} \min\{A(T, s), H(T, s)\} + (2n - 1)(n - |T|) - (2n - |T|)(2n - |T| - 1)/2$
$= \sum_{s \in S} \min\{A(T, s), H(T, s)\} - |T|(|T| - 1)/2 = \alpha(T)$.   □

From here, we abbreviate $\sum_{s \in S} \min\{A(T, s), H(T, s)\}$ to $\min\{A(T), H(T)\}$.

In a pattern set, if a team has the row obtained by substituting "A" for "H" and "H" for "A" of the row of team t, we say that the team is the *complement team* of t and vice versa. We represent the complement team of t as $\tilde{t}$, and say that $\{t, \tilde{t}\}$ is a *complement pair*.

Lemma 1. *For any pattern set, if* $T \subseteq U$ *contains a complement pair* $\{t, \tilde{t}\}$, $|T| \leq n \implies \alpha(T \setminus \{t, \tilde{t}\}) < \alpha(T)$.

Proof. In each $s \in S$, exactly one of t and $\tilde{t}$ contributes to the value of $\min\{A(T, s), H(T, s)\}$.
Hence, $\min\{A(T \setminus \{t, \tilde{t}\}), H(T \setminus \{t, \tilde{t}\})\} = \min\{A(T), H(T)\} - (2n - 1)$.
Thus, $\alpha(T) - \alpha(T \setminus \{t, \tilde{t}\})$
$= + \min\{A(T), H(T)\} - |T|(|T| - 1)/2$
$\quad - \min\{A(T \setminus \{t, \tilde{t}\}), H(T \setminus \{t, \tilde{t}\})\} + (|T| - 2)(|T| - 3)/2$
$= -|T|(|T| - 1)/2 + (2n - 1) + (|T| - 2)(|T| - 3)/2$
$= 2(n - |T|) + 2 > 0$.   □

For $T \subseteq U$, let $\tilde{T} \subseteq U$ be the subset that consists of the complement teams of all teams in T, if exists: $\tilde{T} \overset{\text{def.}}{=} \{\tilde{t}_1, \tilde{t}_2, \ldots, \tilde{t}_{|T|}\}$ where $T = \{t_1, t_2, \ldots, t_{|T|}\}$. The following theorem shows the relationship between the values of $\alpha(T)$ and $\alpha(\tilde{T})$.

Theorem 3. *For any pattern set,* $\forall T \subseteq U$ *such that* $\tilde{T} \subseteq U$, $\alpha(T) = \alpha(\tilde{T})$.

Proof. Since $A(T, s) = H(\tilde{T}, s)$ and $H(T, s) = A(\tilde{T}, s)$ in each $s \in S$,
$\alpha(T) = \min\{A(T), H(T)\} - |T|(|T| - 1)/2$
$= \min\{H(\tilde{T}), A(\tilde{T})\} - |\tilde{T}|(|\tilde{T}| - 1)/2 = \alpha(\tilde{T})$.   □

	1	2	3	4	5			1	2	3	4	5
1 :	A	A̲	A̲	A̲	A̲		1 :	@6	@2	@4	@5	@3
2 :	A	H	A	H	A		2 :	@3	1	@6	4	@5
3 :	H	H̲	A	A̲	H		3 :	2	4	@5	@6	1
4 :	H	A	H	A	A̲		4 :	5	@3	1	@2	@6
5 :	A	A̲	H	H̲	H̲		5 :	@4	@6	3	1	2
6 :	H	H̲	H̲	H̲	H̲		6 :	1	5	2	3	4

Fig. 7. Undesirable feasible pattern set of six teams and a corresponding schedule

4 Pattern Set with a Minimum Number of Breaks and Characterizing Its Feasibility

In practical sports timetabling, organizers of a round-robin tournament often prefer pattern sets satisfying particular properties. This section concerns the feasibility of such pattern sets.

In Section 4.1, we introduce a pattern set with a minimum number of breaks and its well-known properties, described in [3,4,6]. In Section 4.2, we propose a polynomial-time algorithm to check whether a pattern set with a minimum number of breaks satisfies the necessary condition of Theorem 1. In Section 4.3, we perform computational experiments that show that the necessary condition is highly efficient for finding feasible pattern sets with a minimum number of breaks.

4.1 Pattern Set with a Minimum Number of Breaks

Although the pattern set shown in Figure 7 is feasible, it is not desirable for most of organizers, because team 1 plays five consecutive away games. In general, a pattern set in which a particular team has many consecutive away games is not preferred because the team is at a disadvantage compared to the others. Similarly, many consecutive home games are not desirable, such as team 6 in Figure 7.

If a team has two "A"s or two "H"s in slots s and $s+1$, we say that the team has a *break* at slot $s+1$. In this paper, a break is expressed with an underline below the letter "A" or "H" in a pattern set. For instance, in the pattern set of Figure 7, team 4 has consecutive "A"s in slots 4 and 5, and we say that team 4 has a break at slot 5.

Generally, a pattern set with a few breaks is preferred to a pattern set with many breaks. The following theorem [3] shows the minimum number of breaks in a feasible pattern set of $2n$ teams.

Theorem 4. *For any feasible pattern set of $2n$ teams, the number of breaks is greater than or equal to $2n - 2$.*

Proof. There are at most two teams without any breaks, "AHAH...AHA" and "HAHA...HAH". Hence, the number of breaks is greater than or equal to $2n-2$. □

	1	2	3	4	5	6	7
1 :	A	H	A	H	A	H	A
2 :	H	A	A̲	H	A	H	A
3 :	H	A	H	A	A̲	H	A
4 :	H	A	H	A	H	H̲	A
5 :	H	A	H	A	H	A	H
6 :	A	H	H̲	A	H	A	H
7 :	A	H	A	H	H̲	A	H
8 :	A	H	A	H	A	A̲	H

Fig. 8. PSMB of eight teams

	$\cdots$	$s-1$	s	$\cdots$
t :	$\cdots$	A	A̲	$\cdots$
$\cdots$:	$\cdots$	H	A	$\cdots$
$\cdots$:	$\cdots$	H	A	$\cdots$
$\cdots$:	$\cdots$	A	H	$\cdots$
$\cdots$:	$\cdots$	A	H	$\cdots$
$\cdots$:	$\cdots$	A	H	$\cdots$

Fig. 9. Contradiction to condition (i)

If a pattern set of $2n$ teams has exactly $2n - 2$ breaks, we call the pattern set a *pattern set with a minimum number of breaks* (PSMB). In a PSMB, two teams have no break and other $2n - 2$ teams have just one break. Figure 8 is an example of a PSMB.

For any PSMB, the following theorem holds [3].

Theorem 5. *At each slot of any PSMB, there are exactly two breaks or no breaks.*

Proof. In a PSMB, each team has at most one break. Thus, there are at most two teams with a break at slot s, as "AH...AH̲H̲A...A" and "HA...HAA̲H...H". Assume that only team t has a break at slot s. Then the number of "A"s and that of "H"s are different in slot $s - 1$ because all teams except t have no break at slot s (see Figure 9). This contradicts condition (i). □

From Theorem 5, a PSMB of $2n$ teams consists of n complement pairs, and is characterized by the slots at which breaks occur [3].

When each team in a pattern set has exactly one break, the pattern set is called an *equitable pattern set* [3]. Figure 10 is an example of an equitable pattern set. Although each PSMB is an optimal pattern set in terms of minimizing the number of breaks, some organizers prefer an equitable pattern set because each team evenly has one break in it.

In fact, an equitable pattern set is essentially equivalent to a PSMB when we consider the feasibility, because every equitable pattern set is obtained by

	1	2	3	4	5	6	7
1 :	A	A̲	H	A	H	A	H
2 :	A	H	A	A̲	H	A	H
3 :	A	H	A	H	A	A̲	H
4 :	A	H	A	H	A	H	H̲
5 :	H	H̲	A	H	A	H	A
6 :	H	A	H	H̲	A	H	A
7 :	H	A	H	A	H	H̲	A
8 :	H	A	H	A	H	A	A̲

Fig. 10. Equitable pattern set of eight teams

	1	2	3	4	5	6	7
1 :	H	A	A̲	H	A	H	A
2 :	H	A	H	A	H	H̲	A
3 :	H	A	H	A	A̲	H	A
4 :	A	H	H̲	A	H	A	H
5 :	A̲	H	A	H	A	H	A
6 :	H̲	A	H	A	H	A	H
7 :	A	H	A	H	H̲	A	H
8 :	A	H	A	H	A	A̲	H

		1	2	3	4	5	6	7
(6 →)	1 :	H̲	A	H	A	H	A	H
(4 →)	2 :	A	H	H̲	A	H	A	H
(7 →)	3 :	A	H	A	H	H̲	A	H
(8 →)	4 :	A	H	A	H	A	A̲	H
(5 →)	5 :	A̲	H	A	H	A	H	A
(1 →)	6 :	H	A	A̲	H	A	H	A
(3 →)	7 :	H	A	H	A	A̲	H	A
(2 →)	8 :	H	A	H	A	H	H̲	A

Fig. 11. PSMB before and after sorting of its rows

cyclic rotation of the slots of a PSMB. For example, the equitable pattern set of Figure 10 is constructed from the PSMB of Figure 8, by letting the slot s of the PSMB be slot $s + 1$ ($s = 1, 2, \ldots, 6$), and slot 7 be slot 1. Thus, it is natural to count a team with no break for having a break at slot 1, as "A̲HAHAHA".

For a PSMB, let function σ take team t and return the slot at which t has a break. From Remark 1, without loss of generality, we may assume that a PSMB satisfies the following conditions:

- $\sigma(1) < \sigma(2) < \cdots < \sigma(n)$ and $\sigma(1) = 1$;
- teams $1, 2, \ldots,$ and n have "H"s in slot $2n - 1$;
- team $t + n$ is the complement team of t ($t = 1, 2, \ldots, n$).

If a given PSMB does not satisfy the above, we sort its rows first to meet them (see Figure 11). The sorting procedure takes $O(n^2)$ steps.

In the remainder of this paper, we regard a team with no break as a team having a break at slot 1, and assume that the rows of a PSMB have already been sorted.

4.2   Necessary Condition for Feasible Pattern Sets with a Minimum Number of Breaks

In Section 3, we proposed the necessary condition for feasible pattern sets: $\forall T \subseteq U$, $\alpha(T) \geq 0$. However, to check the inequality for all $T \subseteq U$ takes exponential steps. In this section, we consider this condition in the case of a PSMB.

The objective is to show that the following proposition holds.

Proposition 1. *For any PSMB, whether the PSMB satisfies $\forall T \subseteq U$, $\alpha(T) \geq 0$ or not can be checked in polynomial steps.*

To show that Proposition 1 is true, we define several terms and show a number of lemmas. Note that we now consider a PSMB whose rows have been already sorted.

On the assumption that the next team of $2n$ is 1, if we can construct a consecutive sequence of all elements in $T \subseteq U$, we say that T is *consecutive*. For example, when $2n = 8$, both $\{2, 3, 4\}$ and $\{7, 8, 1, 2\}$ are consecutive, while $\{1, 3, 4\}$ is not.

For $T \subseteq U$, if there exists P such that $T \subseteq P \subseteq U$, $|P| = n$, and P is consecutive, we say that T is *narrow*. For example, when $2n = 8$, both $\{3, 4, 5, 6\}$ and $\{7, 8, 2\}$ are narrow, while $\{1, 3, 6\}$ is not. In particular, if T contains a complement pair, T is not narrow.

Lemma 2. *For any PSMB, a subset $T \subseteq U$ is narrow if and only if there exists a slot at which all teams in T have the same characters ("A" or "H").*

Proof. Let subsets $P_1, P_2, \ldots, P_{2n} \subseteq U$ be $\{1, 2, \ldots, n\}$, $\{2, 3, \ldots, n+1\}, \ldots,$ $\{2n, 1, \ldots, n-1\}$, respectively. For each $i \in \{1, 2, \ldots, 2n\}$, P_i is consecutive and $|P_i| = n$.

If T is narrow, there exists P such that $T \subseteq P \subseteq U$, $|P| = n$, and P is consecutive. Then, $P \in \{P_1, P_2, \ldots, P_{2n}\}$. For $P_1, P_2, \ldots, P_{n+1}, P_{n+2}, \ldots, P_{2n}$, all teams in T have the same characters in slot $\sigma(n)$, $\sigma(n+1), \ldots, \sigma(2n)$, $\sigma(1), \ldots, \sigma(n-1)$, respectively (see Figure 11, right).

If there exists a slot at which all teams in T have the same characters, let the slot be s. Then, the following holds:

- if there exists $t \in \{1, 2, \ldots, n-1\}$ such that $\sigma(t) \leq s < \sigma(t+1)$,
  then $T \subseteq P_{t+1}$ or $T \subseteq P_{t+n+1}$;
- if $\sigma(n) \leq s$, then $T \subseteq P_1$ or $T \subseteq P_{n+1}$.

Thus, there exists a subset P such that $T \subseteq P \subseteq U$, $|P| = n$, and P is consecutive, i.e. T is narrow. $\quad\square$

Next, we define partition of 2^U as below:

$\mathcal{T}_1 = \{T \subseteq U : |T| \leq n \text{ and } T \text{ is consecutive}\};$
$\mathcal{T}_2 = \{T \subseteq U : |T| \leq n \text{ and } T \text{ is not consecutive but narrow}\};$
$\mathcal{T}_3 = \{T \subseteq U : |T| \leq n \text{ and } T \text{ is neither consecutive nor narrow}\};$
$\mathcal{T}_4 = \{T \subseteq U : |T| > n\}.$

Table 1. Partition of 2^U

	Consecutive	Not consecutive	
$|T| \leq n$	$\mathcal{T}_1$	$\mathcal{T}_2$	narrow
	–	$\mathcal{T}_3$	not narrow
$|T| > n$		$\mathcal{T}_4$	not narrow

From the definition, $\mathcal{T}_1 \cup \mathcal{T}_2 \cup \mathcal{T}_3 \cup \mathcal{T}_4 = 2^U$ and $\mathcal{T}_i \cap \mathcal{T}_j = \emptyset$ for $i \neq j$. Table 1 illustrates the partition of 2^U.

Here we give a precise description of Proposition 1. It consists of the following proposition.

Proposition 2. *For any PSMB,* $\forall T \in \mathcal{T}_1, \alpha(T) \geq 0 \iff \forall T \subseteq U, \alpha(T) \geq 0$.

For each $T \subseteq U$ of any pattern set, we can check whether T satisfies $\alpha(T) \geq 0$ or not in polynomial steps. Since $|\mathcal{T}_1| = O(n^2)$, Proposition 2 directly implies Proposition 1. In order to prove that Proposition 2 is true, we show several lemmas in the remainder of this section.

Lemma 3. *For any PSMB,* $\forall T \in \mathcal{T}_1, \alpha(T) \geq 0 \implies \forall T \in \mathcal{T}_2, \alpha(T) \geq 0$.

Proof. We show that $\forall T \in \mathcal{T}_2, \exists T^* \in \mathcal{T}_1, \alpha(T^*) \leq \alpha(T)$.

For $T \in \mathcal{T}_2$ of a given PSMB, find P' such that $T \subseteq P' \subseteq U$, $|P'| = n$, P' is consecutive, and $P' \cap \{2n, 1\} \leq 1$. If there is no such P', consider $\widetilde{T}$ instead of T in the rest of this proof. Theorem 3 justifies this replacement.

We represent T as a general form $T = \{t_1, t_2, \ldots, t_{|T|}\}$ where $t_1 < t_2 < \cdots < t_{|T|}$. By cyclically rotating the slots of the given PSMB, let t_1 be having a break at slot 1. Since T is narrow, now $1 = \sigma(t_1) < \sigma(t_2) < \cdots < \sigma(t_{|T|}) \leq 2n - 1$ holds. By the following procedure, we construct T^* such that $T^* \in \mathcal{T}_1$, $|T^*| = |T|$ and $\alpha(T^*) \leq \alpha(T)$.

Let $m \in T$ be the element satisfying $m = t_{\lceil \frac{\cdot T \cdot}{2} \rceil}$. Define $T^* \subseteq U$ as follows:

$$T^* = \left\{ m - \left\lceil \frac{|T|}{2} \right\rceil + 1, m - \left\lceil \frac{|T|}{2} \right\rceil + 2, \ldots, m, \ldots, m + \left\lfloor \frac{|T|}{2} \right\rfloor \right\}.$$

We also represent T^* as $T^* = \{t_1^*, t_2^*, \ldots, t_{\lceil \frac{\cdot T \cdot}{2} \rceil}^*, \ldots, t_{|T|}^*\}$ where $t_1^* < t_2^* < \cdots < t_{\lceil \frac{\cdot T \cdot}{2} \rceil}^* < \cdots < t_{|T|}^*$.

The subset T^* belongs to $\mathcal{T}_1$ because T^* is consecutive and $|T^*| = |T| \leq n$. Thus, it is sufficient to show $\min\{A(T^*), H(T^*)\} \leq \min\{A(T), H(T)\}$ instead of $\alpha(T^*) \leq \alpha(T)$.

In order to make the proof easy, we introduce another expression of a PSMB. For a PSMB, replace each "A" with 0 and each "H" with 1 in all odd slots, and each "A" with 1 and each "H" with 0 in all even slots, respectively. For example, the PSMB of Figure 12 changes to Figure 13. In the new expression of a PSMB,

	1	2	3	4	5	6	7	8	9	10	11	12	13	14	15
1 :	H	A	H	A	H	A	H	A	H	A	H	A	H	A	H
2 :	A	A	H	A	H	A	H	A	H	A	H	A	H	A	H
3 :	A	H	A	A	H	A	H	A	H	A	H	A	H	A	H
4 :	A	H	A	H	A	A	H	A	H	A	H	A	H	A	H
5 :	A	H	A	H	A	H	A	A	H	A	H	A	H	A	H
6 :	A	H	A	H	A	H	A	H	A	A	H	A	H	A	H
7 :	A	H	A	H	A	H	A	H	A	H	H	A	H	A	H
8 :	A	H	A	H	A	H	A	H	A	H	A	H	A	A	H
9 :	A	H	A	H	A	H	A	H	A	H	A	H	A	H	A
10 :	H	H	A	H	A	H	A	H	A	H	A	H	A	H	A
11 :	H	A	H	H	A	H	A	H	A	H	A	H	A	H	A
12 :	H	A	H	A	H	H	A	H	A	H	A	H	A	H	A
13 :	H	A	H	A	H	A	H	H	A	H	A	H	A	H	A
14 :	H	A	H	A	H	A	H	A	H	H	A	H	A	H	A
15 :	H	A	H	A	H	A	H	A	H	A	A	H	A	H	A
16 :	H	A	H	A	H	A	H	A	H	A	H	A	H	H	A

Fig. 12. PSMB of A–H expression

counting the number of "A"s and "H"s in each slot corresponds to counting the number of "0"s and "1"s.

This 0–1 expression clarifies that the following equalities hold (see Figures 14 and 15):

$$\min\{A(T), H(T)\} = \sum_{i=1}^{\lceil \frac{\cdot T\cdot}{2} \rceil} \left(\sigma(m) - \sigma(t_i)\right) + \sum_{i=\lceil \frac{\cdot T\cdot}{2} \rceil+1}^{|T|} \left(\sigma(t_i) - \sigma(m)\right);$$

$$\min\{A(T^*), H(T^*)\} = \sum_{i=1}^{\lceil \frac{\cdot T\cdot}{2} \rceil} \left(\sigma(m) - \sigma(t_i^*)\right) + \sum_{i=\lceil \frac{\cdot T\cdot}{2} \rceil+1}^{|T|} \left(\sigma(t_i^*) - \sigma(m)\right).$$

From the definition of T^*, if $i \leq \lceil \frac{|T|}{2} \rceil$, then $\sigma(t_i) \leq \sigma(t_i^*)$, otherwise $\sigma(t_i) \geq \sigma(t_i^*)$. Note that $t_{\lceil \frac{\cdot T\cdot}{2} \rceil} = m = t_{\lceil \frac{\cdot T\cdot}{2} \rceil}^*$.

Hence, $\min\{A(T), H(T)\} - \min\{A(T^*), H(T^*)\}$

$$= +\sum_{i=1}^{\lceil \frac{\cdot T\cdot}{2} \rceil} \left(\sigma(m) - \sigma(t_i)\right) + \sum_{i=\lceil \frac{\cdot T\cdot}{2} \rceil+1}^{|T|} \left(\sigma(t_i) - \sigma(m)\right)$$

$$- \sum_{i=1}^{\lceil \frac{\cdot T\cdot}{2} \rceil} \left(\sigma(m) - \sigma(t_i^*)\right) - \sum_{i=\lceil \frac{\cdot T\cdot}{2} \rceil+1}^{|T|} \left(\sigma(t_i^*) - \sigma(m)\right)$$

$$= \sum_{i=1}^{\lceil \frac{\cdot T\cdot}{2} \rceil} \left(\sigma(t_i^*) - \sigma(t_i)\right) + \sum_{i=\lceil \frac{\cdot T\cdot}{2} \rceil+1}^{|T|} \left(\sigma(t_i) - \sigma(t_i^*)\right) \geq 0. \qquad \square$$

Example 2. Figures 14 and 15 are subsets of Figure 13. For $T = \{1, 3, 5, 6, 8\} \in \mathcal{T}_2$ of Figure 14, we construct $T^* = \{3, 4, 5, 6, 7\} \in \mathcal{T}_1$ of Figure 15, and $\alpha(T) = \alpha(\{1, 3, 5, 6, 8\}) = 9 > 1 = \alpha(\{3, 4, 5, 6, 7\}) = \alpha(T^*)$.

Proposition 3. *For any PSMB,*

$$\forall T \in \mathcal{T}_1 \cup \mathcal{T}_2, \ \alpha(T) \geq 0 \implies \forall T \in \mathcal{T}_3, \ \alpha(T) \geq 0.$$

	1	2	3	4	5	6	7	8	9	10	11	12	13	14	15
1 :	1	1	1	1	1	1	1	1	1	1	1	1	1	1	1
2 :	0	1	1	1	1	1	1	1	1	1	1	1	1	1	1
3 :	0	0	0	1	1	1	1	1	1	1	1	1	1	1	1
4 :	0	0	0	0	0	1	1	1	1	1	1	1	1	1	1
5 :	0	0	0	0	0	0	0	1	1	1	1	1	1	1	1
6 :	0	0	0	0	0	0	0	0	0	1	1	1	1	1	1
7 :	0	0	0	0	0	0	0	0	0	0	1	1	1	1	1
8 :	0	0	0	0	0	0	0	0	0	0	0	0	0	1	1
9 :	0	0	0	0	0	0	0	0	0	0	0	0	0	0	0
10 :	1	0	0	0	0	0	0	0	0	0	0	0	0	0	0
11 :	1	1	1	0	0	0	0	0	0	0	0	0	0	0	0
12 :	1	1	1	1	1	0	0	0	0	0	0	0	0	0	0
13 :	1	1	1	1	1	1	1	0	0	0	0	0	0	0	0
14 :	1	1	1	1	1	1	1	1	1	0	0	0	0	0	0
15 :	1	1	1	1	1	1	1	1	1	1	0	0	0	0	0
16 :	1	1	1	1	1	1	1	1	1	1	1	1	1	0	0

Fig. 13. PSMB of 0–1 expression

	1	2	3	4	5	6	7	8	9	10	11	12	13	14	15
1 :	1	1	1	1	1	1	1	1	1	1	1	1	1	1	1
3 :	0	0	0	1	1	1	1	1	1	1	1	1	1	1	1
5 :	0	0	0	0	0	0	0	1	1	1	1	1	1	1	1
6 :	0	0	0	0	0	0	0	0	0	1	1	1	1	1	1
8 :	0	0	0	0	0	0	0	0	0	0	0	0	0	1	1

Fig. 14. $T = \{1, 3, 5, 6, 8\} \in \mathcal{T}_2$, $\alpha(T) = 9$

	1	2	3	4	5	6	7	8	9	10	11	12	13	14	15
3 :	0	0	0	1	1	1	1	1	1	1	1	1	1	1	1
4 :	0	0	0	0	0	1	1	1	1	1	1	1	1	1	1
5 :	0	0	0	0	0	0	0	1	1	1	1	1	1	1	1
6 :	0	0	0	0	0	0	0	0	0	1	1	1	1	1	1
7 :	0	0	0	0	0	0	0	0	0	0	1	1	1	1	1

Fig. 15. $T^* = \{3, 4, 5, 6, 7\} \in \mathcal{T}_1$, $\alpha(T^*) = 1$

To show that Proposition 3 is correct, we divide $\mathcal{T}_3$ into two subsets. Let $\mathcal{T}_{3\mathrm{nc}}$ consist of the subsets that contain no complement pairs, and $\mathcal{T}_{3\mathrm{wc}} = \mathcal{T}_3 \setminus \mathcal{T}_{3\mathrm{nc}}$. Note that if $T \subseteq U$ contains a complement pair and $|T| \leq n$, T belongs to $\mathcal{T}_{3\mathrm{wc}}$.

Lemma 4. *For any PSMB,*

$$\forall T \in \mathcal{T}_1 \cup \mathcal{T}_2,\ \alpha(T) \geq 0 \implies \forall T \in \mathcal{T}_{3\mathrm{nc}},\ \alpha(T) \geq 0.$$

Proof. We show that $\forall T \in \mathcal{T}_{3\mathrm{nc}}$, $\exists T^* \in \mathcal{T}_1 \cup \mathcal{T}_2$, $\alpha(T^*) \leq \alpha(T)$. In this proof, we have much help from graphical expression of a subset of a PSMB defined below.

Draw a regular $2(2n-1)$-gon, and name the vertices of the polygon $\bar{1}, \bar{2}, \dots,$ $\overline{2n-1}, \underline{1}, \underline{2}, \dots, \underline{2n-1}$ clockwise. For a given $T \in \mathcal{T}_1 \cup \mathcal{T}_2 \cup \mathcal{T}_{3\text{nc}}$ of 0–1 expression, we colour some vertices black or grey. Recall that $|T| \leq n$ and T includes no complement pairs. If T contains a team that has "1" in slot $2n-1$ and a break at slot s, we colour $\bar{s}$ black and $\underline{s}$ grey. If T contains a team that has "0" in slot $2n-1$ and a break at slot s, we colour $\underline{s}$ black and $\bar{s}$ grey. Then, $2|T|$ vertices of the polygon are coloured, while the remaining vertices are still uncoloured. Note that each pair of vertices $\{\bar{v}, \underline{v}\}$ is a pair of black and grey vertices, or of two uncoloured vertices. For $T \in \mathcal{T}_1 \cup \mathcal{T}_2 \cup \mathcal{T}_{3\text{nc}}$, $f(T)$ denotes the corresponding set of all black vertices generated by the colouring method. For any $V' \subseteq f(T)$, there exists the subset $T' \subseteq T$ satisfying $f(T') = V'$.

Next, we define a *black run*. A black run is a maximal subset of neighbouring vertices of the polygon satisfying that all contained vertices are black or uncoloured, and both end vertices are black. We also define a *grey run* in a similar way. Black runs and grey runs are called runs for simplicity. The size of a run is the number of coloured vertices in the run.

In the following, we introduce some functions that take a vertex of the $2(2n-1)$-gon or a subset of the vertices:

$$shift^+(\bar{v}) \stackrel{\text{def.}}{=} \begin{cases} \underline{1} & (\text{if } \bar{v} = \overline{2n-1}), \\ \overline{v+1} & (\text{otherwise}), \end{cases} \qquad shift^+(\underline{v}) \stackrel{\text{def.}}{=} \begin{cases} \bar{1} & (\text{if } \underline{v} = \underline{2n-1}), \\ \underline{v+1} & (\text{otherwise}), \end{cases}$$

$$shift^-(\bar{v}) \stackrel{\text{def.}}{=} \begin{cases} \underline{2n-1} & (\text{if } \bar{v} = \bar{1}), \\ \overline{v-1} & (\text{otherwise}), \end{cases} \qquad shift^-(\underline{v}) \stackrel{\text{def.}}{=} \begin{cases} \overline{2n-1} & (\text{if } \underline{v} = \underline{1}), \\ \underline{v-1} & (\text{otherwise}), \end{cases}$$

$$vshift^+(V) \stackrel{\text{def.}}{=} \{shift^+(v) : v \in V\}, \quad vshift^-(V) \stackrel{\text{def.}}{=} \{shift^-(v) : v \in V\},$$

$$opp(\bar{v}) \stackrel{\text{def.}}{=} \underline{v}, \; opp(\underline{v}) \stackrel{\text{def.}}{=} \bar{v}, \; vopp(V) \stackrel{\text{def.}}{=} \{opp(v) : v \in V\},$$

$$del(\bar{v}) \stackrel{\text{def.}}{=} v, \; del(\underline{v}) \stackrel{\text{def.}}{=} v.$$

For a vertex subset V, we construct a 0–1 matrix $M = g(V)$ whose rows are indexed by V, columns are indexed by S. An element of $M = (m_{vs})$ is indexed by $(v, s) \in V \times S$. Each element m_{vs} is defined by

$$m_{vs} \stackrel{\text{def.}}{=} \begin{cases} 0 & \left(\text{if } v \in \{\bar{1}, \bar{2}, \dots, \overline{2n-1}\} \text{ and } s < del(v)\right), \\ 0 & \left(\text{if } v \in \{\underline{1}, \underline{2}, \dots, \underline{2n-1}\} \text{ and } s \geq del(v)\right), \\ 1 & (\text{otherwise}). \end{cases}$$

For $M = g(V)$, we define function γ as follows:

$$\gamma(M) \stackrel{\text{def.}}{=} \sum_{s \in S} \min \left\{ \sum_{v \in V} m_{vs}, \sum_{v \in V} (1 - m_{vs}) \right\} - \frac{|V|(|V| - 1)}{2}.$$

Also, function β is defined by $\beta(V) = \gamma(M)$ where $M = g(V)$. The definition of γ directly implies that, for all $T \in \mathcal{T}_1 \cup \mathcal{T}_2 \cup \mathcal{T}_{3\text{nc}}$, $\alpha(T) = \beta(V)$ holds when $V = f(T)$.

Now we construct $T^* \in \mathcal{T}_1 \cup \mathcal{T}_2$ such that $\alpha(T^*) \leq \alpha(T)$ for a given $T \in \mathcal{T}_{3nc}$. Input the set of black vertices $V = f(T)$ into the following procedure, which consists of outermost, middle and innermost loops. This procedure outputs $V^* \subseteq V$ such that $V^* = f(T^*)$:

begin

> $V' := V$
> let all the vertices of the $2(2n-1)$-gon uncoloured
> colour all the vertices in V' black and all the vertices in $vopp(V')$ grey
> **repeat**
>> choose a minimum size black run R' from the $2(2n-1)$-gon
>> put $V'_{\min}$ be the set of black vertices in R'
>> **repeat**
>>> **if** $\beta\big((V' \setminus V'_{\min}) \cup vshift^-(V'_{\min})\big) \leq \beta(V')$ **then**
>>>> **repeat**
>>>>> $V' := (V' \setminus V'_{\min}) \cup vshift^-(V'_{\min})$
>>>>> $V'_{\min} := vshift^-(V'_{\min})$
>>>>> **until** $vopp(V'_{\min}) \cap (V' \setminus V'_{\min}) \neq \emptyset$
>>> **else**
>>>> **repeat**
>>>>> $V' := (V' \setminus V'_{\min}) \cup vshift^+(V'_{\min})$
>>>>> $V'_{\min} := vshift^+(V'_{\min})$
>>>>> **until** $vopp(V'_{\min}) \cap (V' \setminus V'_{\min}) \neq \emptyset$
>>> **endif**
>>> $V'_{\text{temp}} := vopp(V'_{\min}) \cap (V' \setminus V'_{\min})$
>>> $V' := V' \setminus V'_{\text{temp}}$
>>> $V'_{\min} := V'_{\min} \setminus vopp(V'_{\text{temp}})$
>> **until** $V'_{\min} = \emptyset$
>> let all the vertices of the $2(2n-1)$-gon uncoloured
>> colour all the vertices in V' black and all the vertices in $vopp(V')$ grey
> **until** the number of runs in $2(2n-1)$-gon is less than or equal to two
> $V^* := V'$
> **output** V^*

end

Before showing that the above procedure correctly runs for any subset $T \in \mathcal{T}_{3nc}$, we describe how it works with a running example. Let $T \in \mathcal{T}_{3nc}$ be $T = \{1, 2, 3, 6, 7, 8, 12, 13\}$, which is a subset of the PSMB of Figure 13. Accordingly, $V = f(T) = \{\overline{1}, \overline{2}, \overline{4}, \underline{6}, \underline{8}, \overline{10}, \overline{11}, \overline{14}\}$ and $\alpha(T) = \beta(V) = 20$. Inputted V, the procedure returns $V^* = \{\overline{1}, \overline{10}, \overline{11}, \overline{14}\}$ as follows:

1. In Figure 16 (before the first iteration of middle loop),
   $V' = V = \{\overline{1}, \overline{2}, \overline{4}, \underline{6}, \underline{8}, \overline{10}, \overline{11}, \overline{14}\}$, $V'_{\min} = \{\underline{6}, \underline{8}\}$ and $\beta(V') = 20$;

2. In Figure 17 (after the first iteration of the innermost loop),
   $V' = \{\overline{1}, \overline{2}, \overline{4}, \underline{5}, \underline{7}, \overline{10}, \overline{11}, \overline{14}\}$, $V'_{\min} = \{\underline{5}, \underline{7}\}$ and $\beta(V') = 20$;

	1	2	3	4	5	6	7	8	9	10	11	12	13	14	15
1 :	1	1	1	1	1	1	1	1	1	1	1	1	1	1	1
2 :	0	1	1	1	1	1	1	1	1	1	1	1	1	1	1
3 :	0	0	0	1	1	1	1	1	1	1	1	1	1	1	1
6 :	0	0	0	0	0	0	0	0	0	1	1	1	1	1	1
7 :	0	0	0	0	0	0	0	0	0	0	1	1	1	1	1
8 :	0	0	0	0	0	0	0	0	0	0	0	0	0	1	1
12 :	1	1	1	1	1	0	0	0	0	0	0	0	0	0	0
13 :	1	1	1	1	1	1	1	0	0	0	0	0	0	0	0

Fig. 16. $T = \{1,2,3,6,7,8,12,13\} \in \mathcal{T}_{3nc}$, $\alpha(T) = 20$, $V = \{\bar{1}, \bar{2}, \bar{4}, \underline{6}, \underline{8}, \overline{10}, \overline{11}, \overline{14}\}$

	1	2	3	4	5	6	7	8	9	10	11	12	13	14	15
$\bar{1}$:	1	1	1	1	1	1	1	1	1	1	1	1	1	1	1
$\bar{2}$:	0	1	1	1	1	1	1	1	1	1	1	1	1	1	1
$\bar{4}$:	0	0	0	1	1	1	1	1	1	1	1	1	1	1	1
$\overline{10}$:	0	0	0	0	0	0	0	0	0	1	1	1	1	1	1
$\overline{11}$:	0	0	0	0	0	0	0	0	0	0	1	1	1	1	1
$\overline{14}$:	0	0	0	0	0	0	0	0	0	0	0	0	0	1	1
$\underline{5}$:	1	1	1	1	0	0	0	0	0	0	0	0	0	0	0
$\underline{7}$:	1	1	1	1	1	1	0	0	0	0	0	0	0	0	0

Fig. 17. $V' = \{\bar{1}, \bar{2}, \bar{4}, \underline{5}, \underline{7}, \overline{10}, \overline{11}, \overline{14}\}$, $\beta(V') = 20$

3. In Figure 18 (after the second iteration of the innermost loop), $V' = \{\bar{1}, \bar{2}, \bar{4}, \underline{4}, \underline{6}, \overline{10}, \overline{11}, \overline{14}\}$, $V'_{min} = \{\underline{4}, \underline{6}\}$ and $\beta(V') = 20$;

4. In Figure 19 (after the first iteration of the middle loop), $V' = \{\bar{1}, \bar{2}, \underline{6}, \overline{10}, \overline{11}, \overline{14}\}$, $V'_{min} = \{\underline{6}\}$ and $\beta(V') = 18$;

5. In Figure 20 (after the innermost loop in the second iteration of the middle), $V' = \{\bar{1}, \bar{2}, \underline{2}, \overline{10}, \overline{11}, \overline{14}\}$, $V'_{min} = \{\underline{2}\}$ and $\beta(V') = 14$;

6. In Figure 21 (at the end of the procedure), $V^* = V' = \{\bar{1}, \overline{10}, \overline{11}, \overline{14}\}$, $V'_{min} = \emptyset$ and $\beta(V') = 8$.

The procedure returns $V^* = \{\bar{1}, \overline{10}, \overline{11}, \overline{14}\}$ for $V = \{\bar{1}, \bar{2}, \bar{4}, \underline{6}, \underline{8}, \overline{10}, \overline{11}, \overline{14}\}$. Since $V^* \subseteq V = f(T)$, the subset T^* is a subset of T. In this example, $T^* = \{1,6,7,8\} \subseteq \mathcal{T}_1 \cup \mathcal{T}_2$ and $\alpha(T^*) = 8 < 20 = \alpha(T)$.

To prove the correctness of the procedure for a general case, we need to show the following:

- the finiteness of the procedure;
- $\beta(V^*) \leq \beta(V)$;
- $V^* \subseteq V$, and $T^* \in \mathcal{T}_1 \cup \mathcal{T}_2$ where $V^* = f(T^*)$.

The iterations in the innermost and middle loops of the procedure terminate finitely because V'_{min} corresponds to a minimum size black run. At the end of each iteration of the outermost loop, the number of black and grey runs decreases

	1	2	3	4	5	6	7	8	9	10	11	12	13	14	15
$\overline{1}$:	1	1	1	1	1	1	1	1	1	1	1	1	1	1	1
$\overline{2}$:	0	1	1	1	1	1	1	1	1	1	1	1	1	1	1
$\overline{4}$:	0	0	0	1	1	1	1	1	1	1	1	1	1	1	1
$\overline{10}$:	0	0	0	0	0	0	0	0	0	1	1	1	1	1	1
$\overline{11}$:	0	0	0	0	0	0	0	0	0	0	1	1	1	1	1
$\overline{14}$:	0	0	0	0	0	0	0	0	0	0	0	0	0	1	1
$\underline{4}$:	1	1	1	0	0	0	0	0	0	0	0	0	0	0	0
$\underline{6}$:	1	1	1	1	1	0	0	0	0	0	0	0	0	0	0

Fig. 18. $V' = \{\overline{1}, \overline{2}, \overline{4}, \underline{4}, \underline{6}, \overline{10}, \overline{11}, \overline{14}\}$, $\beta(V') = 20$

	1	2	3	4	5	6	7	8	9	10	11	12	13	14	15
$\overline{1}$:	1	1	1	1	1	1	1	1	1	1	1	1	1	1	1
$\overline{2}$:	0	1	1	1	1	1	1	1	1	1	1	1	1	1	1
$\overline{10}$:	0	0	0	0	0	0	0	0	0	1	1	1	1	1	1
$\overline{11}$:	0	0	0	0	0	0	0	0	0	0	1	1	1	1	1
$\overline{14}$:	0	0	0	0	0	0	0	0	0	0	0	0	0	1	1
$\underline{6}$:	1	1	1	1	1	0	0	0	0	0	0	0	0	0	0

Fig. 19. $V' = \{\overline{1}, \overline{2}, \underline{6}, \overline{10}, \overline{11}, \overline{14}\}$, $\beta(V') = 18$

	1	2	3	4	5	6	7	8	9	10	11	12	13	14	15
$\overline{1}$:	1	1	1	1	1	1	1	1	1	1	1	1	1	1	1
$\overline{2}$:	0	1	1	1	1	1	1	1	1	1	1	1	1	1	1
$\overline{10}$:	0	0	0	0	0	0	0	0	0	1	1	1	1	1	1
$\overline{11}$:	0	0	0	0	0	0	0	0	0	0	1	1	1	1	1
$\overline{14}$:	0	0	0	0	0	0	0	0	0	0	0	0	0	1	1
$\underline{2}$:	1	0	0	0	0	0	0	0	0	0	0	0	0	0	0

Fig. 20. $V' = \{\overline{1}, \overline{2}, \underline{2}, \overline{10}, \overline{11}, \overline{14}\}$, $\beta(V') = 14$

by four. Since the number of the runs is finite, this procedure terminates after a finite number of steps.

In each iteration of the innermost loop in a iteration of the middle loop, the value of $\beta(V')$ decreases by a non-negative constant, because only $V'_{\min}$ affects the decrement. At the end of each iteration of the middle loop, where two vertices are removed from V', the value of $\beta(V'_{\min})$ decreases; this can be shown a similar way to the proof of Lemma 1. Thus, the value $\beta(V')$ is non-increasing throughout this procedure, and therefore $\beta(V^*) \leq \beta(V)$.

At the end of each iteration of the middle loop, V' is contained in V because $V'_{\min}$ corresponds to a minimum size black run throughout this procedure. Hence $V^* \subseteq V$ and there exists $T^* \subseteq T$ such that $f(T^*) = V^*$. After running the procedure, the number of runs becomes two; one is black and the other is grey. Thus T^* is narrow, i.e. $T^* \in \mathcal{T}_1 \cup \mathcal{T}_2$.

	1	2	3	4	5	6	7	8	9	10	11	12	13	14	15
1 :	1	1	1	1	1	1	1	1	1	1	1	1	1	1	1
6 :	0	0	0	0	0	0	0	0	0	1	1	1	1	1	1
7 :	0	0	0	0	0	0	0	0	0	0	1	1	1	1	1
8 :	0	0	0	0	0	0	0	0	0	0	0	0	0	1	1

Fig. 21. $V^* = \{\overline{1}, \overline{10}, \overline{11}, \overline{14}\}$, $T^* = \{1, 6, 7, 8\} \in \mathcal{T}_2$, $\alpha(T^*) = 8$

From the above consideration, we conclude $\alpha(T^*) = \beta(V^*) \leq \beta(V) = \alpha(T)$ and $T^* \in \mathcal{T}_1 \cup \mathcal{T}_2$. □

Lemma 5. *For any PSMB,*

$$\forall T \in \mathcal{T}_1 \cup \mathcal{T}_2 \cup \mathcal{T}_{3\mathrm{nc}}, \, \alpha(T) \geq 0 \implies \forall T \in \mathcal{T}_{3\mathrm{wc}}, \, \alpha(T) \geq 0.$$

Proof. We prove that $\forall T \in \mathcal{T}_{3\mathrm{wc}}, \exists T^* \in \mathcal{T}_1 \cup \mathcal{T}_2 \cup \mathcal{T}_{3\mathrm{nc}}, \alpha(T^*) \leq \alpha(T)$.

For a given $T \in \mathcal{T}_{3\mathrm{wc}}$, remove a complement pair $\{t, \tilde{t}\}$ from T. If $T \setminus \{t, \tilde{t}\}$ still contains another complement pair, repeat this procedure until there exists no complement pairs. Let T^* be the finally obtained set of teams. By applying Lemma 1 repeatedly, we have $\alpha(T) > \alpha(T \setminus \{t, \tilde{t}\}) > \cdots > \alpha(T^*)$. Since T^* includes no complement pairs and $|T^*| < n$, $T^* \in \mathcal{T}_1 \cup \mathcal{T}_2 \cup \mathcal{T}_{3\mathrm{nc}}$. □

Lemma 6. *For any PSMB,*

$$\forall T \in \mathcal{T}_1 \cup \mathcal{T}_2, \, \alpha(T) \geq 0 \implies \forall T \in \mathcal{T}_3, \, \alpha(T) \geq 0.$$

Proof. From Lemma 4, $\forall T \in \mathcal{T}_1 \cup \mathcal{T}_2, \alpha(T) \geq 0 \implies \forall T \in \mathcal{T}_{3\mathrm{nc}}, \alpha(T) \geq 0$. Lemma 5 shows $\forall T \in \mathcal{T}_1 \cup \mathcal{T}_2 \cup \mathcal{T}_{3\mathrm{nc}}, \alpha(T) \geq 0 \implies \forall T \in \mathcal{T}_{3\mathrm{wc}}, \alpha(T) \geq 0$. Since $\mathcal{T}_3 = \mathcal{T}_{3\mathrm{nc}} \cup \mathcal{T}_{3\mathrm{wc}}$, the proposition holds. □

Lemma 7. *For any PSMB,*

$$\forall T \in \mathcal{T}_1 \cup \mathcal{T}_2 \cup \mathcal{T}_3, \, \alpha(T) \geq 0 \iff \forall T \in \mathcal{T}_4, \, \alpha(T) \geq 0.$$

Proof. Theorem 2 gives direct proof of this lemma. □

Theorem 6. *For any PSMB, $\forall T \in \mathcal{T}_1, \alpha(T) \geq 0 \iff \forall T \subseteq U, \alpha(T) \geq 0$.*

Proof. From Lemmas 3, 6 and 7, $\forall T \in \mathcal{T}_1, \alpha(T) \geq 0 \implies \forall T \subseteq U, \alpha(T) \geq 0$ holds. The converse is trivial. □

Theorem 7. *For any PSMB, whether the PSMB satisfies $\forall T \subseteq U, \alpha(T) \geq 0$ or not can be tested in $O(n^4)$ steps.*

Proof. For each $T \subseteq U$, we can judge whether $\alpha(T) \geq 0$ or $\alpha(T) < 0$ in $O(n^2)$ steps (see Figure 6). Theorem 6 states that we need to check only subsets $T \in \mathcal{T}_1$. Since $|\mathcal{T}_1| = O(n^2)$, the overall steps take $O(n^4)$.   □

Remark 5. In fact, we do not need to examine all subsets $T \in \mathcal{T}_1$. Since Theorem 3 shows $\alpha(T) = \alpha(\widetilde{T})$, it is sufficient to observe one of the values $\alpha(T)$ and $\alpha(\widetilde{T})$ for each $T \in \mathcal{T}_1$.

Remark 6. From Remark 2, the proposition $\forall T \in \mathcal{T}_1, \alpha(T) \geq 0 \iff \forall T \subseteq U, \alpha(T) \geq 0$ holds for any equitable pattern set, and whether an equitable pattern set satisfies $\forall T \subseteq U, \alpha(T) \geq 0$ or not can be tested in $O(n^4)$ steps.

Remark 7. In Section 4.1, we have already seen that each PSMB is characterized by the slots at which breaks occur [3]. By making effective use of this fact, we can check whether each $T \in \mathcal{T}_1$ satisfies $\forall T \subseteq U, \alpha(T) \geq 0$ or not in $O(n)$ steps. This leads an $O(n^3)$ algorithm to examine whether a PSMB satisfies the condition.

4.3   Computational Experiment

In the previous section, we showed that whether a PSMB satisfies the condition of Theorem 1 can be checked in polynomial steps. If we find $T \subseteq U$ such that $\alpha(T) < 0$ for a given PSMB, we conclude that the PSMB is infeasible. However, if the PSMB satisfies the condition, we have no additional information about the feasibility. If there are many infeasible PSMBs satisfying the condition, the condition is not strong enough for finding feasible PSMBs. In this section, we show the strength of the condition by computational experiment in the following.

First, we enumerated all PSMBs, including both feasible and infeasible ones. The number of PSMBs is $\binom{2n-2}{n-1}$, by choosing the slots with breaks. Next, we examined whether each PSMB satisfies the condition $\forall T \subseteq U, \alpha(T) \geq 0$, by checking $\forall T \in \mathcal{T}_1, \alpha(T) \geq 0$. Then, for each PSMB satisfying the condition, we decided its feasibility by solving an integer programming problem with ILOG CPLEX 7.0 [9]. (See [12,11] for formulation of the pattern set feasibility problem with integer programming.)

Table 2 is the results of the experiment. The experiment shows that, when the number of teams is less than or equal to 26, the number of PSMBs satisfying the condition and that of feasible PSMBs are equal.

Corollary 1. *When the number of teams is less than or equal to 26, a PSMB is feasible if and only if $\forall T \subseteq U, \alpha(T) \geq 0$.*

The number of PSMBs that satisfy the condition is surprisingly small compared with the number of all PSMBs. We did not perform computational experiment for more than 26 teams because it would take too much time to solve the integer programming problems.

Table 2. Results of computational experiment. #candidates: the number of PSMBs satisfying $\forall T \subseteq U$, $\alpha(T) \geq 0$; #feasible: the number of feasible PSMBs

#teams	#PSMBs	#candidates	#feasible
4	2	2	2
6	6	3	3
8	20	8	8
10	70	10	10
12	252	30	30
14	924	49	49
16	3 432	136	136
18	12 870	216	216
20	48 620	580	580
22	184 756	1045	1045
24	705 432	2772	2772
26	2 704 156	5122	5122

Remark 8. In fact, we did not solve all the integer programming problems in Table 2. Since cyclic rotation of the slots does not change the feasibility of a pattern set, we solved (#candidates/n) integer programming problems. Also, it is sufficient to check the necessary condition for $(\binom{2n-2}{n-1}/n)$ PSMBs.

Remark 9. From Remark 2 and Corollary 1, when the number of teams is less than or equal to 26, $\forall T \subseteq U$, $\alpha(T) \geq 0$ is a necessary and sufficient condition for feasible equitable pattern sets.

5   Future Work and Conclusions

There are many extensions of the pattern set feasibility problem.

The computational experiment showed that, when the number of teams is less than or equal to 26, the proposed condition $\forall T \subseteq U$, $\alpha(T) \geq 0$ is a necessary and sufficient condition for feasible PSMBs. We have the following conjecture.

Conjecture 1. A PSMB of an arbitrary number of teams is feasible if and only if $\forall T \subseteq U$, $\alpha(T) \geq 0$.

Although there is a possibility of this conjecture failing, our results are practical enough in terms of the number of teams for a real timetabling problem of a PSMB.

We proposed a necessary condition for feasible pattern sets. In the case of a PSMB, we proved that the necessary condition can be checked in polynomial steps, and showed the strength of the condition by the computational experiment. Then how does the condition work for a general pattern set? For a general pattern set, it is conjectured that pattern set feasibility problem is NP-complete [12]. To the best of our knowledge, this conjecture is still open.

	1	2	3	4	5	6	7
1 :	H	2	H	A	A	A	H
2 :	A	@1	H	H	A	@8	H
3 :	H	A	A	A	H	H	H
4 :	7	H	A	H	A	A	H
5 :	H	A	H	@6	H	H	A
6 :	H	A	H	5	H	A	A
7 :	@4	H	A	A	A	H	A
8 :	A	H	A	H	H	2	A

Fig. 22. Pattern set with partial assignment of games

	1	2	3	4	5	6	7
1 :	A	@4		2		H	8
2 :	H	A		@1		7	
3 :			8	A		H	
4 :	A	1		A		H	A
5 :		A	H	A	@6		H
6 :	@8	H	H		5		@1
7 :			A			@2	H
8 :	6	H	@3	H		A	H

Fig. 23. Incomplete pattern set with partial assignment of games

The pattern set feasibility problem with partial assignment of games is also a challenging one. In the pattern set of Figure 22, some games have already fixed. Is the pattern set feasible? Furthermore, there is a feasibility problem of an incomplete pattern set with or without fixed games (Figure 23). Such situations often appear in the field of real timetabling [12]. With a few modification, the proposed condition is applicable to these problems.

In this paper, we considered the pattern set feasibility problem. We proposed a new necessary condition for feasible pattern sets. For a pattern set with a minimum number of breaks (PSMB), we proved a theorem leading a polynomial-time algorithm to check the proposed condition. The computational experiment showed that the condition is a necessary and sufficient condition for feasible PSMBs of up to 26 teams. We conjecture that the condition is a necessary and sufficient condition for feasible PSMBs of an arbitrary number of teams. Considering the feasibility of pattern sets yields various interesting problems, and is still significant in sports timetabling.

References

1. Armstrong, J.R., Willis, R.J.: Scheduling the Cricket World Cup – a Case Study. J. Oper. Res. Soc. **44** (1993) 1067–1072
2. Blest, D.C., Fitzgerald, D.G.: Scheduling Sports Competitions with a Given Distribution of Times. Discr. Appl. Math. **22** (1988/89) 9–19
3. de Werra, D.: Geography, Games and Graphs. Discr. Appl. Math. **2** (1980) 327–337
4. de Werra, D.: Minimizing Irregularities in Sports Schedules Using Graph Theory. Discr. Appl. Math. **4** (1982) 217–226
5. de Werra, D.: On the Multiplication of Divisions: the Use of Graphs for Sports Scheduling. Networks **15** (1985) 125–136
6. de Werra, D.: Some Models of Graphs for Scheduling Sports Competitions. Discr. Appl. Math. **21** (1988) 47–65
7. Ferland, J.A., Fleurent, C.: Computer Aided Scheduling for a Sport League. INFOR **29** (1991) 14–25
8. Henz, M.: Scheduling a Major College Basketball Conference—Revisited. Oper. Res. **49** (2001) 163–168
9. ILOG: ILOG CPLEX 7.0 User's Manual. ILOG, Gentilly, 2000
10. Miyashiro, R., Matsui, T.: Note on Equitable Round-Robin Tournaments. Proc. 6th KOREA–JAPAN Joint Workshop on Algorithms and Computation (2001) 135–140
11. Miyashiro, R., Matsui, T.: Notes on Equitable Round-Robin Tournaments. IEICE Trans. Fund. Electron. Commun. Comput. Sci. **E85-A** (2002) 1006–1010
12. Nemhauser, G.L., Trick, M.A.: Scheduling a Major College Basketball Conference. Oper. Res. **46** (1998) 1–8
13. Russell, K.G.: Balancing Carry-Over Effects in Round Robin Tournaments. Biometrika **67** (1980) 127–131
14. Russell, R.A., Leung, J.M.Y.: Devising a Cost Effective Schedule for a Baseball League. Oper. Res. **42** (1994) 614–625
15. Schreuder, J.A.M.: Combinatorial Aspects of Construction of Competition Dutch Professional Football Leagues. Discr. Appl. Math. **35** (1992) 301–312
16. Trick, M.A.: A Scheduling-Then-Break Approach to Sports Timetabling. In: Burke, E, Erben W. (Eds.): Practice and Theory of Automated Timetabling III (PATAT 2000, Konstanz, Germany, August, selected papers). Lecture Notes in Computer Science, Vol. 2079. Springer-Verlag, Berlin Heidelberg New York (2001) 242–253
17. van Weert, A., Schreuder, J.A.M.: Construction of Basic Match Schedules for Sports Competitions by Using Graph Theory. In: Burke, E, Carter, M. (Eds.): Practice and Theory of Automated Timetabling II (PATAT 1997, Toronto, Canada, August, selected papers). Lecture Notes in Computer Science, Vol. 1408. Springer-Verlag, Berlin Heidelberg New York (1998) 201–210
18. Willis, R.J., Terrill, B.J.: Scheduling the Australian State Cricket Season Using Simulated Annealing. J. Oper. Res. Soc. **45** (1994) 276–280
19. Wright, M.: Timetabling County Cricket Fixtures Using a Form of Tabu Search. J. Oper. Res. Soc. **45** (1994) 758–770

Solving the Travelling Tournament Problem: A Combined Integer Programming and Constraint Programming Approach

Kelly Easton[1], George Nemhauser[1], and Michael Trick[2]

[1] School of Industrial and Systems Engineering,
Georgia Institute of Technology,
Atlanta, GA 30332, USA
{keaston, george.nemhauser}@isye.gatech.edu
[2] Graduate School of Industrial Administration,
Carnegie Mellon, Pittsburgh, PA 15213, USA
trick@cmu.edu

Abstract. The Travelling Tournament Problem is a sports timetabling problem requiring production of a minimum distance double round-robin tournament for a group of n teams. Even small instances of this problem seem to be very difficult to solve. In this paper, we present the first provably optimal solution for an instance of eight teams. The solution methodology is a parallel implementation of a branch-and-price algorithm that uses integer programming to solve the master problem and constraint programming to solve the pricing problem. Additionally, constraint programming is used as a primal heuristic.

1 Introduction

The Travelling Tournament Problem (TTP) represents the fundamental issues involved in creating a schedule for sports leagues where the amount of team travel is an issue. For many of these leagues, the scheduling problem includes a myriad of constraints based on thousands of games and hundreds of team idiosyncrasies that vary in their content and importance from year to year, but at its heart are two basic requirements. The first is a feasibility issue in that the home and away pattern must be sufficiently varied so as to avoid long home stands and road trips. The second is the goal of preventing excessive travel. For simplicity, we state this objective as minimize total travel distance.

While each issue has been addressed by either the integer programming or constraint programming communities (and sometimes both), their combination is a relatively new problem for both groups. Constraint programming has been successfully used to solve complex sets of home and away pattern constraints. Integer programming methods have been used to solve large travelling salesman and vehicle routing problems that require minimizing travel distance. It might seem that insights from these solution methodologies would make the TTP relatively easy to solve. However, even very small instances of the TTP have proven

E. Burke and P. De Causmaecker (Eds.): PATAT 2002, LNCS 2740, pp. 100–109, 2003.

difficult for traditional methods. Thus the TTP seems to be a good problem for a combined approach.

The TTP was introduced in [6], where we gave results for some basic formulations. Here, we outline an approach that combines integer and constraint programming methods to solve larger problems much more quickly. This approach is appealing not just for its application to the TTP, but for the possibilities inherent for other difficult combinatorial problems.

In the following section, we give a formal definition for the Travelling Tournament Problem. In Section 3, we discuss the single-team problem, show its complexity, and discuss how that problem can be used to bound solutions on the TTP. In Section 4, we present our solution methodology. We then conclude with some computational results and a discussion of future research.

2   The Travelling Tournament Problem

Given n teams, n even, a double round-robin tournament (DRRT) is a set of games in which each team plays each other team once at home and once away. A schedule for a DRRT is a mapping of games to slots, or time periods, such that each team plays exactly once in each slot. A DRRT schedule covers exactly $2(n-1)$ slots. The distances between the team venues are given by an n by n matrix D. For the distance calculations, it is important to note that each team starts and finishes the tournament at its home venue.

A road trip, or trip, is defined as a series of consecutive away games. Similarly, a home stand is defined as a series of consecutive home games. The length of a road trip or home stand is the number of games in the series (not the travel distance).

The TTP is defined as follows:

Input. A set of n teams $T = \{t_1, \dots, t_n\}$ with n even; D a symmetric n by n integer distances matrix with elements d_{ij}; l, u integer parameters, $l \leq u$.

Output. A double round-robin tournament on the teams in T such that

- the length of every home stand and road trip is between l and u inclusive and
- the total distance travelled by the teams is minimized.

For $u = n - 1$, the maximum value for u, a team may visit every opponent without returning home, which is equivalent to a travelling salesman tour. For small u, a team must return home often, and consequently, its travel distance increases. For $u = 1$, the objective becomes constant, and the problem is solely one of feasibility. In practice, $l = 1$ and $u = 3$ or $u = 2$ are most commonly used. Additional background on these parameters and a description of other instances appear in [6].

Although we do not have a formal proof, we strongly believe that the TTP is NP-hard. Indeed, computationally, it appears to be much harder than the Travelling Salesman Problem (TSP). As shown in the next section, some simple relaxations of TTP are provably hard.

3   The Single Team Problem

Fundamental to our solution methodology is the idea of a tour. A tour is a vector that describes the travel of a single team. Each element in the vector is associated with a slot in the schedule and gives the venue for the game in that slot. For example, in an instance with $n = 6$, a tour for team 1 could be (1, 2, 3, 4, 1, 1, 5, 1, 6, 1). Note that each opponent's venue appears exactly once and the home venue appears exactly $n - 1$ times.

A team's optimal tour minimizes the travel distance for that team exclusive of the other teams in the schedule. The sum over n teams of the distances associated with their optimal tours provides a simple but strong lower bound on the TTP. We call this the Independent Lower Bound (ILB).

For most values of l and u, simply generating an optimal tour for a single team is NP-hard, although there exist some polynomial solvable special cases. Here we study the complexity of generating a minimum distance tour for each (feasible) combination of l and u. We will assume that the triangle inequality holds for the distances. The decision form of the Single Team Problem (STP) is formally defined as

Input. A set of n teams $T = \{t_1, \ldots, t_n\}$ with n even; a designated team $t_r \in T$; D a symmetric n by n integer distance matrix with elements d_{ij} satisfying the triangle inequality; l, u integer parameters; integer threshold k.
Output. Does there exist a tour for t_r such that

- the length of every home stand and road trip is between l and u inclusive and
- the sum of the distance travelled by t_r is less than or equal to k?

For specified l and u, we use the notation STP(l, u).

At the extremes of l and u, the complexity of STP(l, u) is obvious. For STP(1, 1), the objective is constant, and a solution can be obtained by alternating away and home games. For STP($l, n - 1$), the problem is a TSP and is NP-complete. In fact, as long as u is proportional to n^α for a fixed constant α, there is an immediate reduction from the TSP to prove completeness.

The STP is also NP-complete for any constant $u > 2$ and $l \le u$. The reduction is from the NP-complete problem Partition into Isomorphic Subgraphs Restricted to Paths [9]. The proof of this is given in [5].

In the special case where $u = 2$ ($l = 1$ is the only feasible assignment for l since n is even), an $O(n^3)$ algorithm can be obtained using weighted matchings to find the optimal tour. To see this first note that there will always be an optimal tour with the minimum number of trips. Two single team trips may be combined into one two team trip with a distance no more than the sum of the two single team trips by the triangle inequality. Secondly, since we are working with even n, a minimum set of trips will always include exactly one single team trip. Given these observations, the following algorithm finds a minimum distance tour.

Algorithm STP(1, 2).

1. Given team t_r, generate a complete graph $G = (V, E)$ with n nodes. Let vertex i represent team t_i for all $i \in \{1, \ldots, n\}$. Assign weights to the edges as follows: Let $w_{pq} = d_{rp} + d_{pq} + d_{qr}$ for all $p, q \in \{1, \ldots, n\} \setminus \{r\}, p \neq q$. Set $w_{rp} = 2 * d_{rp}$ for all $p \in \{1, \ldots, n\} \setminus \{r\}$.
2. Find a minimum weight perfect matching on G.
3. Create a minimum distance tour from the matching. For each matched pair of nodes (p, q) such that $p, q \in \{1, \ldots, n\} \setminus \{r\}$, create a two game road trip with t_r playing at t_p then at t_q. For the node p that is matched to r, create a single team trip in which t_r plays at t_p. In addition, create a set of home stands mimicking the trips, i.e. one home stand with one slot and $n/2 - 1$ home stands with two trips. Finally, schedule the tour by alternately selecting trips and home stands and scheduling them consecutively in the slots of the tournament.

Theorem 1. *Algorithm STP(1,2) yields an optimal solution.*

Proof. Every feasible tour with a minimum number of trips corresponds to a perfect matching in G with the distance of the tour equal to the weight of the matching. Therefore, finding a minimum weight perfect matching in G yields an optimal solution to STP(1,2).   □

The running time of this algorithm is dominated by the work required to solve the matching. A weighted perfect matching on a general graph can be found in $O(|E||V| + \log(|V|)|V|^2) = O(n^3)$ [7]. Therefore, the algorithm runs in $O(n^3)$. This algorithm can be easily generalized in the case the triangle inequality does not hold by using b-matchings instead of perfect matchings. This use of matchings in the case where trips consist of at most a pair of teams is fundamental to the approaches of [2] and [11].

4   Solution Methodology

The TTP has feasibility elements (the home and away pattern) and optimization elements (the travel distance). Constraint programming has been successfully used to solve complex timetabling feasibility problems [8] while integer programming has been successfully used to solve difficult optimality problems like the TSP [1]. The combination of feasibility and optimality in a timetabling problem seems to be difficult for either method, thus we chose a combined approach.

Our solution methodology for the TTP is a branch-and-price (column generation) algorithm in which individual team tours are the columns. In branch-and-price, the linear programming (LP) relaxation at the root node of the branch-and-bound tree includes only a small subset of the columns. To check the LP objective, a subproblem, called a pricing problem, is solved to determine whether there are any additional columns available to enter the basis. If the pricing problem returns one or more columns, the LP is reoptimized. If no more columns can be found to enter the basis and the LP solution is fractional, the algorithm branches. Branch-and-price is a generalization of branch-and-bound with LP relaxations. In our combined integer programming-constraint programming approach, we use constraint programming to solve the pricing problem.

Although the numbers of teams in the instances of the TTP that we have been able to solve seem small, these problems are actually quite large in terms of the numbers of possible columns. For an instance with eight teams there are roughly a total of 4.5 million tours. In order to solve these instances in a reasonable amount of time, we run a parallel version of our branch-and-price algorithm. We use the master–slave paradigm. After generating an initial set of nodes, the master processor passes one node to each slave along with the best known solution (if any). Each slave evaluates a fixed number of nodes and passes back its best solution and all the nodes it has generated, specifying which nodes have been evaluated and which have not. If a slave completely fathoms its portion of the tree, no nodes are returned to the master. The master maintains the best solution reported by the slaves and a list of all unevaluated nodes. A single node from this list and the current best solution are passed to the now idle slave for processing. When the master's list is empty and all the processors are idle, the algorithm is complete and the current best solution is an optimum.

At the root node, we start with a small set of columns $P = \{1, 2, \ldots, m\}$ composed of the optimal tours for each team. Let the vector c represent the distances associated with the tours, T be the set of teams in the tournament and S be the set of slots. Let P_t be the indices of tours for team $t \in T$ and x be a binary decision variable that represents the tours. This leads to an integer program:

Minimize $\sum_{i \in P} c_i x_i$

subject to

$\sum_{i \in P_t} x_i = 1, \quad \forall t \in T,$

$\sum_{\{i \in P : i \notin P_t \text{ and } t \text{ is opponent in slot } s \text{ in } i\}} x_i$
$\quad + \sum_{\{i \in P_t : t \text{ is away in slot } s \text{ in } i\}} x_i = 1, \quad \forall s \in S, \forall t \in T,$

x_i binary for $i \in P$.

The first set of constraints forces exactly one tour to be selected for each team in the tournament. The second set of constraints requires each team to play exactly once in each slot.

In some branch-and-price algorithms, columns are generated until no more negative reduced cost columns remain. In others, columns are generated only if a node is about to be cut off. The branch-and-price routine described here falls into the latter category. One reason for this is that if fewer columns are generated, the master problem will solve more quickly. More importantly, the pricing problem is easier to solve in the latter case. So generating columns only if necessary reduces the time it takes to solve both the master and the pricing problem. By not always generating all columns, however, we are unable to use a best bound node selection strategy. With parallel programming, however, it is difficult to assess the best bound at any given time, so this disadvantage is not key.

Prior research using branch-and-price algorithms indicates that it is most efficient to generate a pool of columns, select from this pool at each node, then generate a new pool when necessary [3]. Thus at each node in our search tree, we first check to see if there are any negative reduced-cost columns in the pool. If so, we select up to 10 columns for each team. If not, we refill the pool. The pricing problem is solved using constraint programming.

When invoked, the constraint program for the subproblem is run once for each team. It generates all negative reduced-cost tours. The variables are the venues at which the team's games are played in each slot. The domains of these variables are the teams in the tournament. Two dummy variables are added for slots 0 and $2(n-1)+1$. These variables are both set equal to the home team and are used for distance calculations. The distances between consecutive venues are also variables but are used strictly for calculating the objective or goal. Before the CP is run, the variable domains are reduced according to prior branching.

The constraints in this model are as follows (for the $l = 1, u = 3$ case):

- each opponent venue appears exactly once,
- the home venue appears $n - 1$ times,
- no more than three home venues may appear consecutively,
- no more than three non-home venues may appear consecutively,
- the sum of the distances and the dual values associated with the games in the tour must be negative (equivalently, the reduced cost of the resulting tour must be negative).

Variables are selected in order of domain size. In the case that all uninstantiated variables have domains of equal size, the variable closest to an end of the tournament (either the first or the last uninstantiated slot) is selected. Ties in this second criteria are broken arbitrarily. Once a variable is selected, it is instantiated with the team in its domain that will make the most negative (or least positive) contribution to the total cost of the tour which includes the distances between consecutive venues and the reduced costs associated with opponents and slots.

A variable instantiation triggers domain reduction for the remaining uninstantiated variables. If an away game is being scheduled, the algorithm first eliminates that opponent from further consideration. Otherwise, a home game is being scheduled, and the home venue is eliminated from further consideration only if the home game tally reaches $n - 1$. Secondly, the pattern constraints are considered. Any members of the remaining domains that would create road trips or home stands longer than three slots are eliminated. Finally, the contribution to the objective of potential assignments is considered. Generating new columns is a fairly simple procedure, and we have observed no runtime savings from adding a degree of look ahead to our algorithm. If a domain is reduced to no remaining elements, we backtrack to the last decision point. If a domain is reduced to one, the variable is marked. At the completion of an iteration of constraint propagation, the algorithm selects a marked variable (if there are any), instantiates it with the single member of its domain, and begins another iteration of constraint propagation. We use our own code to solve the pricing problem, but a model in OPL is given below for clarity.

```
int home = 1;
int  nbTeams = 8;
range Teams 1..nbTeams;
int nbSlots = 14;
range Slots 1..nbSlots;
float EPS = 0.0001;
int Distance[Teams,Teams] = ...;
float rc[Teams,Slots] = ...;
var int travel[0..nbSlots] in 0..1380;
var int venue[0..nbSlots+1] in 1..nbTeams;
var float totcost;

solve {

/* Venue variables */
    forall (i in Teams: i <> home) sum (s in Slots) (venue[s]=i)=1;
      sum (s in Slots) (venue[s]=home)=0.5*nbSlots;
      venue[0]=home;
      venue[nbSlots+1]=home;
      forall (s in 1..nbSlots-4) sum (j in 0..3) (venue[s+j]=home) <= 3;
      forall (s in 1..nbSlots-4) sum (j in 0..3) (venue[s+j] <> home) <= 3;

/* Distance calculations */
    forall (s in 0..nbSlots) travel[s]=Distance[venue[s],venue[s+1]];

/* Generate only negative reduced cost columns */
    totcost = sum (s in 0..nbSlots) travel[s] +sum (s in Slots) rc[venue[s],s];
    totcost <= -EPS;

};
```

We have found that it is helpful to populate the column pool with "good" tours (relatively small distances) in addition to tours with negative reduced costs. Thus we add tours to the pool with distances within an increment of the ILB, increasing this increment iteratively. This reduces the number of times we need to refill the column pool.

As discussed above, parallel programming makes it difficult to determine the current best bound at any given time. For this reason, we use a simple depth first node selection strategy.

Rather than branching on a tour, we use higher-order branching variables. Our higher-order variables are the patterns indexed by team and slot. In other words, at any given node in which the master LP value is less than our current best solution, we divide the solution space into schedules in which team t is home in slot s and schedules in which team t is away in slot s. In order to select a higher order variable on which to branch, we use a strategy known as strong branching. We create a candidate list by selecting the 10 higher order variables with total weight closest to 0.5. For each variable in the candidate list, we do 50 iterations of the simplex method, setting the variable equal to 0 in one case and equal to 1 in a second case. We then select the variable with the largest total change in objective value summed over both cases. A similar strong branching strategy was used successfully in [10].

The final component of our solution methodology is a primal heuristic that makes use of an expanded version of the constraint program for the pricing problem. The primal heuristic works on the whole schedule, as opposed to just one tour, and the model is modified accordingly. Specifically, the variable arrays are expanded to include n teams, and the following constraints are added:

- No more than two teams can play at one venue in one slot;
- If a team plays away, its opponent must play at home.

The objective is to minimize the distances summed over all teams and slots.

Simply running this model does not produce "good" solutions quickly enough; however, by fixing certain elements we can generate solutions within 5–6% of the ILB in a reasonable amount of time. Specifically, we select $n/2$ teams that are each forced to play one of their optimal tours. Note this does not mean that we select $n/2$ columns to fix. A team's set of optimal tours includes all tours with minimum distance. "Fixing" a team amounts to adding a set of constraints to the model regarding trips the selected team may play. Once we have chosen a set of "fixed" teams, we run the constraint program for a specific time interval, outputting solutions and improved solutions as they are generated. Variable and value selection strategies are similar to those used in the pricing problem. We instantiate the variables associated with the fixed teams first. Domain reduction considers the inter-team constraints after the intra-team constraints and before considering objective contribution.

At the end of the set time interval, we chose a new set of $n/2$ teams and run the model again. Fixed teams are chosen in order of proximity to the estimated center of the geographical region defined by the teams participating in the tournament. One processor is dedicated to running the primal heuristic. The master processor checks the primal heuristic output file at intervals that depend on the size of the instance. If a better integer solution has been found, it is sent to the slave processors along with the next nodes they are given to process.

5   Computational Results

The TTP was created to capture the essence of real sports leagues, Major League Baseball (MLB) in particular. Unfortunately, MLB has far too many teams for the current state-of-the-art for finding optimal solutions. MLB is divided into two leagues: the National League and the American League. Almost all of the games each team plays are against teams in its own league, so it is reasonable to limit analysis to an individual league.

Table 1. Results for TTP Instances

Name	l	u	ILB	Optimal solution	Processors	Time (s)
NL4	1	3	8 276	8 276	1	30
NL6	1	3	22 969	23 552	20	912
NL8	1	3	38 670	39 479	20	362 630

We have generated the National League distance matrices by using "air distances" from the city centres. To generate smaller instances, we simply take subsets of the teams. In doing so, we create instances NL4, NL6, NL8, NL10, NL12, NL14 and NL16, in which the number indicates the number of teams in the instance. All of these instances are on the challenge page associated with this work: `http://mat.gsia.cmu.edu/TOURN`. In this work, however, we do not impose the "no back-to-back" games constraint, so the values here are not directly comparable to those on the web page.

Instances with $n = 4$ are nearly trivial to solve. Instances with $n = 6$ are more challenging. We have found several models that can solve these instances in a reasonable amount of time without parallel programming. When 20 processors are used to solve instances with $n = 6$, the computation time is on the order of minutes. Finally, we have found that it is necessary to use parallel programming to solve instances with $n = 8$ teams. On 20 processors, these problems take approximately four days.

The first table gives wall-clock times for finding and proving optimal solutions to NL4, NL6 and NL8. These problems were solved on a network of PCs with 300 MHz Pentium II processors and 512 MB RAM running Redhat Linux 7.1.

Another approach for combining constraint programming and integer programming for this problem was presented in [4]. Their work was able to solve the size 6 instance but could not solve the size 8 instance.

Additionally, we have obtained the bounds for NL16 by running the constraint program from our primal heuristic on a single processor for 24 hours of computation time. We get a lower bound of 248 852 and an upper bound of 312 623.

6   Future Research

While the large increase in solution time between NL6 and NL8 leads us to believe that we will not be able to solve NL16 to optimality, we do hope to solve at least NL10. More broadly, however, we believe the framework underlying our solution methodology may be used to develop algorithms for solving a wide range of difficult timetabling and other scheduling problems with competing feasibility and optimality features.

Acknowledgements. The authors would like to thank CPLEX, a division of ILOG, for its software support. This work was supported, in part, by NSF grants DMI-0101020 and DMI-0121495 to the Georgia Institute of Technology.

References

1. Applegate, D., Bixby, R., Chvatal, V., Cook, W.: On the Solution of Traveling Salesman Problems. Documenta Mathematica (Journal der Deutschen Mathematiker-Veringung, International Congress of Mathematicians) (1998) 645–656
2. Ball, B.C., Webster D.B.: Optimal Scheduling for Even-Numbered Team Athletic Conferences. AIIE Trans. (1977) **9** 161–169
3. Barnhart, C., Johnson, E.L., Nemhauser, G.L., Savelsbergh, M.W.P., Vance, P.H.: Branch-And-Price: Column Generation for Huge Integer Programs. Oper. Res. (1998) **46** 316–329
4. Benoist, T., Laburthe, F., Rottembourg, B.: Lagrange Relaxation and Constraint Programming Collaborative Schemes for Traveling Tournament Problems. CPAI-OR, Wye College (2001) 15–26
5. Easton, K.: Using Integer and Constraint Programming to Solve Sports Scheduling Problems. Doctoral Dissertation, Georgia Institute of Technology, Atlanta (2002)
6. Easton, K., Nemhauser, G.L., Trick M.A.: The Traveling Tournament Problem: Description and Benchmarks, Principle and Practices of Constraint Programming. In: Proc. CP'01. Lecture Notes in Computer Science, Vol. 2239. Springer-Verlag, Berlin Heidelberg New York (2001) 580–585
7. Gabow, H.N.: Data Structures for Weighted Matching and Nearest Common Ancestors with Linking. In: Proc. 1st Annu. ACM–SIAM Symp. Discr. Algorithms. SIAM, Philadelphia, PA (1990) 434–443
8. Henz, M.: Scheduling a Major College Basketball Conference: Revisted. Oper. Res. (2001) **49** 163–168
9. Kirkpatrick, D.G., Hell, P.: On the Complexity of a Generalized Matching Problem. In: Proc. 10th Annu. ACM Symp. Theory Computing. Association for Computing Machinery, New York (1978) 240–245
10. Klabjan, D., Johnson, E.L., Nemhauser G.L.: Solving Large Airline Crew Scheduling Problems: Random Pairing Generation and Strong Branching. Comput. Optim. Appl. (2001) **20** 73–91
11. Russell, R.A., Leung, J.M.: Devising a Cost Effective Schedule for a Baseball League. Oper. Res. (1994) **42** 614–625

Employee Timetabling

Personnel Scheduling in Laboratories

Philip Franses[1] and Gerhard Post[1,2]

[1] ORTEC Consultants BV,
Osloweg 131, 9723 BK Groningen, The Netherlands
[2] Department of Mathematical Sciences,
University Twente, PO Box 217, 7500 AE Enschede, The Netherlands

Abstract. We describe an assignment problem particular to the personnel scheduling of organisations such as laboratories. Here we have to assign tasks to employees. We focus on the situation where this assignment problem reduces to constructing maximal matchings in a set of interrelated bipartite graphs. We describe in detail how the continuity of tasks over the week is achieved to suit the wishes of the planner. Finally, we discuss the implementation of the algorithm in the package IPS. Its main characteristic is the introduction of profiles, which easily allows the user to steer the algorithm.

1 Introduction

Our topic is the scheduling of tasks or workplaces of workers in laboratories, chemists, x-ray departments and operation departments. In such laboratories the employees usually have fixed working hours (between 8AM and 5PM); irregular shifts do not play an important role. During working hours, a number of tasks are to be carried out by skilled employees during a period of at least half a day. To do this planning by hand for 50 employees is very time-consuming. The system IPS originally provided a priority-based algorithm for the task scheduling. It turned out to be hard to steer this algorithm to an acceptable result. The principal idea of the generator we will describe is the use of profiles. In a profile a subset of employees and tasks is fixed, and the generator tries to find as good as possible match between them. However, even in this stage we take into account the other tasks. A final wish is that the generator should provide a quick result (in one or two minutes).

2 Problem Description and Model

2.1 Basic Model

We assume that we have a set of W workers, numbered by $i = 1, \ldots, W$, with T tasks indexed by $j = 1, \ldots, T$ and P periods, $k = 1, \ldots, P$. In each period k a worker i can be available or not for task j. There are two principal reasons for unavailability: first, it can be that a worker is not present (most of the workers

E. Burke and P. De Causmaecker (Eds.): PATAT 2002, LNCS 2740, pp. 113–119, 2003.

are part-time workers). Another reason for unavailability for a task, is that a worker is not skilled for it. We introduce the 0–1 variables a_{ijk} to denote this:

$$a_{ijk} = \begin{cases} 1 & \text{if } i \text{ is available for task } j \text{ in period } k, \\ 0 & \text{otherwise}. \end{cases}$$

Our assumption is that a_{ijk} is known in advance.

Next we introduce the 0–1 decision variables x_{ijk}, by

$$x_{ijk} = \begin{cases} 1 & \text{if } i \text{ is assigned to task } j \text{ in period } k, \\ 0 & \text{otherwise}. \end{cases}$$

Our first condition is then

$$x_{ijk} \leq a_{ijk} \qquad \text{for all } i, j, k \tag{1}$$

which represents the availabilities. We assume that a worker can do only one task in each period, leading to the condition

$$\sum_{j=1}^{T} x_{ijk} \leq 1 \qquad \text{for all } i, k. \tag{2}$$

Finally, for each task j and period k we have demands, such as the number of employees b_{jk} that are needed to work on task j in period k. These can be formulated as

$$\sum_{i=1}^{W} x_{ijk} \geq b_{jk} \qquad \text{for all } j, k. \tag{3}$$

We put $\geq$ here and not equality; it is easy to remove some assignments to obtain equality.

Equations (1–3) constitute our basic model. Note that for each period k we essentially search for a maximal matching in a bipartite graph. The bipartite graph has "task nodes": we have b_{jk} copies for task j. Moreover, there are "resource nodes": i.e. all workers i that are available in period k; those with $a_{ijk} \neq 0$ for a j. We connect a task node with a resource node when the worker (corresponding to the resource node) is available for the task (corresponding to the task node).

2.2   Objectives

In practice, there are a number of factors influencing which task is done by which employee. An important aspect we have to cope with is the "history". By this we mean how often a person has worked on each of the tasks in the previous period. The tasks for the week we want to plan are usually aimed to provide variation from the previous weeks, but one can also aim to maintain workers in the same tasks. Moreover, in certain professions there are legal regulations that require an employee to maintain his skills. This means that in the period of, say, a year an employee must perform a certain task a number of times. The history can be incorporated in the bipartite graph formulation. To each edge we add costs which are related to the history: in our implementation we simply count the number of times a task was executed by an employee; the corresponding edge gets this number as its cost. If we denote the costs that employee i works on task j by c_{ij}, the cost function $f(x)$ becomes

$$f(x) = \sum_{i,j,k} c_{ij} x_{ijk} . \tag{4}$$

The Hungarian method for finding a maximal matching can be adjusted in order to minimise this cost function: see for example [1].

This does not, however, take into account the practice of how these tasks are assigned. Specifically the continuity of the planning is built upon how the tasks are shared over the week. At present three types of tasks are implemented in IPS:

half-day task: This kind of task is required to circulate as much as possible among all people that are qualified to do it. Hence in the morning and the afternoon different people should do this task.

day task: This task should be assigned to the same person during the morning and the afternoon, but the next day a different task for this person is required.

week task: This task should be assigned for the whole week to one person.

The individual situation determines which type of task "day task" or "week task" is allocated to each task: there is sometimes a preference to give the employees the same task during the day, or in other situations for the whole week. The first type, the "half-day task", is usually connected with the boring tasks. As an example one can think of the task of answering the telephone.

The introduction of the types of tasks complicates our problem. If we have week tasks, an idea would be to require that $x_{ijk_1} = x_{ijk_2}$ for all i, j, k_1, k_2 (worker i is doing the same task in all periods). Our IP problem (1)–(4) is now in most cases infeasible: as we have said, most workers are part-time workers, and it is very unlikely that the demands can be fulfilled with workers assigned to a single task. If by chance a feasible solution exists, it is probably one of few: consequently the costs are hardly taken into account. Note that this problem is a binpacking problem: the tasks during the week can be seen as the bins, while each employee is an item of size s, where s is the number of half-days, that the employee of available. Binpacking is an NP-hard problem, see for instance [2].

2.3 Profiles

There are usually further limitations in finding an acceptable planning. We mention two important ones, though the mechanism of "profiles" (see below) is general enough to incorporate additional limitations.

1. *Priorities.* Some tasks are more important than other tasks. This means that these tasks should be assigned first to cater for cases where not all tasks can be assigned.
2. *Supervisors.* There are tasks that require the presence of extra qualified employees. The task is accomplished by a team, one member of which acts as a supervisor without the task "supervisor" being separately specified.

To incorporate the extra requirements we work with "profiles". A profile consists of

- a number of tasks;
- a number of employees;
- type of tasks (half-day, day or week);
- relation to history (same tasks as before or not);
- next profile.

The chain of profiles is treated consecutively, thus creating a multi-objective function. At first the top profile is active. We select all tasks and employees in it, and give to the combination of task and employee the cost depending on history, and the relation to the history, as specified in the profile. The tasks and employees in the lower profiles are also incorporated; however, these get very high costs, equal for all of them. This expresses the fact that, at the moment, we are not interested *where* these employees are scheduled, but only *that* it is possible to schedule them. The actual scheduling of these lower profiles will be done at a later stage.

The profiles give the user an excellent possibility to steer the algorithm. For instance, suppose we have a week task 1 and a day task 2 that can be assigned to employee A. If we first assign task 1 (thus filling the whole week), it will not be possible to assign employee A to task 2. This might be undesirable. This can be avoided by putting task 2 and employee A in the first profile, and task 1 (and once again employee A) in the next profile. In this way we defer the assignment of task 1 to employee A.

3   The Algorithm

The algorithm we have designed consists of two steps: the first step is finding a maximal matching in all periods, the second phase consists of improving this matching. The working of the algorithm is described below. In practice, a period corresponds to a half-day, and the set of all periods corresponds to a week.

3.1   Phase I. Matching

We use the bipartite graph model as starting point for our algorithm, and improve it by local search. In each period we use the weighted Hungarian method to find the maximal matching with minimal costs. At this point it is good to recall some properties of the Hungarian method.

- We can make the Hungarian method work from top to bottom. So we arrange the tasks as they occur in the profiles (top profile first), and start the Hungarian method. In this way we first assign the most important tasks. Hence, the maximal matching we obtain contains the most important tasks (if this is possible).
- What is assigned once, always stays assigned. In the local search phase we can remove some assignments to obtain a better solution. We can be sure that all other assignments remain: i.e. all tasks (employees) remain assigned, though not necessarily to the same employee (task).

program
```
    initialize costs due to history;
    profile := top_profile;
    while non-empty(profile) do
        type := type_of(profile);
        make all costs outside profile high;
        for k := 1 to P do
            calculate maximal matching;
            adjust costs in next periods depending on type;
        end for;
        improve matching within profile according to type;
        fix assignments within profile;
        profile := next(profile);
    end while
end program
```

If the top profile contains half-day tasks, we adjust the costs belonging to this profile after a matching for a half-day is made. Hence, after the Monday morning is matched (task A to employee 1), we put the Monday afternoon cost for this combination A–1 at a very high level. Moreover, the costs for the next days are also increased. After the matching is completed for the whole week, we check whether any tasks are scheduled in the morning and afternoon to the same employee. If so, we try to change the morning assignment, by increasing the corresponding cost.

If the top profile consists of day tasks we proceed more or less the same as with half-day tasks. However, the afternoon should be the same as the morning, so that we decrease the afternoon cost. The next days' costs are raised as above.

3.2 Phase II. Improvement

The algorithm in phase I traverses each of the half-days, from beginning to the end of the week. Some costs for future possibilities can be adjusted during this process when dealing with half-day and day tasks. In phase II we aim to improve the matching, depending on the type of task. Hence, the original costs are discarded. For half-day tasks and day tasks, we try to adjust morning assignments, so that morning and afternoon tasks are the same, by adjusting the costs for the morning.

The procedure for week tasks is more intricate and interesting. In phase II a new score is considered; we call it score to distinguish it from the costs of phase I (moreover, a high score is better than a low one). This score is composed of two parts: the first part is the number of half-days that the employee is available for tasks. The second part is the number of times that the task is assigned to the employee. We proceed then in a greedy way: first the combination task–employee with the highest score is found. We check if it is possible, without losing maximal matching, to assign this combination during the whole week. If so, we assign the combination permanently (we fix it), and we proceed with the next best task–employee combination:

```
procedure improve(profile, weektask)
    matrix S(i: worker; j: task);
    for all (i, j) in profile do
        if SkilledForTask(i, j)
            then S(i, j) := #(available periods of i) + #(j assigned to i)
            else S(i, j) := −∞;
        end if;
    end for all;
    while max_{i,j} S(i, j) > −∞ do
        Find i*, j* with S(i*, j*) = max_{i,j} S(i, j);
        Remove availabilities and assignments of i* for other tasks;
        if Can match all tasks
            then Fix assignments of i*
            else Restore original matching
        end if;
        S(i*, j*) := −∞;
    end while;
end procedure.
```

In practical situations this procedure works quite well: usually the tasks for several persons are already the same or "almost" the same during the week. Employees that work the most during the week we are trying to plan, are assigned the week tasks first, due to the score we introduced. This seems quite reasonable, since these employees are the most difficult to fit in at the end.

When all task–employee combinations have been tried, we come to the final step. We repeat the procedure above, but now without the requirement of assignment for the whole week. So we calculate the scores in the same way as above, discarding all that was already fixed in the previous round, and find the best task–employee combination. We assign this combination, and lower all other scores of this employee. This forces the algorithm to consider first other employees, i.e. those for which no tasks have been found so far. Once all employees have been considered, we continue with the tasks that are left over. In this way we try to diminish the number of different tasks per employee.

4   Conclusion

The algorithm has been rigorously tested (and is now in daily use over several departments) at Canisius Wilhelmina Ziekenhuis Nijmegen administration and x-ray departments and in the Medisch Laboratorium Noord, Groningen in two laboratory departments. Critical points on which the algorithm has been judged are the sharing of unwanted shifts such as telephone duty and the continuity in the allocation of standard tasks, workers doing where possible the same task over a week. Use is made of the profiles (up to six) to generate a planning over the various task-sorts.

The algorithm has become an integral part of the planning process. In the summary written by MLN about their method of working, item 6 is "generator plans tasks in per week". By an ingenious use of the possibilities of the profiles, MLN have tailored the generator to routinely solve a part of their planning puzzle.

The success of the algorithm lies in the fact that the generator improves the success of task planning, while providing an enormous saving in the time previously needed to make the plan. The planner is able to trust that the algorithm provides a solution that both planner and workers can happily accept. Following on the success of the above-mentioned test-sites, the generator is also in increasing use in the growing number (now over 30) of hospital departments where IPS is used.

References

1. Bondy, J.A., Murty, U.S.R.: Graph Theory with Application. Macmillan, London (1976)
2. Coffman Jr., E.G., Garey, M.R., Johnson, D.S.: Approximation Algorithms for Bin Packing: A Survey. In: Hochbaum, D. (Ed.): Approximation Algorithms for NP-Hard Problems. PWS Publishing, Boston, MA (1997) 46–93

Scheduling Doctors for Clinical Training Unit Rounds Using Tabu Optimization

Christine A. White[1] and George M. White[2]

[1] Department of Internal Medicine,
University of Ottawa,
Ottawa K1N 6N5, Canada
[2] School of Information Technology and Engineering,
University of Ottawa,
Ottawa K1N 6N5, Canada
white@site.uottawa.ca

Abstract. Hospitals must be staffed 24 hours a day, seven days a week by teams of doctors having certain combinations of skills. The construction of schedules for these doctors and the medical students who work with them is known to be a difficult NP-complete problem known as *personnel scheduling, employee timetabling, labour scheduling* or *rostering*. We have constructed a program that uses a constraint logic formalism to enforce certain scheduling rules followed by a tabu search heuristic optimizing algorithm to produce a call schedule that is used at the Ottawa Hospital. This call schedule can be later changed by the chief resident to accommodate last-minute personnel changes by means of a spreadsheet-based program.

1 Introduction

At the Ottawa Hospital Clinical Teaching Unit, the chief resident is responsible for casting and posting the call schedule that assigns medical staff to their overnight duty rosters. A poll of former chief residents reported that the single most unpleasant duty they faced was the creation of these schedules by hand and the reaction of the staff when they were posted. It was suggested that a scheduling program might be able to provide better-quality solutions than can be done manually, and with a good deal less time and effort.

Schedules that are intended to be used in a real setting have to be solved in their entirety, without omission of anything relevant. This includes the so-called *soft constraints* which take into account religious holidays, family affairs, vacations and training sessions. These soft constraints or side constraints can determine whether the schedule is widely acceptable to both staff and management or whether it will be damaging to institutional morale. Therefore, the common practice of eliminating awkward side constraints is not permissible. Because of the difficulty of finding solutions that are acceptable to those concerned, a number of heuristic techniques have been brought to bear on these problems by

E. Burke and P. De Causmaecker (Eds.): PATAT 2002, LNCS 2740, pp. 120–128, 2003.
© Springer-Verlag Berlin Heidelberg 2003

several groups of researchers, and results have been reported for several similar problem instances in both medical and non-medical settings.

Carter and Lapierre [5] have listed the constraints that govern the scheduling of physicians in an emergency ward. Cowling et al. [4] have employed a *hyper-heuristic* approach to personnel scheduling by means of a mechanism to guide the specific heuristic optimization method used depending on the properties of the solution space being examined at a given time. Chan and Weil [8] have generated solutions to problems similar to ours using a constraint logic programming (CLP) approach implemented in CHIP. Harald Meyer auf'm Hofe [18] investigated the incorporation of fuzzy constraints and branch & bound procedures into a constraint logic approach in his generation of solutions for nurse rostering in German hospitals. Similar approaches using constraint logic have described by Abdennadher and Schlenker [1,2]. Cheng et al. [6] used what they termed *redundant modelling* within a CLP formulation to reduce search time. A genetic algorithm approach was used by Kragelund [16] and an evolutionary algorithm approach has been studied by Jan et al. [13].

Franses and Post [10] have recently constructed a system for the staffing of health care laboratories that creates an optimal schedule for certain criteria and then improves it by local search. The optimal schedule uses the weighted Hungarian method and subsequent treatment implements local search in a two-part process. Solutions generated by networks of scheduling agents have been investigated by Meisels and Kaplansky [17]. The agents use an exhaustive search on a portion of the solution space and send messages to a central agent that looks for global constraint violations. Case-based reasoning has been used by Petrovic et al. [19] for nurse rostering. The possibility of relaxing "hard" constraints has been investigated by De Causmaecker and Vanden Berghe [3]. Sometimes the specific scheduling instance is found to be overconstrained and when this is discovered, some of the hard constraints are relaxed. This appears to increase the satisfaction of personnel.

2   The Rostering Problem

There are many possible schedules for the rounds of medical personnel in a hospital that do not violate any "hard" constraints, for example that a person cannot be scheduled to do two different tasks simultaneously. These are called *feasible* schedules. However, feasible schedules can have very pronounced differences that can lead to one being considered much better than another. Some of these schedules may be considered quite bad. A schedule that required someone to work two consecutive rounds or to work every weekend would not be well regarded. Some of the constraints are imposed by conservation laws, others by union rules and others by cherished traditions that flourish within the unit. It seems that no two hospitals have the same set of constraints [5,11]. Basic circadian rhythms must also be considered. Costa [7] and Knauth [14,15] summarize some useful guidelines for casting duty rosters.

Table 1. Conditions and their penalties

Condition	Penalty
One student missing	5
Student replaces missing junior; senior is on student's team	10
Two students missing	20
Junior and student missing;	
student takes junior's place; senior is on student's team	40
Student replaces missing junior; senior is not on student's team	80
Junior and student missing;	
student takes junior's place; senior is not on student's team	100
Two juniors missing; replaced by two students	300
Anything else	500

In the end, the "best" schedule is the one that pleases most of the people involved the most of the time.

For our purposes, medical personnel are experienced doctors, newer doctors and medical students, classified by *rank*, seniors, juniors and students, and by *unit*, team A, team B, GM-consult, Ambulatory, Float, second year, and third year (for seniors), team A and team B (for juniors and students). During any round, medical duties are supported by a senior and two teams consisting of a junior (if available) and a student (if available). The most desirable situation occurs when the unit is staffed by a senior, two juniors and two students; unfortunately this occurs rarely because of the continuing shortage of medical staff. Usually, the round is supported by a mandatory senior and either two juniors and a student, a junior and two students or two students with an additional Float senior. If a student is on duty without a junior from the same team, the senior's unit should be the same as that of the junior to provide additional support. Each of these possible combinations is assigned a *penalty* ranging from 0 for the best case, a senior, two juniors and two students, to 300 for the worst case, a senior, no juniors and two students, accompanied by an additional Float senior. The penalties and the condition that causes them to be applied are shown in Table 1.

These penalties can be adjusted at run time by the operator to represent the perceived seriousness of personnel deficiencies at various times. These may vary depending on the mix of available staff.

The medical staff may take vacations at random times and are required to work a maximum of seven calls in any 28 day scheduling block. This maximum number of calls may be reduced depending on the number of vacation days taken, and the rank of the person involved.

Calls should be spaced out evenly over the period and weekends should be assigned fairly to all personnel respecting a pattern that groups a Friday and a Sunday call in one weekend and a Saturday call in another. Staff must not work two consecutive shifts and the number of patterns of two shifts in three days should be minimized. The penalty for consecutive shifts is set to 500 currently but can be adjusted at run time. Some other rules are in effect dealing with long weekends and certain other personnel matters.

The penalties can be divided into two broad classes. The composition of staff on duty during a given call is governed by rank and team considerations that are referred to as *horizontal* constraints. The penalties that are incurred because the weekend shifts are not as they should be or because a member has to work two shifts in three days are due to *vertical* constraints. These two sets conflict directly with each other and are the chief obstacle to casting perfect schedules. Depending on the number and the mix of the medical staff on duty during a block the operator may choose to favour the vertical constraints over the horizontal constraints or vice versa. A constraint can be effectively eliminated by setting its penalty to zero.

3   The Algorithm

An automatic scheduler has been developed that uses tabu search [12] to perform an heuristic optimization of the scheduling space in an attempt to find the least objectionable schedule. An initial call schedule is generated by a combination of constraint logic that satisfies the weekend requirement followed by a simple bin packing procedure that satisfies the requirements of the rest of the week and considers vacation and rank factors. The weekend requirement specifies that a Friday and the following Sunday plus a Saturday from a different weekend be assigned to staff where possible. This would be very simple to do if it were not for holidays, vacations and days off which can occur any time. The weekend requirement is enforced by a simple chronological backtracking algorithm. The weekday slots are then filled in turn by finding a suitable staff member who is not already working, is available, and has not yet completed the required number of calls.

Then a multiphase tabu search is performed that heuristically reduces a penalty function by considering the seniors, juniors and students in sequence. A similar strategy has been used to obtain heuristally optimal examination schedules [9,20]. The first of two basic moves that define the neighbourhood is a *swap*, exchanging the places of two seniors, juniors or students as appropriate. As there are five persons on duty during a call, there are five neighbourhoods generated: one for the seniors, one for the juniors on team A, one for the juniors on team B, one for the students on team A and one for the students on team B. This is complicated by the requirement that when a junior is not available, the place must be taken by a student from the missing junior's team. This implies that a second junior from the same team should not be placed on the call. The entire space was partitioned into five sub-spaces to reduce the amount of time required to complete an evaluation of the neighbourhood. Since there are 28 days in a schedule, there are $28 \times 27/2 = 378$ possible swaps, each of which must be evaluated for each tabu move. If one larger neighbourhood is used, there are $378^5 = 7.7 \times 10^{12}$ possible swaps.

	Senior	Junior A	Junior B	Student A	Student B
Tue:	Narayan	Seidler	Holden*	Firoz*	------
Wed:	Schneider_A	Trottier*	Amhalal	------	Dickie*
Thu:	Al_Qassabi	Aggarwal*	Charlebois	------	Davidson*
Fri:	Davis	Seidler	Kay	Matar*	Pinto*
Sat:	Schneider_A	Stewart	Amhalal	Lund*	Davidson*
Sun:	Davis	Seidler	Kay	Aggarwal*	Holden*
Mon:	Al_Qassabi	Lund*	Davidson*	------	------
Tue:	Ling_B	Abdelgader	Holden*	Matar*	------
Wed:	White	Firoz*	Kay	------	Pinto*
Thu:	Al_Qassabi	Aggarwal*	Amhalal	------	Holden*
Fri:	Ling_B	Abdelgader	Dickie*	Firoz*	------
Sat:	Al_Qassabi	Seidler	Kay	Aggarwal*	Davidson*
Sun:	Ling_B	Abdelgader	Dickie*	Lund*	------
Mon:	White	Stewart	Pinto*	Trottier*	------
Tue:	Schneider_A	Firoz*	Amhalal	------	Dickie*
Wed:	Davis	Matar*	Charlebois	------	Davidson*
Thu:	Narayan	Seidler	Pinto*	Trottier*	------
Fri:	Schneider_A	Lund*	Charlebois	------	Holden*
Sat:	Davis	Trottier*	Dickie*	------	------
Sun:	Schneider_A	Aggarwal*	Charlebois	------	Pinto*
Mon:	White	Abdelgader	Davidson*	Matar*	------
Tue:	Narayan	Stewart	Amhalal	Aggarwal*	Dickie*
Wed:	Ling_B	Seidler	Holden*	Matar*	------
Thu:	Al_Qassabi	Lund*	Dickie*	------	------
Fri:	Narayan	Stewart	Amhalal	Trottier*	Davidson*
Sat:	Ling_B	Abdelgader	Charlebois	Firoz*	Pinto*
Sun:	Narayan	Stewart	Amhalal	Matar*	Holden*
Mon:	Davis	Trottier*	Kay	------	Pinto*

Final penalty is: 4976

Fig. 1. An example of a call schedule

A tabu minimization is performed using each of the five neighbourhoods in turn. When this is completed, the schedule is examined to discover whether the solution could be improved by *removing* one of the students. Curiously, in this environment, more is not always better. In spite of the chronic understaffing of the hospital, if there are too many students on duty and too few seniors and juniors to supervise their work, the situation is deemed worse than if there were fewer students. Therefore, during this part of the algorithm, surplus students are removed from the schedule if this would reduce the total schedule penalty.

At this point, the neighbourhood is again redefined. The juniors and students from team A and the juniors and students of team B are joined to produce a larger neighbourhood created by redefining the *move* to be a *rotation* across two lines of the schedule. This involves changing the "slots" of four persons. This is done for the A team followed by the B team.

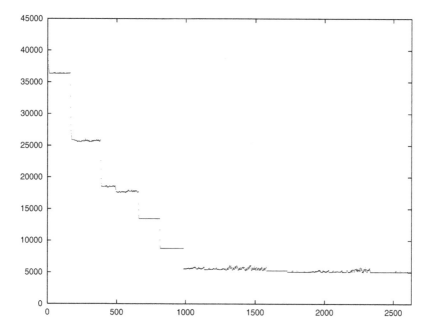

Fig. 2. Total penalty versus move number

Finally, the original moves and neighbourhoods are restored and the tabu optimization is recycled through the five neighbourhoods until no further improvement can be found. An example of a call schedule is presented in Figure 1. An asterisk shown after a name indicates that the person is a student. This has no computational significance and is used here only to aid in interpreting the results.

In this instance, there are five seniors, six juniors and nine students. The block extends over 28 days.

The effectiveness of the algorithm and the influence of neighbourhood redefinition is shown in Figure 2 where the total penalty is reduced from its initial value of 43 715 to 5075. The sharp reduction in penalty resulting from neighbour redefinition and the emptying of tabu lists is clearly visible. Approximately 20 sets of data have been examined so far and they all exhibit the same kind of behaviour.

4   Post-optimization Processing

When the tabu algorithm has terminated, the optimized call schedule can be further processed by hand. This may seem counter-productive since manual processing at this stage is likely to make the call schedule less, rather than more, optimized. However, there are several reasons why post-optimization processing may be desirable:

(a) Illness or some other reason may make one of the medical staff unavailable to work at the required time.
(b) Other considerations that are not reflected in the tabu optimization program may be enforced at this point: for example, some staff members may not work well with other staff members.
(c) New staff members may arrive and must be inserted in the schedule with minimum disruption to the calls of the other staff.

A spreadsheet, augmented with add-ins, has been created to assist in this processing. The tabu-optimized call schedule is loaded into the spreadsheet where specialized functions, *swap*, *triple swap* and *add new member* functions permit the call schedule to be manipulated. When the calls of the medical staff are being changed in this way, certain statistics are continually being monitored and displayed to the user. Thus the total number of calls and the weekend placements are always visible to the user who can then ensure that the changes do not violate the corresponding rules.

The add-ins were coded in Visual Basic and appended to the Excel spreadsheet. Although the end result is never more optimal than the solution generated by the tabu procedure, on-the-spot modifications can be made by the chief resident as required. Somewhat to our surprise, this facility has proven to be the most popular feature of the system.

5   Conclusions

Many experiments were done during the construction of this system to evaluate the effectiveness of certain strategies. Special attention was paid to the construction of the initial schedule. A simple bin packing algorithm that avoided the obvious vacation clashes was tried in the hope that the tabu optimization phase would be powerful enough to produce good schedules. It was found that although the tabu phase did improve the initial schedules greatly, the results were still not good enough for use in the hospital. The use of constraint logic to enforce the complicated weekend requirements and vacation requests was tried later and found to give much better results when combined in the same way with tabu search. It appears that the shape of the optimization surface is such that tabu alone cannot find regions of great improvement when started from regions that are not fairly good already with moderate run times. Our run times are about one minute in length. Work in this area is continuing.

The length of the tabu list was also investigated. It was found that results improved with increasing tabu list fixed length up to a value of about 40. Beyond this value, run lengths increased while the quality of the solution levelled off. A value of 40 was used for all the results reported here.

Our most important conclusions are

(a) the initial solution has a great influence on the quality of the final solution,
(b) modifying the neighbourhood definition by modifying the *move* leads to improved solutions,

(c) unifying two disjoint neighbourhoods (a form of *ejection chain*) by redefining the basic move improves the solution quality,

(d) a fixed tabu tenure of 40 results in good behaviour of the algorithm.

We would like to compare our algorithm with others using similar constraints and data but we have been unable to find a sufficiently equivalent problem.

A working implementation has been developed in Java using the abstract window toolkit (AWT) to build the user interface. Specially written *tabu classes* have been developed to implement evaluation functions, moves and tabu lists. The program runs in a Windows 2000 environment. The call schedule shown above took about one minute to cast on an IBM ThinkPad, model A21e. The program has been used on two campuses of the Ottawa Hospital for about a year and a half. Work is continuing on

(a) improving the quality of the initial solution,

(b) changing the user interface toolkit from AWT to SWT,

(c) compiling the Java code for increased security and execution speed.

References

1. Abdennadher, S., Schlenker, H.: Interdip – an Interactive Constraint Based Nurse Scheduler. In: Proc. Int. Conf. Practical Applic. Constraint Technol. Logic Program. (PACLP'99) (1999)
2. Abdennadher, S., Schlenker, H.: Nurse Scheduling Using Constraint Logic Programming. In: Proc. 11th Annu. Conf. Innov. Applic. Artif. Intell. (1999)
3. De Causmaecker, P., Vanden Berghe, G.: Relaxation of Coverage Constraints in Hospital Personnel Rostering. In: Proc. 4th Int. Conf. Pract. Theory Timetabling (21–23 August, 2002, Gent, Belgium) (2002) 187–206
4. Cowling, P., Kendall, G., Soubeiga, E.: A Hyperheuristic Approach to Scheduling a Sales Summit. In: Burke, E, Erben W. (Eds.): Practice and Theory of Automated Timetabling III (PATAT 2000, Konstanz, Germany, August, selected papers). Lecture Notes in Computer Science, Vol. 2079. Springer-Verlag, Berlin Heidelberg New York (2001) 176–190
5. Carter, M.W., Lapierre, S.: Scheduling Emergency Room Physicians. In: INFORMS Fall 1996 Meeting. Publication 99–23 (1996)
6. Cheng, B.M.W., Lee, J.H.M., Wu, J.C.K.: A Nurse Rostering System Using Constraint Programming and Redundant Modeling. IEEE Trans. Inf. Technol. Biomed. **1** (1997) 44–54
7. Costa, G.: The Impact of Shift and Night Work on Health. Appl. Ergonomics **27** (1996) 9–16
8. Chan, P., Weil, G.: Cyclical Staff Scheduling Using Constraint Logic Programming. In: Burke, E, Erben W. (Eds.): Practice and Theory of Automated Timetabling III (PATAT 2000, Konstanz, Germany, August, selected papers). Lecture Notes in Computer Science, Vol. 2079. Springer-Verlag, Berlin Heidelberg New York (2001) 159–175
9. Di Gaspero, L., Schaerf, A.: Tabu Search Techniques for Examination Timetables. In: Burke, E, Erben W. (Eds.): Practice and Theory of Automated Timetabling III (PATAT 2000, Konstanz, Germany, August, selected papers). Lecture Notes in Computer Science, Vol. 2079. Springer-Verlag, Berlin Heidelberg New York (2001) 104–117

10. Franses, P., Post, G.: Personnel Scheduling in Laboratories Using IPs. In: Proc. 4th Int. Conf. Pract. Theory Timetabling (21–23 August, 2002, Gent, Belgium) (2002) 175–178

11. Gabow, H.N., Kohno, T.: A Network-Flow-Based Scheduler: Design, Performance History and Experimental Analysis. In: ALENEX – Algorithm Engineering and Experiments (Jan. 7–8, 2000)

12. Glover, F., Laguna, M.: Tabu search. Kluwer, Dordrecht (1997)

13. Jan, A., Yamamoto, M., Ohuchi A.: Evolutionary Algorithms for Nurse Scheduling Problem. In: 2000 Congress Evolut. Comput. (2000)

14. Knauth, P.: The Design of Shift Systems. Ergonomics **36** (1993) 15–28

15. Knauth, P.: Design Better Shift Systems. Appl. Ergonomics **27** (1996) 39–44

16. Kragelund, L.V.: Solving a Timetable Problem Using Hybrid Genetic Algorithms. Softw. Practice and Experience **27** (1996) 1121–1134

17. Meisels, A. Kaplansky, E.: Scheduling agents – Distributed Timetabling Problems (DisTTP). In: Proc. 4th Int. Conf. Pract. Theory Timetabling (21–23 August, 2002, Gent, Belgium) (2002) 182–184

18. Meyer auf'm Hofe, H.: Solving Rostering Tasks as Constraint Optimization. In: Burke, E, Erben W. (Eds.): Practice and Theory of Automated Timetabling III (PATAT 2000, Konstanz, Germany, August, selected papers). Lecture Notes in Computer Science, Vol. 2079. Springer-Verlag, Berlin Heidelberg New York (2001) 191–212

19. Petrovic, S., Beddoe, G., Vanden Berghe, G.: Storing and Adapting Repair Experiences in Personnel Rostering. In: Proc. 4th Int. Conf. Pract. Theory Timetabling (21–23 August, 2002, Gent, Belgium) (2002) 185–186

20. White, G.M., Xie, B.S.: Examination Timetables and Tabu Search with Longer-Term Memory. In: Burke, E, Erben W. (Eds.): Practice and Theory of Automated Timetabling III (PATAT 2000, Konstanz, Germany, August, selected papers). Lecture Notes in Computer Science, Vol. 2079. Springer-Verlag, Berlin Heidelberg New York (2001) 85–103

Relaxation of Coverage Constraints in Hospital Personnel Rostering

Patrick De Causmaecker and Greet Vanden Berghe

KaHo St-Lieven, Information Technology,
Gebr. Desmetstraat 1, 9000 Gent, Belgium
{Patrick.DeCausmaecker,greetvb}@kahosl.be

Abstract. Hospital personnel scheduling deals with a large number of constraints of a different nature, some of which need to be satisfied at all costs. It is, for example, unacceptable not to fully support patient care needs and therefore a sufficient number of skilled personnel has to be scheduled at any time. In addition to personnel coverage constraints, nurse rostering problems deal with time-related constraints arranging work load, free time, and personal requests for the staff.

Real-world nurse rostering problems are usually over-constrained but schedulers in hospitals manage to produce solutions anyway. In practice, coverage constraints, which are generally defined as hard constraints, are often relaxed by the head nurse or personnel manager.

The work presented in this paper builds upon a previously developed nurse rostering system that is used in several Belgian hospitals. In order to deal with widely varying problems and objectives, all the data in the model are modifiable by the users.

We introduce a set of specific algorithms for handling and even relaxing coverage constraints, some of which were not supposed to be violated in the original model. The motivation is that such practices are common in real scheduling environments. Relaxations enable the generation of better-quality schedules without enlarging the search space or the computation time.

1 Introduction

Automating the personnel scheduling process of large organisations increases the quality of the personal schedules [2,3,4,14]. Compared to manual scheduling approaches, timetabling heuristics reduce the computation time considerably [6, 8,15].

The relaxation of coverage constraints is part of the development of a nurse rostering package which was designed to meet the requirements of personnel rostering in Belgian hospitals [4,6,8]. Cyclical scheduling is very unusual in these practical health care environments for several reasons. The round-the-clock work is irregular, part-time contracts are common, nurses want the freedom to choose days off and holiday periods freely, and the personnel requirements fluctuate more than in most other personnel scheduling problems (such as employee scheduling in banks, shops, and crew scheduling in public transport).

E. Burke and P. De Causmaecker (Eds.): PATAT 2002, LNCS 2740, pp. 129–147, 2003.
© Springer-Verlag Berlin Heidelberg 2003

The modifiable constraints and objectives allow for a wide applicability among very diverse hospital teams, while the model offers a general approach for calculating solutions. Drawbacks are the possibilities for users to define parts of their problems in such a way that they cause difficulties for the algorithms to find good-quality or even feasible solutions.

In real rostering environments, feasible schedules are very rare and are often less attractive than under- or overstaffed schedules with lower soft-constraint violations. A careful observation of how experienced planners deal with infeasibilities in practice lies at the basis of this research. We present a set of flexible algorithms for increasing the quality of the solutions by relaxing appropriate hard constraints. Some of the algorithms are interactive, others co-operate with the meta-heuristics that generate the solutions. The methodology can be adapted to different practical contexts of hospitals and it can be extended towards other timetabling and scheduling approaches.

Section 2 schematically introduces the sort of problems tackled in this research. The constraints and their contribution to the quality of the schedule are explained briefly. The scheduling options and planning algorithms contributing to better satisfying the constraints on personal schedules, provided that some hard constraints can be relaxed, are explained in Section 3. Section 4 presents test results of the relaxation algorithms for real-world problems. We draw conclusions on the approaches in Section 5.

2   Problem Description

Different personnel scheduling approaches can have widely varying objectives. The nurse rostering problem in this paper concerns finding a good-quality schedule that satisfies the coverage constraints. The quality of a schedule is expressed by its value of the cost function, counting the violations of time-related constraints.

We briefly explain the terminology used in this nurse rostering approach:

- A *shift type* is a personnel task with a fixed start and end time. Unlike in other organisations, hospitals require more shift types than the regular early, late, and night shift. The number of different shift types is S. In the representation of the problem, we define a different assignment unit t for each shift type. For more details about the representation we refer to [6].
- *Skill category*: Personnel members in a ward belong to skill categories that are based upon the level of qualification, the experience, and the responsibility of the nurses. The number of different skill categories in the model is Q. Examples of skill categories are head nurse, regular nurse, junior nurse, caretaker, etc. When personnel members are allowed to replace people belonging to other skill categories, they are said to have alternative skills. This allows for a more complex replacement model than the hierarchical skill categories (as in [2,11,14]).
- *Work regulation*: Hospital personnel have work regulations or contracts with their employer. Most healthcare organisations allow several job descriptions

such as part-time work, night nurses, weekend workers, etc. Each of these
regulations is subject to a different set of constraints.

- A *schedule* is a two-dimensional structure in which the rows p $(1, \ldots, P)$
  represent the timetables for all the personnel members. The columns denote
  the assignment units. There is one assignment unit per shift type and per
  day t $(1, \ldots, T)$, where $T = S \times D$ and D is the number of days in the
  planning period.

We define the coverage constraints as the required number, minimum or
preferred, of people of each skill category at any time in the planning period.
Coverage constraints belong to the category of hard constraints so they must
always be satisfied, i.e. at least the minimum number of personnel and at most
the preferred number should be scheduled at any time.

All the time-related constraints on personal schedules are soft constraints.
They will preferably be satisfied but violations can be accepted to a certain
extent. Examples of such constraints are overtime, undertime, maximum number
of assignments per week, minimum number of consecutive working days and
free days, personal requests for days or shifts off, personal requests for working
a specific shift on a particular day, cyclical patterns, etc. An extended list of
such constraints can be found at [19]. In our approach, planners can set the cost
parameters of the soft constraints and even modify the constraint definition. The
modular cost function sums all the violations (of soft constraints) multiplied by
the corresponding cost parameter. For details about the evaluation function, we
refer to [6].

In Figure 1, we briefly introduce a few examples of soft constraints that,
in our approach, play an important role in the relaxation of hard constraints.
The set contains mainly constraints for which the point in time is relevant, such
as special personal requests for a duty or absence on specific days. Although
these constraints are soft, violations often considerably reduce the quality of the
schedule "sensed" by individual nurses. For the search algorithms that attempt
to generate schedules within the feasible part of the search space, such violations
might be unavoidable. The dominant influence of this category of soft constraints
on the satisfaction of coverage constraints, and vice versa, makes it worthwhile
pre-evaluating them. Other soft constraints for evaluating maximum values or
consecutiveness cannot be easily pre-evaluated. The constraints on overtime and
undertime are exceptional. They only determine one of the post-planning options
in our model (see "add hours" in Figure 2).

When comparing a set of nurse rostering approaches in the literature, we
can distinguish two main objectives: solving coverage constraints and the time-
related constraints on personal schedules. Some models allow the scheduling algorithms to make coverage decisions by incorporating them as soft constraints [9,
15,16,17,21]. Other approaches consider a fixed number of staff, which is supposed to be sufficient for satisfying the coverage constraints [2,4,6,8,11,13,14,21].
Approaches that require a strict application of the time-related constraints are
rare and solve smaller-size problems than those tackled in this paper. Berrada
et al. [3], for example, define fewer time-related constraints than we do, but some

– *Day off*: This constraint is a personal request for a day off. Personnel members can have a list of such requests.
– *Shift off*: Unlike the previous constraint, this one only forbids the assignment of a particular shift type on a certain day.
– *Assignment*: Nurses can list their personal requests for carrying out certain shifts on particular days.

For each of the previous constraints, a high or low priority can be set. It determines the penalty for violating the constraint.

– A *pattern* is a complex cyclical constraint that is built with a combination of different pattern types. Each pattern has a predefined length, equal to a number of days. Per pattern day, one of the following restrictions holds:

PAT-1 no free day (at least one shift type must be assigned),
PAT-2 assignment of a particular shift type,
PAT-3 assignment of a shift type of a certain duration (a small deviation of that duration, generally 15 minutes, is allowed),
PAT-4 no restriction on that day,
PAT-5 free day,
PAT-6 day off,
PAT-7 forbidden shift types.

Although both PAT-5 and PAT-6 do not allow assignments on the corresponding days, there is a difference in interpretation. PAT-6 represents a day off that influences the evaluation of some other constraints (e.g. when it is a compensation day, holiday, refresher course, etc.) by contributing to the working hours.

– *Overtime*: According to their work agreement or contract, personnel members have a different maximum working time per planning period. The constraint is cumulative. For the evaluation, access hours in the previous planning period are added to the work in the current period.
– *Undertime*: Similarly, a minimum number of working hours can be defined per work agreement. Schedules with fewer hours assigned are penalised. However, holiday periods and illness leave will induce a recalculation of the minimum number of hours that a member of personnel should work. Note that undertime is not related to undercoverage, which is a shortage in personnel.

Fig. 1. Set of time-related soft constraints that can easily be pre-evaluated

of them are hard (e.g. weekend working patterns and the number of weekly working days). Miller et al. [16] define the personnel requirements as a minimum and preferred number of people per day (and not per shift type, as in our approach). They divide the time-related constraints into two categories: a feasibility set (maximum number of assignments, minimum/maximum number of consecutive working days) and non-binding constraints (examples are number of working weekends, working complete weekends, consecutive free days, in addition to stricter formulations of the constraints in the feasibility set). The total number of constraints is very low compared to the set used in this research.

There is a range of different ways to tackle coverage constraints. Isken and Hancock [12] allow variable starting times for the personnel members instead of working strictly defined shifts. Over- and understaffing can occur in solutions, but they are penalised. The method proposed by Ozkarahan [18] aims at minimising over- and understaffing. Although there is a defined range of feasible personnel attendance, solutions generated by Miller et al. [16] do not necessarily satisfy all the coverage constraints. Warner [20] allows the scheduling of more nurses than strictly required in order to compensate for unwanted personal schedules.

The goal function in Warner and Prawda's work [21] aims at minimising the difference between a given lower limit for the number of nurses, and their actual number. The difference must not be under zero. Ahmad et al. [1] only search solutions in the feasible domain, which includes the satisfaction of some time-related constraints in addition to personnel coverage. Scheduling too many personnel members is not penalised. Meyer auf'm Hofe [15] defines minimum and standard staffing levels which are treated as fuzzy constraints. Understaffing is considerably more penalised than overstaffing in [15].

The observation of the behaviour of manual planners revealed that certain circumstances can permit the relaxation of some hard constraints. Undertime is often considered as a worse violation than overtime, whereas overcoverage is generally less wanted than understaffing, etc. It is not possible to incorporate this knowledge in the problem description because it reflects implicit dissatisfaction rather than countable violations. Moreover, it differs strongly from one hospital to another. It is therefore important to provide interactive scheduling relaxation tools.

3   Algorithms for Relaxing the Hard Constraints

3.1   Overview of the Algorithms

The relaxation algorithms are pre- and post-planning heuristics, which can be combined with different meta-heuristic approaches. Examples of such meta-heuristics for the nurse rostering problem are variable neighbourhood search [7], tabu search [8], and memetic algorithms [4].

The relaxation heuristics are presented as separate planning options, in order to isolate the meta-heuristics from typical shortcomings of the model when applied in practice. Figure 2 schematically demonstrates the order in which the relaxation heuristics (presented in bold) appear in the total planning process.

Certain circumstances require a reduced search space, as is schematically presented in the box between the consistency check and the initialisation. It is the case when a previously generated schedule can or must be partly reused. The large box "per skill category" denotes that the nurse rostering problem in our approach can be divided into sub-problems that are related to the skill

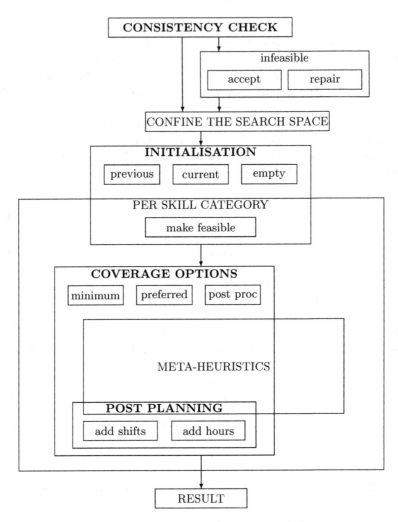

Fig. 2. Overview of the relaxation algorithms which are applied in combination with the meta-heuristics

categories. Planners define a hierarchy determining the planning order of the skill categories. The large solution space can thus be split into smaller regions for each separate skill category. Search spaces for different categories are not necessarily completely disjunctive (see Section 2).

The meta-heuristics are not specified in Figure 2 because any search algorithm maintaining the coverage is allowed. Boxes presented on the same level denote exclusive options. Examples are "accept" and "repair" in case the consistency check discovers infeasibilities. The "post-planning" options start after fin-

- the days off in personal schedules
- the assignment units corresponding to shifts off in personal schedules
- the assignment units corresponding to other than requested assignments on the same day in personal schedules
- in case a pattern is defined in a person's work regulation:
  - the assignment units that do not correspond to the specified shift type in case of PAT-2
  - the assignment units corresponding with shift types having a duration that differs from the specified duration in case of PAT-3
  - the entire day corresponding to PAT-5 and PAT-6
  - the assignment units that correspond to the specified shift types in case of PAT-7

Fig. 3. Schedule positions that should remain empty according to the list of soft constraints presented in Figure 1. The set of schedule positions is denoted by SE

ishing the meta-heuristics but they often involve the need to reiterate the meta-heuristics. Additional assignments made during these options will be "marked" in order to highlight them for later possible search actions (Sections 3.3 and 3.4).

3.2   Consistency Check on Available People

In order not to expect too much insight from the hospital planners who set up the data, we developed a simple consistency check that is performed before the planning starts. The procedure warns users for infeasible problems, by checking relevant constraints.

In the pre-planning process, the number of available people of a certain skill category is compared with the number of requested people for that skill category at any time during the planning period. Apart from this obvious check on the hard constraints, planners in practice expect an extra check on some "precedence" soft constraints (Figure 1).

The consistency check respects the planning order of the skill categories. For every personnel member, a list of available time slots (shift types or time intervals) is constructed. Time slots that are not available when satisfying these soft constraints are presented in Figure 3. The personal requests for assignments and the requested shifts in patterns should be satisfied. Figure 4 presents a list of these necessary assignments.

The term "requirements" will be used instead of the more specific coverage constraints called minimum or preferred personnel requirements. We explain in Section 3.4 how the (hard) coverage constraints for the scheduling algorithm are derived from the personnel requirements and the planning options.

The entire consistency check procedure is presented schematically in Figure 5. The variable *occupied* has been introduced in the model in order to keep track the assignment units corresponding to:

- personal requests for the assignment of shifts

and in case a pattern is defined in a person's work regulation:

- one of the assignment units corresponding to the day in case of PAT-1
- the assignment units corresponding to the specified shift type in case of PAT-2
- one of the assignment units corresponding to shift types having the specified duration in case of PAT-3

Fig. 4. Schedule positions that should have assignments, according to the list of soft constraints presented in Figure 1. This set of schedule positions is denoted by SA

of areas in the schedule that are not free for assignments. All the free days in patterns (type PAT-5), for example, render the corresponding days in the schedule occupied. When estimating the number of assignments in a schedule, the algorithm starts with calculating all the personal requests (and requested assignments in patterns) for particular shifts. Afterwards, the algorithm searches positions where extra assignments can be made. All the assignments in this step render the other schedule positions on the corresponding days occupied. When there is still an excess of requirements, the algorithm continues by assigning the shifts to personal schedules which are still unoccupied on the particular day. If there are not enough such locations, the value $inconsistent_{q,t}$ equals the number of assignments minus the number of requirements.

Blocking the entire day on which an assignment is made is often more strict than necessary. However, in real-world situations, it rarely occurs that a personnel member in healthcare environments is assigned to more than one shift per day. We found it better to mark such problems as inconsistent.

Hospital planners are free to accept or ignore the diagnosis. The more flexibility the algorithm allows for doing the imaginary assignments, the more accurate the diagnosis will be. In our model, alternative skill categories are also taken into account. Assignments for alternative skill categories can often help to satisfy the pattern constraint (PAT-1, PAT-2 and PAT-3) when the problem is inconsistent for the main skill category.

At the start of a planning, the algorithm only informs the planner about inconsistencies in the coverage constraints and/or preceding soft constraints. The consistency check does not make any actual assignments. It reports whether the requirements should be increased or decreased by a certain amount. Either users agree to relax the personnel requirements or they accept unavoidable violations of the checked soft constraints. These options determine whether the search

$\forall q\ (1 \leq q \leq Q) : \forall p, t\ (1 \leq p \leq P, 1 \leq t \leq T)$:

    IF $\{p, t\} \in SE$ *(see Figure 3)*
        $schedule_{\{p,t\}} =$ '*occupied*'
    IF $\{p, t\} \in SA$ *(see Figure 4)*
        $schedule_{\{p,t\}} =$ '*assignable*'
        stop = false
        WHILE $(assignable_{q,t} < required_{q,t} \land\ !stop)$
            search, in the following order, a schedule position that is not
            occupied and not yet assignable
                - p with skill category q, PAT-1 type for the day,
                  and not yet assignable for t
                - p with skill category q and not yet assignable for t
                - p with alternative skill category q, PAT-1 type for the day, and
                  not yet assignable for t
                - p with alternative skill category q and not yet assignable for t
            IF (found)
                $\forall i\ (1 \leq i \leq S) : schedule_{\{p,(t/S)*S+i\}} =$ '*occupied*'
                $schedule_{\{p,t\}} =$ '*assignable*'
            ELSE
                stop = true
        $inconsistent_{q,t} = assignable_{q,t} - required_{q,t}$

RESULT

    IF $(\sum_{q,t} |inconsistent_{q,t}| > 0)$
        inconsistent = true
    ELSE
        inconsistent = false

WHERE $assignable_{q,t} = \sum |\{p\ \xi\ 1 \leq p \leq P \land p_q \land schedule_{\{p,t\}} = assignable\}|$
and p_q reflects that p can carry out work for skill class q

Fig. 5. Procedure for consistency check

algorithms will respect the original or the relaxed coverage constraints. Both
options are mentioned in the overview of Figure 2.

3.3   Initialisation

The initialisation of the scheduling algorithm consists of constructing a feasible
initial solution. It suffices for a schedule to satisfy all the hard constraints (or
the relaxed hard constraints proposed by the consistency check) to be called
feasible. We have split the initialisation into two phases. After an input option
has been selected, the first phase loads the input. The second phase creates a
feasible schedule.

Input options for the initialisation. Three possible strategies are introduced for practical planning problems.

Current schedule. The initialisation starts from the currently available schedule, which can either be the result of a previous attempt to generate a solution, or the planning that existed before certain extra restrictions appeared.

Schedule of the previous planning period. This option is only useful when the schedule for the previous planning period is of a very high quality and when the current and the previous planning period have similar constraints. It is not recommended to take the previous schedule as an input when the number of personnel is not the same in both periods (when the planning periods includes bank holidays, or when, for some personnel members, the pattern period differs from the planning period).

Empty schedule. The simplest input option starts the initialisation from an empty schedule.

Although the first two initial schedule constructors may seem very attractive, our experiments show that it is not too difficult for good meta-heuristics to produce schedules of comparable quality starting from a random initial schedule. Indeed, it is often the case that with the two latter initialisation options, the solution represents a local minimum in the search space and the algorithm has problems escaping from it.

Create a feasible solution. In order to satisfy the hard constraints in the initial schedule, an algorithm is applied that adds and/or removes shifts until the personnel requirements (according to the planning options of Section 3.4) are met. The process is mainly randomly driven, although it takes some of the soft constraints into account. Users in practice suggested to initially force satisfaction of the personal requests for days and shifts off, and satisfaction of the patterns. These are precisely the soft constraints that play a role in determining the consistency of the data (Section 3.2). It seems contradictory to the previously explained concept of treating the cost function (which sums violations of all the constraints [6]) as the only evaluation means. However, even though the planners can freely set cost parameters, a schedule with fewer violations of personal constraints will often be preferred to better-quality schedules (with respect to the cost function) containing more violations of these particular constraints.

The main procedure of the initialisation algorithm is comparable to the consistency check. The initialisation is always executed per skill category. The group of people belonging to the skill category is extended with the people who have that skill as an alternative qualification. The initialisation algorithm first makes an attempt, only in the case when the *Empty schedule* option is selected, to satisfy the pattern constraints. When a pattern requires a shift, the algorithm assigns a randomly chosen shift to the corresponding people, provided the assignment never violates the hard constraint on the number of people required.

In case the shift type is specified (building block PAT-2 of the pattern), the shift will be assigned to the person if there is a shortage in the schedule for the shift on the particular pattern day. For the third building block, PAT-3, the algorithm randomly chooses a shift type of the specified duration for which the personnel requirements are not yet met.

The algorithm afterwards moves to an iterative phase that stops when the personnel requirements are fulfilled for every shift type, or when a maximum number of attempts to assign randomly is reached. The maximum number of attempts is a function of the number of people authorised to carry out jobs for the skill category. In case the number of scheduled personnel on a particular day is too low to meet the requirements, the algorithm assigns the lacking shift type randomly to a personal schedule, provided that the person is skilled and available. The people are divided into groups with equal eagerness for the assignment. The algorithm stops if a random assignment is possible in a group, that is if a person of the group has no assignment. If no such person exists, the algorithm moves to the next group. Only when there is nobody available in any of the groups will the initialisation algorithm report a failure. The groups of appropriate people considered for adding assignments are listed below in order.

ADD-1 All the people having a personal request for the shift corresponding to the assignment unit to be scheduled and having no assignment yet.

ADD-2 All the people working according to a predefined pattern, for which a PAT-2 type corresponds to the day, and the detail to the shift to be scheduled, having an empty schedule for that assignment unit.

ADD-3 All the people working according to a predefined pattern, for which a PAT-3 type corresponds to the day and the detail to the duration of the shift to be scheduled (+/- the deviation), and an empty schedule for the corresponding assignment unit.

ADD-4 All the people who belong to the scheduled skill category (main skill) and do not have a personal request (with a high importance) for a day off or shift off at the time to be scheduled, and have an empty schedule at that time.

ADD-5 All the people who have the skill as alternative and do not have a personal request (with a high importance) for a day off or shift off at the time to be scheduled, and have an empty schedule at that time.

ADD-6 All the people who belong to the skill category, and have an assignment for another skill category, which has not been scheduled in this run, i.e. a skill category which is lower in the planning order hierarchy (Section 3.1).

ADD-7 All the people who have the skill as alternative, and have an assignment for another skill category, which has not been scheduled in this run (a skill category that is lower in the planning order hierarchy) and differs from the main skill of these people.

ADD-8 All the people who are authorised for the skill category and have an empty schedule for the shift.

An analogous procedure has been developed for removing shifts when a schedule exceeds the requirements for certain shifts. The hierarchy of the groups to which the removal of a shift is applied is listed below:

REM-1 All the people who have the skill as alternative and have a marked (see Section 3.4) assignment for the skill category being scheduled.

REM-2 All the people who belong to the skill category and have a marked assignment for it.

REM-3 All the people who have the skill as alternative and have an assignment for the skill category being scheduled.

REM-4 All the people who belong to the skill category and have an assignment for it.

In Section 3.4, we explain how the personnel requirements for the initialisation depend on the coverage options.

3.4   Coverage Options

In practice, the number of required personnel on a certain day is not completely strict. Experienced planners know very well when it is reasonable to plan more or less personnel than required. However, there are no clear rules for decisions like this. Therefore our model allows users to choose among different coverage strategies. There is a feasible range between minimum and preferred personnel requirements. Any solution with fewer assignments than the minimum requirements, or with more than the maximum requirements, violates the hard constraints.

Our model provides options to set the coverage constraints, which will function as hard constraints for the rostering algorithms. It is also possible to allow a few post-planning algorithms that modify the coverage after a solution has been generated.

Minimum/Preferred requirements. The hospital scheduler can choose to consider the minimum or the preferred requirements as hard constraints. During the entire meta-heuristic planning process, the number of planned shifts (minimum or preferred) will not change.

Plan towards preferred requirements. Instead of strictly setting the hard constraints, this option allows a range in which the hard constraints are considered satisfied. In Figure 6, the procedure for this option is schematically presented.

The algorithm first takes the minimum requirements as hard constraints. After a result has been computed by the scheduling meta-heuristics, the system tries to add shifts to the schedule wherever this does not involve extra violations of soft constraints. When the actual coverage is lower than the preferred requirements, the algorithm selects the personal schedule for which an additional assignment improves the overall quality of the schedule most. Adding a pair of identical shifts on consecutive days in a personal schedule is often less harmful than adding an isolated shift. In the competition for the best candidates to assign extra shifts to, "twin" assignments are also considered (provided none of

$\forall q \ (1 \leq q \leq Q):$

$$\begin{cases} \text{initialise a solution } schedule_q \text{ that satisfies the minimum personnel} \\ \text{requirements (whether consistent or not) for } q \\ \text{improve the solution } schedule_q \text{ by applying meta-heuristics until} \\ \text{a stop criterion is met} \\ quality(schedule_q) = \sum_{p \, \xi \, p_q = q} quality(schedule_{\{p,q\}}) \end{cases} \quad (1)$$

$\forall q \ (1 \leq q \leq Q):$

> stop = false
> WHILE $(assigned_q < preferred_q \wedge !stop)$
>> select $best$ (the least worsening with respect to the quality) assignment
>> position $schedule_{\{p,t\}}$ (or 'twins' $schedule_{\{p,t\}} + schedule_{\{p,t+S\}}$) for which
>> $schedule_{\{p,t\}} = 0 \ (\wedge \ schedule_{\{p,t+S\}} = 0) \wedge assigned_{q,t} < preferred_{q,t}$
>> IF $(best \leq quality(schedule_q) \vee quality(schedule_p) \leq threshold_{shifts})$
>>> $schedule_{\{p,t\}} = assigned$
>>> $assigned_{q,t} = assigned_{q,t} + 1$
>>> IF (twins)
>>>> $schedule_{\{p,t+S\}} = assigned$
>>>> $assigned_{q,t+S} = assigned_{q,t+S+1}$
>> ELSE
>>> stop = true

WITH

> $assigned_q = |\{t, p \, \xi \, p_q = q \wedge schedule_{\{p,t\}} = assigned\}|$
> $assigned_{q,t} = |\{p \, \xi \, p_q = q \wedge schedule_{\{p,t\}} = assigned\}|$

Fig. 6. Procedure for adding shift types after planning

the assigned shifts causes excess coverage). "Twin" assignments are often more beneficial for the quality of the schedule because they influence many soft constraints (such as constraints on complete weekends, identical shifts in weekends, a minimum number of consecutive shifts or days, etc. [19]). Since the complexity of finding optimal "twins" is exponential, we reduced the search to the selection of the best set of equal shift types on two consecutive days for each personal schedule.

The system allows for a more flexible approach by providing a threshold value for the individual cost function: $threshold_{cost}$. In the case that the threshold is positive, the algorithm will add extra shifts whenever the personal cost function does not exceed that threshold.

Every additional assignment in the post-planning phase is marked in the schedule. The other assignments occurred while considering the entire solution space during the regular search. It is recommended during some planning activities to remember which shift removals will harm the schedule less. We explained the importance of marking in Section 3.3.

$\forall q \ (1 \leq q \leq Q):$

identical to (1) in Figure 6

$\forall q \ (1 \leq q \leq Q):$

$\forall p \ (1 \leq p \leq P):$ FOR each personnel member p belonging to q
    stop = false
    WHILE $(undertime_p > 0 \wedge !stop)$
        select *best* (the least worsening with respect to the quality) assignment
        position $schedule_{\{p,t\}}$ (or twin $schedule_{\{p,t\}} + schedule_{\{p,t+S\}}$) for
        which the preferred personnel requirements at t (and $t + S$) are not 0
        and for which $(best_p < threshold_{hours})$
        IF (found)
            $schedule_{\{p,t\}} = schedule_{\{p,t+S\}} = assigned$
        ELSE
            stop = true

WITH

$undertime_p$ = the difference between the number of hours that p should work
and the number of hours that is scheduled for p

Fig. 7. Procedure for adding hours to a schedule after planning

Add hours. This option attempts to add shifts to personal schedules with undertime. It is a pure post-planning option that does not necessarily respect the (hard) preferred personnel coverage constraints. After a schedule has been generated, an algorithm searches, for every personal schedule, the best points in time for adding extra shifts. A shift cannot be added to a schedule unless such shift already occurs in the personnel requirements for that day and skill category. By applying this option, the coverage can exceed the level of the preferred requirements unless when that level is 0.

By default, this algorithm will not modify the schedule if that generates overtime. As explained in the previous option, there is a possibility for setting a parameter $threshold_{hours}$ for the cost function value of the personal schedule. When a personal schedule has reached this number, the algorithm will not add extra shifts. As explained in the previous section, the extra shifts are marked. Figure 7 schematically demonstrates the procedure.

Since it might be better in terms of the cost function to add a pair of shift types on consecutive days, we also consider twins when searching for the best schedule positions for assigning extra shifts. As explained in Section 3.4, we restrict multiple assignments to "twins" in order to keep the search space small.

4   Test Results

We explain the tests of the relaxation algorithms on a few real-world problems. All the experiments make use of the hybrid tabu search approach introduced in [8]. It consists of a simple tabu search algorithm that moves single shifts between personal schedules while maintaining the coverage. The local search is combined with greedier algorithms concentrating on weekend constraints and improving the worst personal schedule.

We briefly describe the difference between two sample problems. In Problem I, there is a cyclic pattern for every personnel member. Some of the people have loose patterns. The constraints are rather strict for others. We have tested different paths in the planning procedure (Figure 2). The results are presented in Table 1. Problem II is a completely different real-world problem, with fewer strict patterns but more people on leave. The same tests have been carried out and the results are shown in Table 2.

For each experiment, represented by a row in the tables, the planning option is presented in the first column. We distinguish, according to Section 3.4, "minimum" and "preferred" coverage, "shifts" for planning towards the preferred requirements, and "hours" for adding hours. The columns "<M" and ">P" refer to the shortage and excess coverage. ">M" and "<P" denote excess coverage compared to the minimum personnel requirements and shortage compared to the preferred requirements. Violations of the hard coverage constraints are presented in bold. When the planning option "minimum" is selected, excesses of personnel assignments (even if they do not exceed the preferred requirements) are considered violations of the coverage. The same holds for a shortage of assignments when "preferred" is selected. We have carried out experiments with two different cost functions. In the first, undertime does not contribute to the value of the cost function (results are in the upper part of the tables). The second cost function has a penalty of 1 per hour undertime (lower part of the tables).

Obviously, we only present the results of problems that are inconsistent with respect to the hard constraints and precedence soft constraints of Figure 1. Most experiments in which the "repair" option is selected after detecting inconsistencies, produce better-quality results (with respect to the cost function) than the schedules generated with the original hard constraints. This is especially remarkable for the experiments with Problem II, in which many people take days off. However, the solutions for the original requirements (accept option) do not have violations of hard constraints. Relaxing the coverage constraints simplifies solving constraints on personal requests for days off. Exceptions to this finding are the tests with the option "preferred" in Table 1. In these cases, "repair" leads to higher values of the cost function because many other than the soft constraints checked by the consistency procedure are violated due to the increased workload after repairing. The option of planning towards the preferred requirements (denoted by "shifts" in the first column) does not violate hard constraints (unless combined with "repair"). Moreover, it approaches the satisfaction of the preferred personnel coverage better than the option "minimum", and it produces better-quality schedules in all cases. That option is default for most users

Table 1. Test results for problem I

				Coverage			
Options	Threshold	Consistent	Quality	<M	>M	<P	>P
No penalty for undertime							
Minimum		accept	790	0	0	19−	0
Minimum		repair	625	0	7+	16−	4+
Shifts		accept	608	0	6+	13−	0
Shifts	100	accept	608	0	6+	13−	0
Shifts	200	accept	608	0	6+	13−	0
Shifts		repair	550	0	9+	14−	4+
Hours		accept	849	0	20+	0	1+
Hours		repair	709	0	13+	13−	7+
Preferred		accept	938	0	19+	0	0
Preferred		repair	947	0	23+	0	4+
Penalty 1 per hour undertime							
Minimum		accept	1457	0	0	19−	0
Minimum		repair	1249	0	7+	16−	4+
Shifts		accept	1209	0	6+	13−	0
Shifts	200	accept	1209	0	12+	7−	0
Shifts		repair	1138	0	16+	7−	4+
Shifts	200	repair	1138	0	16+	7−	4+
Hours		accept	1298	0	24+	0	5+
Hours	20	accept	1298	0	24+	0	5+
Hours	50	accept	1305	0	25+	0	6+
Hours	100	accept	1348	0	19+	0	7+
Hours	200	accept	1348	0	20+	0	0
Hours		repair	1178	0	18+	11−	10+
Hours	200	repair	1249	0	7+	16−	4+
Preferred		accept	938	0	19+	0	0
Preferred		repair	1291	0	23+	0	4+

in practice. It is very difficult to evaluate the results for the option "hours" without knowing the subjective wishes of the personnel and the hospital. The schedules obtained with this option all have many violations of soft constraints. Attempts to better satisfy the constraint on undertime induce violations of other soft constraints. We have carried out a few tests with a threshold for the cost function value per person. In the case of the "shift" option, only one result was influenced by increasing the threshold (Problem I, threshold 200: more shifts are assigned but the value of the cost function remains unchanged in the end). Higher thresholds have not been considered because they would deteriorate the quality too much. Setting a threshold value for adding hours has a bigger effect

Table 2. Test results for problem II

Options	Threshold	Consistent	Quality	Coverage			
				<M	>M	<P	>P
No penalty for undertime							
Minimum		accept	311	0	0	18−	0
Minimum		repair	85	4−	1+	21−	0
Shifts		accept	261	0	2+	16−	0
Shifts	100	accept	261	0	2+	16−	0
Shifts		repair	85	4−	1+	13−	0
Hours		accept	824	0	18+	0	0
Hours		repair	85	4−	3+	21−	2+
Preferred		accept	836	0	19+	0	0
Preferred		repair	373	4−	15+	7−	0
Penalty 1 per hour undertime							
Minimum		accept	379	4−	1+	21−	0
Minimum		repair	199	0	0	18−	0
Shifts		accept	310	0	4+	14−	0
Shifts		repair	168	4−	5+	17−	0
Hours		accept	878	0	18+	0	0
Hours		repair	180	4−	4+	21−	3+
Preferred		accept	923	0	18+	0	0
Preferred		repair	388	5−	15+	8−	0

than adding shifts towards the preferred requirements. The results in both test sets suffer from undertime, as can be noticed when comparing the results with and without a penalty for undertime.

Relaxing the hard constraints does not necessarily lead to better-quality schedules with respect to the evaluation algorithm, although in most cases it does. It is clear that none of the relaxation algorithms outperforms all the others. They all aim at different goals, and their results strongly depend on the particular needs and wishes of the hospital and its staff.

5   Conclusions

The nurse rostering model developed for application in Belgian hospitals is very flexible. It provides many possibilities for setting constraints and requirements. However, the main drawback of allowing users to define and set their problem is the danger of creating over-constrained and unsolvable problems. Experienced hospital planners often deal with infeasibilities in practice. They know which constraints to relax in difficult circumstances.

By observing the reaction of hospital planners to the results of the scheduling algorithms for nurse rostering, we developed algorithms for addressing over-constrained and even infeasible problems, without drastically changing the objective of giving high priority to coverage constraints. The consistency check algorithm not only considers the hard constraints, but pre-evaluates a few of the most touchy soft constraints, such as personal preferences for days off and for assignments. This interactive tool allows the users to adapt their requirements to the given recommendations in order to create feasible problems.

The other relaxation algorithms are pre- or post-scheduling algorithms which do not necessarily interfere with the meta-heuristic approaches. Depending on the goal chosen by the user (to aim at meeting the minimum or the preferred coverage), these algorithms adapt the overall objective before or after generating the schedule. Adding more hours to some personal schedules is a typical example of a real-world habit, which is not revealed when looking at the formulation of the requirements. We developed interactive ways of adapting the schedule to the soft constraint on "undertime" while ignoring the hard constraint on coverage to a controlled extent.

Experiments have pointed out that relaxing the hard constraints leads to a much higher overall satisfaction for the personnel members. Many infeasible, but not unwanted, solutions are within reach of the modified goals. The interactive software based on the proposed model provides enough feedback for guiding the planners and it allows them to specify their own solution strategy. All the relaxation algorithms described in this paper are applied in practice. They have considerably increased the applicability of our nurse rostering model to various kinds of hospital settings with varying requirements and scheduling habits.

Although the effect of the relaxation procedures is not necessarily clearly visible in terms of the cost function value, they have dramatically contributed to modelling specific wishes and rostering customs, and to solving real-world problems.

Our future work includes applying the relaxation algorithms to automated school timetabling.

References

1. Ahmad, J., Yamamoto, M., Ohuchi, A.: Evolutionary Algorithms for Nurse Scheduling Problem. In: Proc. 2000 Congress Evolut. Comput. (CEC 2000, San Diego, CA) (2000) 196–203
2. Aickelin, U., Dowsland, K.: Exploiting Problem Structure in a Genetic Algorithm Approach to a Nurse Rostering Problem. J. Scheduling **3** (2000) 139–153
3. Berrada, I., Ferland, J., Michelon, P.: A Multi-Objective Approach to Nurse Scheduling with both Hard and Soft Constraints. Socio-Econ. Planning Sci. **30** (1996) 183–193
4. Burke, E.K., Cowling, P., De Causmaecker, P., Vanden Berghe, G.: A Memetic Approach to the Nurse Rostering Problem. Appl. Intell. (special issue on Simulated Evolution and Learning) **15** 2001 199–214
5. Burke, E.K., De Causmaecker, P., Petrovic, S., Vanden Berghe, G.: Floating Personnel Requirements in a Shift Based Timetable, Working Paper, KaHo St-Lieven, 2001

6. Burke, E.K., De Causmaecker, P., Petrovic, S., Vanden Berghe, G.: Fitness Evaluation for Nurse Scheduling Problems. Proc. Congress Evolut. Comput. (CEC 2001, Seoul, South Korea) IEEE Press (2001) 1139–1146

7. Burke, E.K., De Causmaecker, P., Petrovic, S., Vanden Berghe, G.: Variable Neighbourhood Search for Nurse Rostering Problems. Proc. 4th Metaheuristics Int. Conf. (MIC 2001, Porto, Portugal) (2001) 755–760 (accepted for publication in the selected papers volume by Kluwer)

8. Burke, E.K., De Causmaecker, P., Vanden Berghe, G.: A Hybrid Tabu Search Algorithm for the Nurse Rostering Problem. In: McKay, B. et al. (Eds.): Simulated Evolution and Learning 1998. Lecture Notes in Artificial Intelligence, Vol. 1585. Springer-Verlag, Berlin Heidelberg New York (1999) 187–194

9. Chen, J.-G., Yeung, T.: Hybrid Expert System Approach to Nurse Scheduling. Comput. Nursing **11** (1993) 183–192

10. Chiarandini, M., Schaerf, A., Tiozzo, F.: Solving Employee Timetabling Problems with Flexible Workload using Tabu Search. In: Burke, E, Ross, P. (Eds.): Practice and Theory of Automated Timetabling I (PATAT 1995, Edinburgh, Aug/Sept, selected papers). Lecture Notes in Computer Science, Vol. 1153. Springer-Verlag, Berlin Heidelberg New York (1996) 298–302

11. Dowsland K.: Nurse scheduling with Tabu Search and Strategic Oscillation. Eur. J. Oper. Res. **106** (1998) 393–407

12. Isken, M., Hancock, W.: A Heuristic Approach to Nurse Scheduling in Hospital Units with Non-Stationary, Urgent Demand, and a Fixed Staff Size. J. Soc. Health Syst. **2** (1990) 24–41

13. Kawanaka, H., Yamamoto, K., Yoshikawa, T., Shinogi, T., Tsuruoka S.: Genetic Algorithm with the Constraints for Nurse Scheduling Problem, Proc. Congress Evolut. Comput. (CEC 2001, Seoul, South Korea) IEEE Press (2001) 1123–1130

14. Meisels, A., Gudes, E., Solotorevski, G.: Employee Timetabling, Constraint Networks and Knowledge-Based Rules: a Mixed Approach. In: Burke, E, Ross, P. (Eds.): Practice and Theory of Automated Timetabling I (PATAT 1995, Edinburgh, Aug/Sept, selected papers). Lecture Notes in Computer Science, Vol. 1153. Springer-Verlag, Berlin Heidelberg New York (1996) 93–105

15. Meyer auf'm Hofe, H.: Solving Rostering Tasks as Constraint Optimization. In: Burke, E, Erben W. (Eds.): Practice and Theory of Automated Timetabling III (PATAT 2000, Konstanz, Germany, August, selected papers). Lecture Notes in Computer Science, Vol. 2079. Springer-Verlag, Berlin Heidelberg New York (2001) 191–212

16. Miller, M.L., Pierskalla, W., Rath, G.: Nurse Scheduling Using Mathematical Programming. Oper. Res. **24** (1976) 857–870

17. Okada, M.: An Approach to the Generalised Nurse Scheduling Problem – Generation of a Declarative Program to Represent Institution-Specific Knowledge. Comput. Biomed. Res. **25** (1992) 417–434

18. Ozkarahan, I.: A Disaggregation Model of a Flexible Nurse Scheduling Support System. Socio-Econ. Planning Sci. **25** (1991) 9–26

19. Vanden Berghe, G.: Soft constraints in the Nurse Rostering Problem. http://www.cs.nott.ac.uk/ gvb/constraints.ps (2001)

20. Warner, M.: Scheduling Nursing Personnel According to Nursing Preference: a Mathematical Programming Approach. Oper. Res. **24** (1976) 842–856

21. Warner, M., Prawda, J.: A Mathematical Programming Model for Scheduling Nursing Personnel in a Hospital. Manage. Sci. **19** (1972) 411–422

Storing and Adapting Repair Experiences in Employee Rostering

Sanja Petrovic[1], Gareth Beddoe[1], and Greet Vanden Berghe[2]

[1] Automated Scheduling, Optimisation, and Planning Research Group,
School of Computer Science and Information Technology, University of Nottingham,
Jubilee Campus, Nottingham NG8 1BB, UK
`sxp,grb@cs.nott.ac.uk`
[2] KaHo St-Lieven, Information Technology,
Campus Rabot, Gebroeders Desmetstraat 1, B-9000 Gent, Belgium
`greetvb@kahosl.be`

Abstract. The production of effective workforce rosters is a common management problem. Rostering problems are highly constrained and require extensive experience to solve manually. The decisions made by expert rosterers are often subjective and are difficult to represent systematically. This paper presents a formal description of a new technique for capturing rostering experience using case-based reasoning methodology. Examples of previously encountered constraint violations and their corresponding repairs are used to solve new rostering problems. We apply the technique to real-world data from a UK hospital.

1 Introduction

The rostering of employees within an organisation must satisfy operational, legal and management requirements whilst taking into account the conflicting considerations of staff morale and sensible working practice. Manual rostering experts develop strategies for balancing these requirements, drawing on their extensive experience to make rostering decisions. This paper describes a method for capturing this experience for re-use in an automated setting.

Capturing such rostering knowledge in the form of logical rules (e.g. IF THEN rules) is difficult and can lead to incomplete and inflexible domain models [15]. This is because rostering decisions are made by often subjective interpretations of the subtle interactions of a number of parameters. We move away from explicit representations of rostering rules and introduce a system that stores them implicitly in a history of past experience.

Case-based reasoning (CBR) [11] is an artificial intelligence methodology that aims to imitate human style decision making by solving new problems using knowledge about the solutions to similar problems. CBR methodology operates under the premise that similar problems will require similar solutions. Previous problems and solutions are stored in a *case-base* and accessed during reasoning by processes of identification, retrieval, adaptation and storage. The identification and retrieval phases search the case-base for cases containing problems that are

E. Burke and P. De Causmaecker (Eds.): PATAT 2002, LNCS 2740, pp. 148–165, 2003.

most similar to the current problem in terms of a set of characteristic features called *indices*. The solutions from these retrieved cases are then adapted and applied to the context of the current problem. If the new solution might be useful for solving future problems then it is stored as a new case in the casebase. CBR is suited to the problem of capturing rostering knowledge because it enables us to build a history of correspondences between constraint violations and their solutions.

A number of different approaches have been used for solving employee rostering problems in the past including linear and integer programming [3,4,14,18], goal programming [2,5], and constraint satisfaction techniques [1,8,12,13]. CBR was employed by Scott and Simpson [16] by storing shift patterns used for the construction of rosters. A number of meta-heuristic methods have also been developed with some success using tabu search [6,9,10] and memetic algorithms [7]. These methods traverse the search space through neighbourhoods defined using extensive domain knowledge and experimental trial and error.

The knowledge capturing technique described here aims to provide a means to intelligently and dynamically define neighbourhoods tailored to individual problems. The repairs used to solve constraint violations are stored in a case-base of previous experience and are adapted when new violations are encountered. This method is not intended to produce final solutions to rostering problems but instead works on the more *local* level by repairing individual constraint violations. It provides a means by which to store and re-use rostering experience and could in the future be incorporated in some form of metaheuristic or other problem solving framework.

We investigated the problem of rostering nurses in an ophthalmological ward at the Queens Medical Centre University Hospital Trust (QMC) in Nottingham, United Kingdom. Section 2 of this paper will describe the problem in this ward including the manual procedures used at present. Sections 3 and 4 will define the problem mathematically and present the CBR based local repair algorithms that have been developed. An example of a problem solving instance is given in Section 5. The results of some experiments are provided in Section 6 before a discussion on the future directions of this research.

2   Problem Description

The rostering problem at the QMC is rather more complex than many previously investigated in the literature in terms of the levels of detail that must be considered. The descriptions of nurses qualifications and abilities does not lend itself well to the problem subdivision methodologies of [2,10] – we cannot simply divide the nurses up into disjoint "levels" of qualification. This also results in more complex constraint descriptions in the form of appropriate skill mixes and cover requirements for the ward.

The problem consists of assigning shifts to nurses over a set time period (usually 28 days) subject to a number of constraints. Nurses have one of four levels of "qualification" depending on the training they had to become a nurse. These are, in descending order of seniority: Registered (RN), Enrolled (EN), Auxiliary (AN) and Student (SN). RNs are the most qualified and have had extensive training in both the practical and management aspects of nursing whereas ENs have had less training in only practical nursing. ANs are unqualified nurses who can perform basic duties and SNs are training to be either RNs or ENs. Three additional qualification types are used to group these "real" types. Registered and enrolled nurses are grouped together as Qualified Nurses (QN) and RNs, ENs, and ANs, are classified as Employed Nurses (PN). The classification XN groups all of the nurses together.

In addition to these qualifications nurses can receive specialty training specific to the ward they work on (in this ophthalmological ward it is "eye-training" (ET)). A grade is also assigned to each nurse and is determined by a combination of their qualification, specialty training and the amount of practical experience they have. These grades range from A to I with I being the most senior. Two final attributes taken into account during the rostering process are gender (M or F) and international status (I or H). The latter is important in UK hospitals due to an increased reliance on overseas-trained nurses to overcome staff shortages in the public sector. In this paper we refer to all of these attributes of a nurse as their "descriptive features".

At present, roster production in the QMC ward is a three-stage process. The self-rostering planning approach is used to collect shift preference information from all members of staff (see [17] for a comprehensive survey of the use of self-rostering in UK hospitals). This approach recognises that nurses are professionals who will fulfil their responsibilities without excessive administrative intervention.

The three stages of roster production are

1. nurses are assigned to teams (according to a particular skill mix);
2. nurses produce partial rosters (called preference rosters) for the planning period in consultation with other members of their teams;
3. partial rosters are combined to produce the ward roster which is invariably infeasible; any constraint violations are repaired by senior staff members.

These preference rosters indicate when individual nurses would like to work a particular shift, and when they would like a day off. If they have no preference then they can leave a particular day blank. The amount of detail and flexibility for shift assignments varies considerably between nurses.

We wish to automate the third stage of the process. At present, this stage takes a considerable number of hours per month to complete. The constraint violations present in the roster need to be repaired whilst retaining as much preference information as possible.

At this stage of this research we consider only a subset of the constraints that are present in the real-world problem. A number of hard and soft constraints can be identified but here we shall deal only with the two most important hard constraints. It is our conjecture that some of the soft constraints will be satisfied as a consequence of the information stored about the repair of hard constraint violations – or at least that a series of these repairs will minimise the degree of violation of the soft constraint.

The two hard constraints (hereinafter referred to simply as "constraints") considered here are

- *Cover* constraints define the skill mix required for a particular shift. They are described by variables indicating the type and number of nurses needed for a specified shift. For example, the early shift requires 4 QNs;
- *Totals* constraints describe the maximum working hours allowed over a particular period. These can be defined for all nurses of a specific type as well as for individual nurses over any time period. For example, the maximum number of hours that can be (legally) worked by any nurse (XN) within a fortnight is 75.

Three basic rostering actions, or repair types, have been identified and these are representative of the kind of actions carried out by rostering experts. RE-ASSIGN repairs are the simplest and involve reassigning a nurse's shift on a particular day. Two nurses can have their shifts on a particular day swapped by the SWAP repair. The final repair type, SWITCH, interchanges the shifts assigned to one nurse on two different (consecutive or non-consecutive) days. These basic repair types require different data and this will be reflected in the mathematical representation described in the following section.

The method proposed in this paper stores information about individual repairs of constraint violations in rosters. It keeps a history consisting of pairs of problems and solutions from which new repairs can be generated.

3   Mathematical Formulation

We define a rostering problem as an ordered pair

$$R = \langle N, C \rangle,$$

where $N = \{nurse_i : 0 \leq i < n\}$ is the set of nurses to be rostered such that

$$nurse_i = \langle NurseType_i, hours_i, NR_i, NP_i \rangle.$$

$NurseType_i = \{f_{i,1} \ldots f_{i,I}\}$ is an array of descriptive features where

$$f_{i,1} \in \{RN, EN, AN, SN, QN, PN, XN\}$$

is the nurse's qualification and $f_{i,2} \ldots f_{i,I}$ describe gender, international status, specialty training, and grade. $hours_i \in \mathbb{R}^+$ is the number of hours the nurse is contracted to work in a week (normally 37.5).

$$NR_i = \{s_{i,j} : 0 \leq j < period\}$$

is a set of assignment variables $s_{i,j}$ which represent the actual shift assignment for $nurse_i$ on day j over the number of days for which the roster has been constructed, *period*.

$$NP_i = \{p_{i,j} : 0 \leq j < period\}$$

is a set of variables $p_{i,j}$ representing the preferred assignment of $nurse_i$ on day j. The variables $s_{i,j}$ and $p_{i,j}$ can take values from the set {UNASSIGNED, EARLY, LATE, NIGHT, OFF}.

The set C consists of a number of constraints that can take one of the following types:

$$COVERCONSTRAINT(NurseType, shift, minimumCover);$$
$$TOTALSCONSTRAINT(NurseType, period, maximumHours).$$

Cover constraints describe the minimum number $minimumCover$ of nurses of type $NurseType$ who must be assigned $shift$ on every day of the roster. Totals constraints describe the maximum number of hours $maximumHours$ that nurses of type $NurseType$ may work over a number of days $period$.

When constraints are applied to N, they generate a set of violations of a type corresponding to the constraint type. We define the problem instance spaces P_v^R and P_r^R as the set of violations and possible repairs given the current roster R. An element $violation_\alpha \in P_v^R$ details the type of violation and the parameters relevant to it. It may be one of the following types:

$$COVER(NurseType, day, shift);$$
$$TOTALS(nurse_i, startDay, endDay).$$

Cover violations are generated by cover constraints and represent that the number of nurses of type $NurseType$ assigned $shift$ on the day indicated is insufficient. Likewise, totals violations are generated by totals constraints and describe that $nurse_i$ has been assigned too many hours between days $startDay$ and $endDay$.

An element $repair_\beta \in P_r^R$ describes the type of repair and the nurses, days, and shift assignments involved. They can be one of the following types:

$$REASSIGN(nurse_i, day_\beta, shift_\beta);$$
$$SWAP(nurse_i, nurse_j, day_\beta);$$
$$SWITCH(nurse_i, day1_\beta, day2_\beta).$$

Reassign repairs assign $shift_\beta$ to $nurse_i$ on day day_β. Swap repairs interchange the shift assignments of $nurse_i$ and $nurse_j$ on day day_β and switch repairs interchange the shift assignments of $nurse_i$ on the days $day1_\beta$ and $day2_\beta$.

The violation and repair problem spaces represent information relevant to a specific instance of a rostering problem (an instantiation of R). The nurses, days and shifts they describe refer only to those specified by R. In order to store and reuse examples of previous violation/repair experiences we need to define a generalised structure.

We define the case-base CB as a set of previously encountered violations and their corresponding repairs. The case-base $CB = W_v \times W_r$ where W_v is the space of stored violations (the problem history) and W_r is the space of stored repairs (the solution history). Therefore a case $c_\gamma = (v_\gamma, r_\gamma) \in CB$ where $\gamma \in \Gamma$. We define v_γ and r_γ as follows:

$$v_\gamma = \langle ViolationType_\gamma, ViolationIndices_\gamma \rangle,$$
$$r_\gamma = \langle RepairType_\gamma, RepairIndices_\gamma \rangle.$$

Here $ViolationType_\gamma$ and $RepairType_\gamma$ contain the type and necessary parameters describing generalised violations and repairs. $ViolationIndices_\gamma$ and $RepairIndices_\gamma$ store the feature information needed to match problem instances during case retrieval, and to generate repairs during case adaptation. The index sets are arrays of feature values, which are generally integers or real numbers, and in the case of the repair indices store shift pattern information. This index information will be described in more detail later in the paper.

We can now define the generalisation functions $\theta_v^R : P_v^R \to W_v$ and $\theta_r^R : P_r^R \to W_r$ that map instance specific violations and repairs respectively to their generalised case representations. The various types of violation and repair are converted to their generalised equivalents and the indices necessary for retrieval and adaptation are calculated. For example, the $TOTALS$ violation and its parameters will be mapped:

$$\theta_v^R \Big(TOTALS(nurse_i, startDay, endDay) \Big)$$
$$= \langle CBTOTALS(NurseType_i), ViolationIndices \rangle,$$

and similarly for a $SWAP$ repair:

$$\theta_r^R \Big(SWAP(nurse_i, nurse_j, day) \Big)$$
$$= \langle CBSWAP(NurseType_i, NurseType_j, s_{i,day}, s_{j,day}), RepairIndices \rangle.$$

The violation and repair indices used are described in Section 4.

Figure 1 gives a graphical summary of the generalisation functions. Note that here we define only the transformation from problem instance to case-base. The inverses of these functions are not well-defined mathematically and in fact would not make sense from an operational point of view. The method of generating new repairs using a combination of current problem information and historical experience is the subject of the next section.

In order to describe the retrieval and adaptation processes the definitions of equality of some variables need to be defined. Two nurses are considered to have the same type if all of their descriptive feature information is the same. We define equality of the $NurseType$ variable mathematically as follows. Given $NurseType_a = \{f_{a_1}, \ldots, f_{a_I}\}$ and $NurseType_b = \{f_{b_1}, \ldots, f_{b_I}\}$

$$NurseType_a = NurseType_b \text{ iff } f_{a_i} = f_{b_i} \forall i, (1 \le i \le I).$$

Equality of the generalised $ViolationType$ and $RepairType$ information are defined similarly. This information is in the following form:

$$TYPENAME_a(param_{a_1}, \ldots, param_{a_M}).$$

Then equality between variables $ViolationType_a$ and $ViolationType_b$, or $RepairType_a$ and $RepairType_b$ occurs if and only if

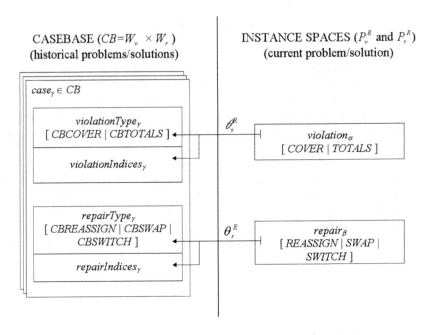

Fig. 1. Relationship between the current instance and the cases from the case-base

$$(TYPENAME_a = TYPENAME_b) \wedge (param_{a_i} = param_{b_i}) \forall i, (1 \leq i \leq M).$$

It remains to define two functions that will be used in the following section. The similarity measure function SIM is a standard nearest-neighbour method common in the CBR literature [11]. We apply it here to the index sets of violations and repairs. Given $IndexSet_a = \{index_{a_1}, \ldots, index_{a_I}\}$ and $IndexSet_b = \{index_{b_1}, \ldots, index_{b_I}\}$, representing either $ViolationIndices$ or $RepairIndices$ sets,

$$\text{SIM}(IndexSet_a, IndexSet_b) = \left(\frac{1}{I} \sum_{i=1}^{I} w_i \times dist(index_{a_i}, index_{b_i}) \right)^{-1},$$

where I is the number of elements in the index sets, w_i are the index weights, and

$$dist(index_{a_i}, index_{b_i}) = \left| \frac{index_{b_i} - index_{a_i}}{index_{max_i} - index_{min_i}} \right|.$$

The values $index_{max_i}$ and $index_{min_i}$ are the maximum and minimum values for the corresponding index recorded in the case-base. The $dist$ function therefore finds the normalised distance between feature values. By weighting each of these distances we assign a relative importance to each feature. In this research these weights are generally flat (all set to 1). The effect of changes to these values will be the subject of future investigation.

During repair adaptation it is necessary to compare the shift patterns of nurses in the current roster with stored patterns in the $RepairIndices$ variables. The shift pattern comparison function $comp$ compares the shift patterns of two different nurses over a five day period (two days before and after the day on which a repair is applied). The difference in shift pattern strings is calculated as follows. Given $nurse_\alpha$ with shift pattern string $x_1x_2x_3x_4x_5$ and a pattern $y_1y_2y_3y_4y_5$ to be compared we define

$$comp_\alpha(y_1y_2y_3y_4y_5) = \sum_{i=1}^{5} \delta_\alpha(y_i),$$

where

$$\delta_\alpha(y_i) = \begin{cases} 0 & y_i = x_i, \\ 1 & y_i \neq x_i. \end{cases}$$

By this definition the shift patterns $EUUNN$ and $EEUUL$ have a difference of $0 + 1 + 0 + 1 + 1 = 3$.

4   Retrieval and Adaptation

The main difference between the classical OR and meta-heuristic approaches to nurse rostering described in the literature and the CBR approach proposed here is that the repairs generated are not optimal in any sense. We have described no measures of the quality of repairs – the aim here is to imitate, as closely as possible, the decisions made by rostering experts without relying on quantified measures of roster quality.

The notion of problem similarity is key to the success of any CBR application. A method must be developed for finding the most similar problems in the case-base to the current problem being solved. Having identified the most similar problems the stored repairs have to be adapted to the context of the current roster. Here again we use the notion of similarity. We must generate a new repair that most closely matches the repair stored in the case.

There are a number of justifications for taking this approach. The aim must not be to replicate exactly the repair from the retrieved case. This would be incorrect in most situations as the nurses and time periods involved would undoubtedly be different. In some instances it may even be impossible if a different set of nurses is defined. It is also clear that simply generating a random repair of a specified type would not be correct. A compromise between these two extreme approaches is reached between by the generation of repairs that are considered *similar* to the retrieved repairs. This emphasis on the similarity of repairs in different instances motivated the generalisations of both violation and repair described in the previous section.

The retrieval process is split into two distinct searches of the case-base. The first search filters the case-base to obtain cases containing violations that match the current problem in terms of violation type and parameters. This is a strict

Table 1. Violation indices

Global problem characteristics	
Number of violations	The level of infeasibility of the roster w.r.t. the (hard) constraints
Nurse satisfaction	The percentage of nurse shift preferences that have remained intact
Utilisation	The percentage of the total assignable hours (w.r.t. nurse contracted hours h) already assigned over the whole roster and within the week of the violation
Local problem characteristics	
Nurse rank	An index assigned based on the qualification level of the nurse or NurseType involved in the violation
Nurse satisfaction	The percentage of shift preferences that have remained intact w.r.t. nurses of the type involved in the violation
Utilisation	The percentage of total assignable hours assigned of the nurses of the type involved in the violation over the whole roster and within the week of the violation

search whereby cases are either accepted or not and no similarities are evaluated. The equality of the *ViolationType* variables described in Section 3 is used to make the case comparisons.

The second search ranks the restricted set of cases using the similarity function SIM applied to the violation indices. These indices are the characteristics of the problem identified as having an influence on the decision making process and are divided into two categories. Global characteristics are the properties of the roster as a whole including the number of violations, levels of staff satisfaction and current utilisation statistics. Local characteristics describe the problem in the location of the violation. They include the magnitude of the violation, and the current satisfaction and utilisation statistics of the types of nurses involved. Table 1 lists all the problem indices and gives a brief description of each.

Formally, the retrieval algorithm is described by the function RETRIEVE ($violation_\alpha, CB, CB'$) (see Figure 2) which from the case-base CB returns a set of cases, CB', sorted in order of similarity, that are compatible to the current problem $violation_\alpha$. Lines 1 and 2 initialise CB' and apply the function θ_v^R to $violation_\alpha$ to get the generalised form. The set of cases CB' is filled, in lines 3 to 7, with all cases in the case-base of the same *ViolationType* as the current problem. By the definition of equality (as defined in Section 3) this includes the parameters of the violation. If there are no such cases then line 8 returns false

RETRIEVE $(violation_\alpha, CB, CB')$
1: $CB' \leftarrow \emptyset$.
2: $\langle ViolationType_\alpha, ViolationIndices_\alpha \rangle \leftarrow \theta_v^R(violation_\alpha)$.
3: **for all** $case_\gamma = \langle\langle ViolationType_\gamma, ViolationIndices_\gamma \rangle, r_\gamma \rangle \in CB$ **do**
4:   **if** $ViolationType_\gamma = ViolationType_\alpha$ **then**
5:     $CB' \leftarrow CB' \cup \{case_\gamma\}$.
6:   **end if**
7: **end for**
8: **if** $|CB'| = 0$ **then return** $false$. **end if**
9: generate an array $score[|CB'|]$.
10: **for all** $case_\gamma = \langle\langle ViolationType_\gamma, ViolationIndices_\gamma \rangle, r_\gamma \rangle \in CB'$ **do**
11:   $score[\gamma] \leftarrow$ SIM$(ViolationIndices_\alpha, ViolationIndices_\gamma)$.
12: **end for**
13: sort CB' according to $score$.
14: **return** $true$.

Fig. 2. Retrieval algorithm

and a manual solution will be required. Lines 9 to 13 generates an array of scores for each of the cases in CB' and then sorts CB' accordingly.

This sorted set of restricted cases generated by RETRIEVE is then passed to another method for repair adaptation. The adaptation process is also separated into two phases. Initially, the method generates, using the data from the current roster, a set of candidate repairs each of the same type as in the retrieved case. The second stage involves ranking these candidate repairs according to their similarity to the repair in the retrieved case. Here the set of repair indices from the retrieved case is compared with the calculated indices of the candidates using the SIM function. The exact indices that are used depend on the type of repair being generated. Table 2 lists the indices used for each of the three different repair types.

Formally, the function GENERATEREPAIR$(R, CB', violation_\alpha, Candidates)$ (see Figure 3) tries to generate a repair as similar as possible to that used in the retrieved cases. The array of possible repairs $Candidates$ is filled by a function DETERMINECANDIDATES. If there are no available candidates given the repair information from the retrieved case then the next case in the $Candidates$ array is considered. Lines 1 to 8 of the algorithm try to produce a set of candidates for each of the cases in CB' starting with the most similar. Then, analogously to the retrieval algorithm, the candidates are ranked according to their similarity to the repair in the retrieved case with respect to their $RepairIndices$.

The DETERMINECANDIDATES $(N, violation_\alpha, RepairType_0)$ function is the key to the generation of repairs. This returns a set of actual repairs that match the retrieved repair in terms of the type of the repair and the types of the nurses and shifts involved. Six sets of rules have been established for determining the candidates – one for each possible pair of violation and repair types. These rules are fairly intuitive and use some very simple domain knowledge to produce.

Table 2. Repair indices

Feature	CBREASSIGN	CBSWAP	CBSWITCH
# nurses assigned $newShift1$ on $day1_\beta$	$\checkmark$	$\checkmark$	$\checkmark$
# nurses assigned $newShift2$ on $day1_\beta$		$\checkmark$	$\checkmark$
# nurses assigned $newShift1$ on $day2_\beta$			$\checkmark$
# nurses assigned $newShift2$ on $day2_\beta$			$\checkmark$
# nurses of type $NurseType1_\beta$ assigned $newShift1$ on $day1_\beta$	$\checkmark$	$\checkmark$	$\checkmark$
# nurses of type $NurseType1_\beta$ assigned $newShift2$ on $day1_\beta$		$\checkmark$	$\checkmark$
# nurses of type $NurseType1_\beta$ assigned $newShift1$ on $day2_\beta$			$\checkmark$
# nurses of type $NurseType1_\beta$ assigned $newShift2$ on $day2_\beta$			$\checkmark$
# nurses of type $NurseType2_\beta$ assigned $newShift1$ on $day1_\beta$		$\checkmark$	
# nurses of type $NurseType2_\beta$ assigned $newShift2$ on $day1_\beta$		$\checkmark$	
# nurses assigned $oldShift$ on $day1_\beta$	$\checkmark$		
# nurses of type $NurseType1_\beta$ assigned $oldShift$ on $day1_\beta$	$\checkmark$		
Assigned/Contract Hours $nurse1_\beta$	$\checkmark$	$\checkmark$	$\checkmark$
Assigned/Contract Hours $nurse2_\beta$		$\checkmark$	
$comp$ value for $nurse1_\beta$ around $day1$	$\checkmark$	$\checkmark$	$\checkmark$
$comp$ value for $nurse1_\beta$ around $day2$			$\checkmark$
$comp$ value for $nurse2_\beta$ around $day2$		$\checkmark$	

GENERATEREPAIR $(N, CB', violation_\alpha, Candidates)$
1: $Candidates \leftarrow \emptyset$
2: $index \leftarrow 0$
3: **while** $|Candidates| = 0$ **do**
4:     $case := \langle v, \langle RepairType, RepairIndices \rangle \rangle \leftarrow CB'[index]$.
5:     $Candidates \leftarrow$ DETERMINECANDIDATES $(N, violation_\alpha, RepairType)$.
6:     $index \leftarrow index + 1$.
7:     **if** $index = |CB'|$ **then return** false. **end if**
8: **end while**
9: generate an array $score[Candidates]$.
10: **for all** $repair_\beta \in Candidates$ **do**
11:     $\langle RepairType_\beta, RepairIndices_\beta \rangle \leftarrow \theta_r^R(repair_\beta)$.
12:     $score[repair_\beta] \leftarrow$ SIM$(RepairIndices_0, RepairIndices_\beta)$.
13: **end for**
14: sort $Candidates$ according to $score[Candidates]$.
15: **return** $true$.

Fig. 3. Adaptation algorithm

Nevertheless, they are quite large when notated and so here we shall only present the following rule, which shall be used in the example in Section 5:

Given $violation_\alpha = COVER(NurseType_\alpha, day_\alpha, shift_\alpha)$ and $repairType = CBREASSIGN(NurseType1, newShift1, oldShift)$, then $\textsc{DetermineCandidates}\left(N, violation_\alpha, repairType\right) = \left\{REASSIGN(nurse1_{i_\beta}, day1_\beta, shift_\beta) | (nurse1_{i_\beta} \in N) \wedge (NurseType1_\beta = NurseType1) \wedge (s_{c,day_\alpha} = oldShift), day1_\beta = day_\alpha, shift_\beta = shift_\alpha\right\}$.

This rule returns a set of repairs all of which have the $REASSIGN$ type with different parameter values. The nurse in each repair must be of the same type as that used in the repair from the retrieved case. In addition the shift currently assigned to each nurse on the day of the violation must be the same as the $oldShift$ parameter of the retrieved repair. The day and $shift$ parameters of the repair must be the same as that of the violation for this rule.

5   Example

In order to give an illustrative example of the method we shall consider a relatively simple problem solving episode. It is assumed that the weighting of all similarity calculations is flat (all weights equal 1). The problem we are attempting to solve is a cover violation of the roster R – that there is no registered nurse (RN) rostered on the early shift ($shift = E$) of the third day ($day = 2$) of the planning period. This is represented as

$$violation_\alpha = COVER(\{RN, 0, 0, 0, 0\}, 2, E).$$

Here the zero-valued elements in the $NurseType$ set indicate that there is no restrictions on these feature values (so a suitable registered nurse could be male or female, specialty trained or untrained, etc.).

This violation is first passed to the $\textsc{Retrieve}$ method. The generalised form of this violation is generated by the θ_v^R as follows:

$$\theta_v^R\left(violation_\alpha\right) = \langle CBCOVER(\{RN, 0, 0, 0, 0\}), ViolationIndices_\alpha\rangle,$$
where $ViolationIndices_\alpha = \{62.00, 99.46, 47.29, 54.32, 1.00, 78.41, 44.63, 50.76\}$.

Each of the values in the $ViolationIndices_\alpha$ array corresponds to one of the violation feature values described in Table 1. We must now find all cases $case_\gamma \in CB$ with $ViolationType = CBCOVER(\{RN, 0, 0, 0, 0\})$ and add them to the set of cases CB'. In this example we find three such cases. The $\textsc{Retrieve}$ method then calculates the similarity between the $ViolationIndices_\gamma$ of each of these cases and the $ViolationIndices_\alpha$ array:

Case	Features	SIM (score)
α	62.00 99.46 47.29 54.32 1.00 78.41 44.63 50.76	NA
$CB'[0]$	57.00 92.31 48.12 52.14 1.00 80.32 74.23 90.32	8.749
$CB'[1]$	31.00 86.32 53.98 78.92 1.00 44.02 70.23 60.12	4.975
$CB'[2]$	80.00 83.24 70.34 80.23 2.00 70.12 80.12 85.23	3.997

Finally, the CB' set is sorted according to the similarity values in the far right column. This set now contains all compatible cases in order of the similarity between their and the current problem's violation features.

Now that a set of cases of similar problems has been identified, their corresponding solutions need to be adapted to the current problem $violation_\alpha$. For simplicity in this example we shall assume that the first case ($CB'[0]$) allows the production of such a set of candidates. The repair part of this case contains the following:

$$r = \langle CBREASSIGN(\{RN, F, H, ET, E\}, E, U), RepairIndices\rangle,$$
$$\text{where } RepairIndices = \langle\{11, 9, 2, 0, 59.1, 0\}, EEUUL\rangle.$$

The GENERATEREPAIR function fills the $Candidates$ array with potential repairs using the DETERMINECANDIDATES function. By applying the conditions given in Section 4 for $CBREASSIGN$ repairs given a $COVER$ violation we get three candidate repairs, each of which uses a different registered nurse with $NurseType = \{RN, F, H, ET, E\}$. These candidate repairs are

$$REASSIGN(nurse_2, 2, E);$$
$$REASSIGN(nurse_8, 2, E);$$
$$REASSIGN(nurse_{15}, 2, E).$$

We apply the generalisation function θ_r^R to each of these candidate repairs and compare their resulting $RepairIndices$:

Repair	Features	SIM (score)
$retrieved$	11 9 2 0 59.1 $COMP_0(EEUUL) = 0$	NA
$\theta_r^R(Candidates[0])$	5 5 3 0 20.3 $COMP_\alpha(EELLO) = 3$	4.090
$\theta_r^R(Candidates[1])$	10 9 3 0 62.7 $COMP_\alpha(OEUUE) = 2$	9.357
$\theta_r^R(Candidates[2])$	10 9 3 0 84.0 $COMP_\alpha(UUUUU) = 3$	5.433

The candidate repair that is closest to the repair in the best case $CB'[0]$ is $Candidates[1]$ – which is a reassignment of $nurse_8$ to the EARLY shift on the third day of the planning period.

The example here is simpler than many encountered for ease of explanation. Other combinations of violation and repair types involve a more complex search for candidate repairs. However, all the principals needed for more complex instances are the same.

6   Results

The results presented in this section illustrate how the method performs at imitating the rostering decisions of humans. We do not compare performance with other rostering methods for two reasons. Firstly, the amount of information used and the way it is presented is incompatible with most existing problem formulations. More significantly, the case-based reasoning method here treats constraint violations as individual *problems* rather than providing solutions to entire rostering problems in the traditional sense.

The method has been implemented and tested on real-world data from the QMC. This data consisted of twelve 28-day rosters and the corresponding preference information for 19 nurses of various qualification and training levels. Nine constraints were defined consisting of eight cover type constraints detailing required skill mixes for the three shifts and one totals constraint limiting the number of hours in a fortnight to 75 per nurse.

Two sets of experiments were defined. The aim of the experiments was to determine the quality of the reasoning process in terms of the agreement between automated decisions and those of the nurse rostering expert. Constraint violations were identified at random and the repairs suggested by this method were compared to the repairs actually made in the final roster. These expert repairs were determined by comparing the final and preference rosters and only those instances where the decision was clearly evident were considered. The "quality" of a generated repair was assessed by comparing it with the expert repair and assigning one the following verdicts:

– Exact match: the generated repair is identical to the expert's repair;
– Equivalent match: the generated repair involves nurses of the same types and the same shifts as those used in the expert's repair;
– Fail: the generated repair is not an exact or equivalent match, or no repair was generated.

The first experiment involved repairing five runs of 120 constraint violations. Three repairs were suggested by the method and compared to the expert repair. The case-base is empty at the start of the run and the expert repair for each of the constraint violations is stored after it is applied to the roster. In this way the method is storing more experience in the case-base as the run progresses. Figure 4 shows the average cumulative number of exact and equivalent matches against the case-base size for each of the three suggested repairs. The bold lines are the first (or best with respect to the reasoning process) repairs for each iteration.

The results in Figure 4 show an increasing gradient of all lines indicating an increasing number of repairs of the given verdict per iteration. It can be seen from this that the case-base learns how to produce more exact or equivalent repairs as its size increases. An increase in the amount of training given to the case-base corresponds to an increase in the quality of the repairs produced. It is particularly encouraging that the first suggestions in general score more exact and equivalent matches than the second and third. The increases in solution quality are made more apparent in Figure 5. This shows the percentage of exact

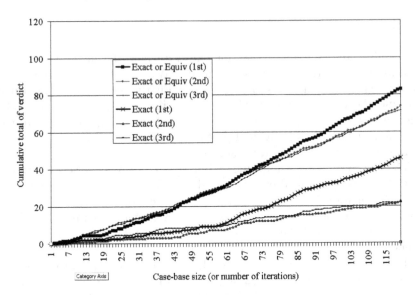

Fig. 4. Average cumulative number of exact and equivalent matches against case-base size over five 120 iteration runs

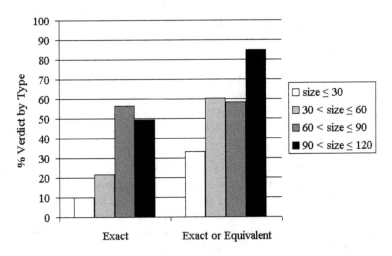

Fig. 5. Effects of case-base size on solution quality

and equivalent repairs at different stages in the runs. In general, in the later stages, when the case-base contains more experience, a larger number of good suggestions are produced.

The second set of experiments was defined to test the influence of different types of indices, namely global and local, on the reasoning process. Figure 6 shows the results of an experiment carried out using case-bases generated during

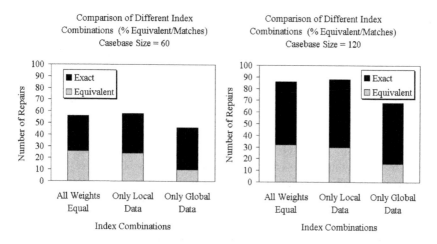

Fig. 6. Effects of different combinations of problem index on the solution quality

the previous test. Case-bases containing 60 and 120 cases were collected during each run. Using each of these 10 case-bases an additional 100 constraint violations were repaired but this time there was no storage of new cases. Three different combinations of problem index weights were set in the similarity function so that all indices, only local indices, and only global indices were used.

The lower-quality results from the only global data set suggest, at first, that the global indices are not useful. The best results were achieved when only local data was used. However, the global data set nevertheless produced a reasonable percentage of good results and this suggests that global information may still be useful for the reasoning process. It is certainly the case that global data are not as important as the local data and this should be reflected in any automated feature weighting that may be carried out in future work.

7   Conclusion

This paper has introduced a new approach to the employee rostering problem. The approach is different to existing methods in the sense that it does not use explicitly defined evaluation functions, which often fail to include all aspects of the rostering problem. The results show that it is possible to capture and imitate the rostering actions of human rostering experts. By storing the rules for repairing constraint violation implicitly in the case information we have created a technique that is both adaptable and flexible.

We intend to increase the number of different types of constraints that are considered including weekend constraints, night shift constraints, and possibly any soft constraints that are not adequately covered by the information in the case-base. This may involve adding additional structural and index elements to the definition of a case. The weighting of indices to reflect their relative importance in the reasoning process must be investigated. A system that dynamically allocates weights under different conditions is being considered. This allocation may depend on the content of the case-base and thus reflect the importance placed on indices by the expert.

The success of this violation/repair model for problem solving will be built on by incorporating the method within an intelligent iterative algorithm. A meta-heuristic could use the technique to decide its next move in the search space at each iteration. Although the processing time for each repair generation is negligible, the cumulative effect of multiple searches of the case-base and the impact on overall algorithm performance will need to be addressed. A hybrid algorithm could be evaluated and compared with existing rostering methods and its design and behaviour will be the subject of future research.

Acknowledgements. This research is supported by the Engineering and Physical Sciences Research Council (EPSRC) in the UK, grant no GR/N35205/01.

References

1. Abdennadher, S., Schlenker, H.: INTERDIP – an Interactive Constraint Based Nurse Scheduler. In: Proc. 1st Int. Conf. Exhib. Pract. Applic. Constraint Technol. Logic Program. (PACLP) (1999)
2. Arthur, J.L., Ravindran, A.: A Multiple Objective Nurse Scheduling Model. AIIE Trans. **13** (1981) 55–60
3. Bailey, J., Field, J.: Personnel Scheduling with Flexshift Models. J. Oper. Manage. **5** (1985) 327–338
4. Beaumont, N.: Scheduling Staff Using Mixed Integer Programming. Eur. J. Oper. Res. **98** (1997) 473–484
5. Berrada, I., Ferland, J.A., Michelon, P.: A Multi-objective Approach to Nurse Scheduling with Both Hard and Soft Constraints. Socio-Econ. Planning Sci. **30** (1996) 183–193
6. Burke, E.K., De Causmaecker, P., Vanden Berghe, G.: A hybrid tabu search algorithm for the nurse rostering problem. In: Proc. 2nd Asia Pacific Conf. on Simulated Evolution and Learning (selected papers). Lecture Notes in Artificial Intelligence, Vol. 1585. Springer-Verlag, Berlin Heidelberg New York (1998) 187–194
7. Burke, E.K., Cowling, P.I., De Causmaecker, P., Vanden Berghe, G.: A Memetic Approach to the Nurse Rostering Problem. Applied Intelligence. Kluwer, Dordrecht (2001) 199–214
8. Cheng, B.M.W., Lee, J.H.M., Wu, J.C.K.: A Constraint-Based Nurse Rostering System Using a Redundant Modeling Approach. Department of Computer Science and Engineering, The Chinese University of Hong Kong (1996)
9. Dowsland, K.A., Thompson, J.M.: Solving a Nurse Scheduling Problem with Knapsacks, Networks and Tabu Search. J. Oper. Res. Soc. **51** (2000) 825–833
10. Dowsland, K.: Nurse Scheduling with Tabu Search and Strategic Oscillation. Eur. J. Oper. Res. **106** (1998) 393–407
11. Kolodner, J.L.: Case-Based Reasoning. Morgan Kaufmann, San Mateo, CA (1993)

12. Meisels, A., Gudes, E., Solotorevski, G.: Employee Timetabling, Constraint Networks and Knowledge-Based Rules: A Mixed Approach. In: Burke, E, Ross, P. (Eds.): Practice and Theory of Automated Timetabling I (PATAT 1995, Edinburgh, Aug/Sept, selected papers). Lecture Notes in Computer Science, Vol. 1153. Springer-Verlag, Berlin Heidelberg New York (1996) 93–105
13. Meyer auf'm Hofe, H.: Solving Rostering Tasks as Constraint Optimization. In: Burke, E, Erben W. (Eds.): Practice and Theory of Automated Timetabling III (PATAT 2000, Konstanz, Germany, August, selected papers). Lecture Notes in Computer Science, Vol. 2079. Springer-Verlag, Berlin Heidelberg New York (2001) 280–297
14. Miller, M.L., Pierskalla, W., Rath, G.: Nurse Scheduling Using Mathematical Programming. Oper. Res. **24** (1976) 857–870
15. Miyashita, K., Sycaras, K., Mizoguchi, R.: Capturing Scheduling Knowledge from Repair Experiences. Int. J. Human-Comput. Stud. **41** (1994) 751–773
16. Scott, S., Simpson, R.: Case-bases Incorporating Scheduling Constraint Dimensions – Experiences in Nurse Rostering. In: Smyth, B., Cunningham, P. (Eds.): Advances in Case-Based Reasoning (EWCBR'98). Lecture Notes in Computer Science, Vol. 1488. Springer-Verlag, Berlin Heidelberg New York (1998) 392–401
17. Silvestro, R., Silvestro, C.: An Evaluation of Nurse Rostering Practices in the National Health Service. J. Adv. Nursing **32** (2000) 525–535
18. Warner, M.: Scheduling Nursing Personnel According to Nurse Preference: a Mathematical Programming Approach. Oper. Res. **24** (1976) 842–856

Scheduling Agents – Distributed Timetabling Problems

Amnon Meisels and Eliezer Kaplansky

Department of Computer Science,
Ben-Gurion University of the Negev,
Beer-Sheva, 84-105, Israel
{am,kaplan_e}@cs.bgu.ac.il

Abstract. Many real-world *Timetabling Problems* are composed of organizational parts that need to timetable their staff in an independent way, while adhering to some global constraints. Later, the departmental timetables are combined to yield a coherent, consistent solution. This last phase involves negotiations with the various agents and requests for changes in their own solutions.
Most of the real-world distributed timetabling problems that fall into this class have global constraints that involve many of the agents in the system. Models that use networks of binary constraints are inadequate. As a result, this paper proposes a new model that contains only one additional agent: the Central Agent that coordinates the search process of all Scheduling Agents (SAs). Preliminary experiments show that a sophisticated heuristic is needed for the CA to effectively interact with its scheduling agents in order to find an optimal solution. The approach and the results reported in this paper are an initial attempt to investigate possible solution methods for networks of SAs.

1 Introduction

Timetabling Problems (TTPs) involve an organization with a set of tasks that need to be fulfilled by a set of resources, with their own qualifications, constraints and preferences. The organization enforces overall regulations and attempts to achieve some global objectives such as lowering the overall cost. This leads naturally to the formulation of TTPs as constraint networks.

Many real-world TTPs are composed of organizational parts that need to produce their timetables in an independent way, while adhering to some global constraints. Later, the departments' timetables are combined to yield a coherent, consistent solution. This last phase involve negotiations with the various agents and requests for changes in their own solutions.

An example of such a real-life problem is the construction of a weekly timetable of nurses in several wards of a large hospital. Each ward needs a weekly timetable for its nurses that satisfies the shift and task requirements, attempting to satisfy also the nurses' personal preferences. The generation of a weekly schedule for the nurses in a ward is typically performed by each ward independently.

E. Burke and P. De Causmaecker (Eds.): PATAT 2002, LNCS 2740, pp. 166–177, 2003.

Based on the department timetables, a transportation plan for all the nurses must be constructed. Since the number of vehicles sent and the distances they cover are limited by hospital saving policies, such a transportation plan generates additional constraints between the weekly schedules of the wards.

Since the separate timetables for wards are generated locally, the constraints among these timetables, imposed by the limits on transportation vehicles, have to be negotiated among the scheduling agents of the wards.

1.1   Distributed Constraint Satisfaction Problem (DisCSP)

In the last decade, Constraint Satisfaction Problem (CSP) techniques have become the method of choice for modelling many types of optimization problems: in particular, those involving heterogeneous constraints and combinatorial search.

In many fields fast response has become one of the most important factors for success, especially in the commercial world of today. In fields that utilize CSP techniques, the running time is a critical factor. This is especially true when we want to use CSP techniques to search for some optimal solution.

Recently, the AI community has shown an increasing interest in solving the distributed problem using the agent paradigm. Different parts of the problem are held by different agents, which behave autonomously and collaborate among themselves in order to achieve a global solution. The World Wide Web offers many opportunities to actually solve real problems through agents.

Several works consider constraint satisfaction in a distributed form (see the book by Yokoo [19] for an introduction). These works are motivated by the existence of naturally distributed constraint problems, for which it is impossible or undesirable to gather the whole problem knowledge into a single agent and to solve it using centralized CSP algorithms. There can be several reasons for that. The cost of collecting all information into a single agent may be too high. This includes not only communication costs, but also the cost of translating the problem knowledge into a common format, which could be prohibitive for some applications. Furthermore, gathering all information into a single agent implies that this agent knows every detail about the problem, which could be undesirable for security or privacy reasons (see [11]). The term for this family of problems is the Distributed Constraint Satisfaction Problem (DisCSP).

The generic model for distributed timetabling is that of distributed search. Agents solve their local assignment problem and interact via messages to make their local schedules compatible with the global constraints of transportation. This scenario falls exactly into the domain of distributed constraint networks (DisCNs) and distributed search algorithms [20]. The agents that generate the weekly schedules of the wards are termed Scheduling Agents (SAs).

1.2   DisTTP Example – University Timetabling

An important example of a real-life Distributed Timetabling Problem (DisTTP) is the construction of university timetables. More specifically, classroom allocation in the context of the course timetabling activities within the university.

Some departments are not able to construct a good timetable, since they are over-constrained by the lack of classrooms. The problem of allocating classrooms dynamically is so severe in some universities that each department is assigned a fixed set of lecture rooms which are managed by the department. For example, this is the case in Imperial College [13]. There are additional complexity factors to this problem:

- The amount and type of lecture space available varies widely between departments. Departments which are short of lecture space need to make arrangements with other departments.
- A better timetable for the students requires each department to try and meet its teaching requirements and hold a maximum number of teaching activities within its buildings.

One popular solution for this class of room allocation problems is to allow the departments to construct a partial timetable using their own resources to a maximum, and then try to allocated a room for the other teaching activities through booking arrangements between departments (see [13]).

Since each department owns different resources, has different teaching requirements, and different preferences, this is a typical DisTTP. Each SA solves the timetabling problem for a single department. Solving the problem separately for each department allows departments to easily set and change their own preferences and constraints independently. Departments can follow different strategies for utilizing their teaching space. Each department needs to solve a different problem and can use a different CSP solver tuned up to its needs. This distributed agent model is in line with the irreversible trend toward decentralized information sources, a trend that makes the current approach of a global, centralized search (based on a central computer) less and less adequate.

Based on the departmental timetables, a room allocation plan for the whole university must be constructed. Since the number of the rooms and their capacities are limited and may be regulated by the university policies, such a room allocation plan generates additional constraints between the schedules of the departments.

The above real-world DisTTP is an instance of a DisCSP. Its DisCN has a typical structure in which the local timetabling problem is quite complex and the global problem includes relatively few constraints. One can think of a DisTTP as a network of SAs.

This paper focuses on real-world problems that by their nature need to be solved by agents. Each agent is responsible to solve a single, local, however very complex, CSP. The local solutions are combined to yield a coherent, globally consistent solution. The last phase involves negotiations between agents and requests for changes in the agents' solutions.

The paper is organized as follows. Section 2 present two methods that a set of scheduling agents can coordinate their activities in order to find a solution that satisfies the problem's global constraints. Section 3 describes a general framework for a DisTTP. This framework consists of two different levels that need to be dealt with in order to solve a DisTTP: Section 3.1 describes the global level that

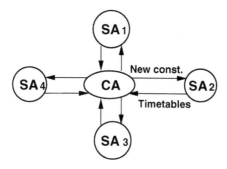

Fig. 1. The Central Agent (CA) adds new constraints to the scheduling agents (SA_i). The SAs solve the modified problem and send back a new timetable

deals with the solution methodology for the central agent (CA) which needs to search for a combination of SA solutions that satisfies the global constraints of the problem and Section 3.2 describes the local level that deals with search algorithms for SAs that can solve a local timetabling problem where constraints are added or removed dynamically. Section 4 presents preliminary results of experiments on Distributed Employee Timetabling Problems (DisETPs) that point in the direction of stochastic search for the SAs. The experiments were performed on the same DisETP scenario and carried out with two different CA control approaches. The overall conclusion is that the model for "intelligent" CA, that controls SAs in a problem-dependent manner, is of value. Section 5 draws some preliminary conclusions.

2   Solving Methods for Scheduling Agents

There are two well known families of algorithms for solving timetabling problems – exhaustive (or complete) search algorithms and stochastic (or local) search algorithms (see [14]). Exhaustive search can go on systematically to find all possible timetables, while stochastic search typically produces one solution, reaching a minimum. It has been shown in the past [10] that for real-world ETPs, stochastic search methods outperform complete search in efficiency terms. Several search algorithms for distributed constraints networks have been proposed in the last decade (see [16,20,4,6,15]). However, existing algorithms for solving DisCNs are all exhaustive. In this paper we show that both methods are useful for solving DisETPs.

One possible approach for solving DisETPs uses exhaustive search for the SAs. In this approach, the SAs perform their exhaustive, systematic, algorithm continuously. Each new local solution they find is sent as a message to the CA. This method is based on the fact that any global solution will always incorporate one local timetable from each SA (i.e. for each ward of the hospital in our example). In order to produce the overall solution, the CA has to search among the set of local solutions, while its domain is constantly enlarged, for one subset that satisfies all the global (inter-agent) constraints. The advantage of the exhaustive search approach is in its high degree of concurrency. In addition, it is guaranteed to find a global solution if such a solution exists. However, for real-world problems, complete search can be very slow [10].

A second, more practical method, uses local search for SAs to generate their local timetables. Each SA can find a local timetable relatively fast, but then it waits for a request, from the central agent, to improve its solution. This request is the result of the CA attempt to resolve some additional global constraint violations. Upon receiving such a message, an SA resumes its local search (i.e. hill climbing) in order to find an improved solution. The advantage of the local search agents approach is that it is likely to converge quickly to some global solution, by using the strategy of improving local solutions. The main drawback of this approach is that the CA may find itself performing cycles in searching for a globally improved solution, since local solutions improve only locally.

This paper investigates combinations of the above two approaches on realistic instances of DisETPs. The local ETPs are real-world instances of wards in a large hospital. The constraints among timetables of the wards are simple limit constraints on the number of nurses that have to be transported, at the end of their shifts, to several residential locations. The induced inter-agent constraint on SAs can be represented very simply by the CA and are also simple to calculate, as partial limits on the number of nurses in a certain group in each of the shifts.

3   A General Framework for DisTTPs

Since the separate timetables for the departments are generated locally, the constraints among these timetables, imposed by the limits on room space, have to be negotiated among departmental scheduling agents. The above generic model falls into the framework of DisCSPs [20,19]. We use the term Scheduling Agent (SA) to refer to the particular case in which an agent has to solve a local CSP (the department TTP in our example) in a global distributed search. In our model, agents solve their local assignment problem and interact via messages to make their local schedules compatible with the global constraints of the problem.

In our model for DisTTPs, shown in Figure 1, an additional agent models the constraints among the agents – the Central Agent (CA). The CA deals with all the intra-SA constraints. All agents communicate by sending and receiving messages from the CA. SAs send their local timetables to the central agent. The CA checks the local timetables for conflicts with global constraints and decides which agents have to be requested to change their local timetables. Based on this model, two families of concurrent algorithms for solving DisTTPs are proposed and compared in Section 4.

There are clearly two different levels (see Figure 2) that need to be deal with in order to solve a DisTTP:

The global level: The solution methodology for the central agent needs to search for a combination of SAs solutions that satisfies the global constraints of the problem. In many cases the search is for an optimal global solution.

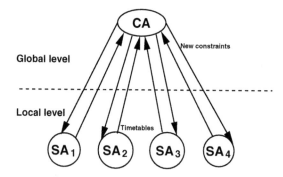

Fig. 2. The different levels of the problem model: The global and the local levels

The local level: Search algorithms for SAs that can solve a local timetable
problem where constraints are added or removed dynamically.

3.1   The Central Agent is an Open CSP (OCSP)

The main difficulty of the CA problem stems from the fact that the values for
assignment in the CA constraint network are acquired incrementally from the
SAs. This transforms the CA CSP into a special class of Dynamic CSP (DCSP).

Preliminary results show that the CA has to have an effective way to interact
with the SAs in order to get an optimal solution in a reasonable time frame. An
important issue here is the ability of the CA to control its SAs in such a way
that the CA can accumulate a promising set of alternative local solutions in a
short time.

Initially, the CA must wait until all SAs return with at least one solution.
Thereafter, the CA starts its own search for a legal combination of SA solutions
satisfying the global constraints of the problem. The CA search space can grow
very fast and usually, quite soon, it becomes a prohibitively large search space:
for example, 10 departments, each with 100 alternative timetables, result in 10^{20}
nodes on the CA's search tree.

Preliminary results on realistic DisETPs (Section 4) have shown that the CA
should start with a complete search, and then move to local search when too
many alternative timetables have arrived. Eventually, the CA performs a local
search that is based on scanning the neighbourhood around the current solution.
One of the basic operations of this scan is a request for an alternative timetable
from some SA. The time that a SA needs for computing the next solution varies
wildly. One possible solution for this difficulty is that the CA will manage a
dynamic buffer of alternative timetables. Another approach for the CA is to
request multiple alternative solutions simultaneously from some selected SAs at
each step of its local search as a form of intelligent improvement moves.

In the paradigm of the experiments presented in this paper, the CA searches
for an optimal global solution. In such a paradigm the basic operation for the CA
is to request an alternative solution from one of its SAs. One way to guide the

SA is by dynamically adding new constraints to the SA's local problem. In our problem, the CA can find out, for example, that on Wednesday, in the time slot of 14:00 to 16:00 there are not enough rooms with capacity range of 100–200. In order to satisfy this global constraint the CA can select the SA with the largest number of requests for such rooms in this time slot and send it a message, asking this SA to search for an alternative local solution. This new local solution has to satisfied additional constraint: "No more than five courses on Wednesday, 14:00 to 16:00, with number of students in range of 100–200". Each new SA solution is an additional new value to one variable of the CA constraint network.

When one tries to solve a DisTTP, in either way, one needs to solve TTPs that are *dynamically* changing. The basic idea, here, is that each SA starts with its original TTP which is an instance of a Local CSP (LCSP). We term the original constraints of the original local problem *original local constraints* (OLCs). Whenever any of the SA finds a solution that complies with all its OLCs, it sends this solution to the CA. When the CA gets at least one solution from each of the SAs, it starts a search for a solution to the *global CSP* (GCSP) which includes the *global constraints*. Each variable of the GCSP represents one SA. The domain of each variable is the set of solutions for the LCSP of its SA. At the start of the search, the domains include only the first solution found by the SAs. Then the domain of each variable of the CA is dynamically increased as the SAs find more solutions that comply with all of its OLCs. In this model the CA must solve a CSP that variable domains are dynamically changing. Until recently, and the increasing use of the Internet, such class of problems has been little treated in the academic research community.

It can be said that the CA operates in an environment that is very similar to the open-world environment of the Internet. Many real-world problems that used to be solved by traditional CSP techniques can be solved more effectively in a distributed setting. For example, for the problem of computer configuration, it is now possible to locate, through the Internet, during the search for a solution, additional suppliers of some needed parts. The term Open CSP (OCSP) was first used for such a search environment by Faltings and Macho-Gonzalez [8] in 2002.

When moving to OCSPs, it immediately becomes obvious that most successful CSP solving methods, in particular constraint propagation, are based on the closed-world assumption that the domains of the variables are completely known and are fixed during the search process. For an OCSP setting, this assumption no longer holds (see [17,2,3,1]). This change requires an adaptation of the current search algorithm.

OCSPs bear some resemblance to the interactive CSP (ICSP) introduced by Lemma et. el. in 1999 [7]. In the ICSP model, domain values are acquired during the solution process only when necessary, and inserted into the variable domains. In this methodology the forward checking algorithm can be modified so that when domains become empty, it launches a specific request for additional values that would satisfy the constraints on that variable. Similarly to our CA search problem, the ICSP acquisition process of new domain values from external agents needs to be guided in order to acquire only new values that satisfy a specific constraint. In the ICSP model the acquisition of new values is an immediate interactive operation and it typically gathers more values than necessary. In the DisTTP model, a request from the CA for a new solution from a SA is usually a very expensive operation and the focus is on minimizing the information gathering activity.

3.2   Scheduling Agents (SA) Solve a Dynamic CSP

For the second approach, where each SA uses local search, the level of DCSP goes deeper into the level of the local ETPs. When the CA gets one solution from each of its SAs, the CA computes the current value of the cost function. In the nurses' transportation problem, this is the cost of transportation of all the nurses at the start and end of each shift. In the next step, the CA tries to reduce the cost by a request for all or some of its SAs to deliver another solution with some new constraints that the CA imposes. For example, suppose that in the first set of solutions that each SA delivered to the CA there is only one nurse, a, that needed transportation to target A, and three sets of four nurses each that needed transportation to three other locations: B, C and D. Furthermore, suppose that there are taxis with capacity of five passengers that are available. In such a case, it is easy to see that a reasonable move of the CA is to ask the SA that makes the schedule for nurse a, SA_a, to replace nurse a with any other nurse that lives in location B, C or D. Now SA_a is facing a CSP with constraints that are changing *dynamically*.

In this model each SA must solve a dynamically changing CSP. This is an instance of a known, albeit little treated, class of CSP problems: the DCSPs (see [5,18,1]).

4   Experimental Study of Interacting SAs

In order to understand the basic behaviour of our model we have conducted preliminary experiments. At the basis of our experiments is a real-world family of problems. One set of problems is the timetabling of nurses in a hospital. There are several wards in the hospital. Each ward schedules its employees independently. The hospital works three shifts a day, seven days a week. The hospital manages transportation for nurses, for each shift. Transportation lines are grouped into destinations. Cab/bus capacities are fixed. To achieve cost effectiveness, the SAs need to coordinate their timetables. SAs are not aware of inter-ward transportation constraints.

In our experiments there are 21 shifts. All wards weekly timetable are similar – 29 employees have to be assigned to three tasks for a total of 105 assignments over seven days. There are five lines of transportation. Cabs have a capacity of 1–7 passengers. There are different costs for lines for different shifts. Since the number of passengers depends on the timetables of all SAs, only the CA can

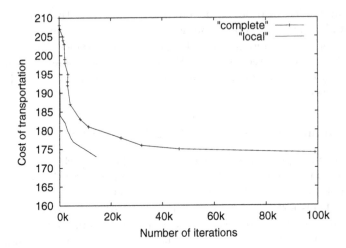

Fig. 3. Comparing complete search to local search; at some point, too many alternative solutions make the complete search ineffective

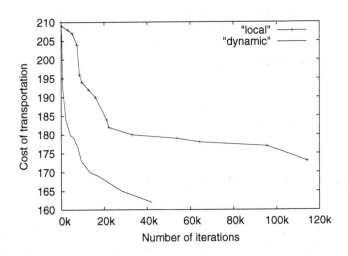

Fig. 4. Using sophisticated interaction between the CA and its SAs enables the search to find a much better solution faster

compute the cost by summing the number of passengers in each shift for each transportation line and then compute how many cabs are needed in each line. The strategy of the CA is to "fill up" the cabs in each line in order to achieve optimality of cost.

In order to explore a wide range of the problem space, many instances of this real-world ETP were generated with varying amounts of binary constraints, by adding or removing conflicts among shifts. Adding conflicts among shifts creates a random sample of problems with the same underlying structure (i.e. shifts, tasks, employees, limits) and different domains of values. This has the additional experimental value of a large random ensemble of problems.

For difficult TTPs the SAs generate only few alternative schedules in a reasonable time frame. For such problem class the CA needs to use a heuristic that will enhance the chance of encountering lower transportation cost for the next SA solution. The CA requests the SAs to generate timetables with needed features in order to avoid cabs with only one passenger. The CA sends messages that add a constraint to selected set of SAs, to limit the number of nurses for some line. In addition, new constraints are needed to prevent the SA from adding a nurse to a line with full cabs. Another option for this heuristic is to add new constraints that request SAs to reduce nurses for lines with an almost empty cab.

In the first set of experiments that is presented in Figure 3 the effect of using local search technique for the CA search is compared with the best complete search algorithm: FC-CBJ [12]. Each point on the graph of Figure 1 represents the average of a set of 100 problem instances. The stopping criterion for all (unsuccessful) runs is based on the total number of assignment attempts by the algorithm and that limit was set to 500 000 assignments. Using the number of iterations performed as the measurement unit for the search effort, we compare the efforts needed to solve the same set of problems by a pure FC-CBJ algorithm and by the simple (*random hill climbing* (RHC) algorithm. It is easy to see that in one-third of the time needed for FC-CBJ, RHC can find a solution that it about 10% cheaper. Each point on the graph of Figure 3 represents the average of a set of 100 problem instances.

The next step of investigation is to check the effectiveness of intelligent interaction between the CA and its SAs. In other words, it is important to have a good strategy for the CA search method. The results shown in Figure 4 shown clearly that the CA must search smartly and coordinate the SAs in order to find better combinations of SA timetables: a combination that enables the CA to find a global solution with a lower transportation cost. In the algorithm marked (*dynamic*, the CA adds and dynamically removes constraints to the problems of the SAs. The best solution achieved is 10% cheaper then the simple local search. From theses experiments we can conclude that the CA needs specific heuristics for directing SAs in improving their timetables.

5   Conclusions

A new, general model for DisTTPs was presented and its behaviour on large and hard constraint networks of TTPs was experimentally tested. The DisTTP can be naturally represented as distributed CSPs, where each of the agents solves a local CSP representing its timetable. Since the local problems of the agents can be quite complex, the resulting DisTTP is very different from the DisCSP that appears in the literature [19]. We termed these agents *scheduling agents* (SAs). SAs are connected by few global constraints, such as classroom capacity in university timetabling. This makes the DisCSP non-binary and these global constraints can naturally be represented by an additional agent that we termed the *central agent* (CA). When the CA performs an exhaustive search for a global solution for the DisTTP, the problem can become huge. This is because each node on the search tree of the CA is a complete solution of the SAs' timetables.

Our model and approaches were implemented for a real-world distributed employee timetabling problem (DisETP) involving several wards of a large hospital. Each of the SAs solves a large ETP with more than 100 needed assignments. The strong result shown in Figure 3 is quite intuitive – when the search space of possible SA timetables become large, local search is essential for the CA (see [9]).

The second set of experiments dealt with the interaction between the CA that tries to satisfy global constraints and the SAs, each searching for a good local timetable. We have tested one simple policy for the CA – improve global transportation cost by dynamically constraining the requested timetables of selected SAs. This policy appears to work much better (Figure 4) than simple hill climbing. This result shows that the heuristics used by the CA have a significant impact on its runtime performance.

References

1. Bellicha, A.: Maintenance of Solution in a Dynamic Constraint Satisfaction Problem. In: Proc. Applic. Artif. Intell. Engng VIII (Toulouse, France) (1993) 261–274
2. Bessière, C.: Arc-Consistency in Dynamic Constraint Satisfaction Problems. Proc. 9th Natl Conf. Artif. Intell. (Anaheim, CA) (1991) 221–226
3. Bessière, C.: Arc-Consistency for Non-binary Dynamic CSPs. In: Proc. 10th Eur. Conf. Artif. Intell. (Vienna, Austria) (1992) 23–27
4. Bessière, C., Maestre, A., Messeguer, P.: Distributed Dynamic Backtracking. In: Workshop Distrib. Constraint (IJCAI'01) (2001)
5. Dechter, R., Dechter, A.: Belief Maintenance in Dynamic Constraint Networks. In: Proc. 7th Annu. Conf. Am. Assoc. Artif. Intell. (1988) 37–42
6. Hamadi, Y., Bessière, C.: Backtracking in Distributed Constraint Networks. In: Proc. ECAI'98 (Brighton, August) (1998) 219–223
7. Lamma, E., Mello, P., Milano, M., Cucchiara, R., Gavanelli, M., Piccardi, M.: Constraint Propagation and Value Acquisition: Why We Should Do It Interactively. In: Dean, T. (Ed.): Proc. Int. Joint Conf. Artif. Intell. (JCAI'99, Stockholm, Sweden). Morgan Kaufmann, San Mateo, CA (1999) 468–477
8. Macho-Gonzalez, S., Faltings, B.: Open Constraint Satisfaction. In: Proc. AAMAS 2002 Workshop Distrib. Constraint Satisfaction (2002)
9. Meisels, A., Schaerf, A.: Modelling and Solving Employee Timetabling Problems. Ann. Math. Artif. Intell. (to appear)
10. Meisels, A., Schaerf, A.: Solving employee timetabling problems by generalized local search. Proc. Ital. AI-1 (Bologna, Italy, May, 1999) (1999) 281–331
11. Meseguer, P., Jimenez, M.A.: Distributed Forward Checking. In: Proc. CP 2000 Workshop Distrib. Constraint Satisfaction (Singapore, September) (2000)
12. Prosser, P.: Hybrid Algorithms for the Constraint Satisfaction Problem. Comput. Intell. **9** (1993) 268–299
13. Richards, E.B., Das, S., Choi, H., El-Kholy, A., Liatsos, V., Harrison, C.: Distributed Optimisation: a Case study of Utilising Teaching Space in a College. In: Proc. Expert Syst. '96 Conf. (Cambridge). SGES Publications (1996) 153–161

14. Meisels, A., Schaerf, A.: Solving employee timetabling problems by generalized local search. In: Proc. Ital. AI-1 (Bologna, Italy, May, 1999) (1999) 493–502
15. Silaghi, M.C., Stefan, S., Sam-Haroud, D., Faltings, B.: Asynchronous Search for Numeric DisCSPs. In: CP'01 (Paphos, Cyprus) (2001)
16. Solotorevsky, G., Gudes, E., Meisels, A.: Modeling and Solving Distributed Constraint Satisfaction Problems (DCSPs). In: Constraint Processing 1996 (New Hampshire, October) (1996)
17. Nguyen, T., Deville, Y.: A Distributed Arc-Consistency Algorithm. Sci. Comput. Program. **30** (1998) 227–250
18. Verfaillie, G., Schiex, T.: Solution Reuse in Dynamic Constraint Satisfaction Problems. In: Proc. 12th Natl Conf. Artif. Intell. (Seattle, WA) (1994)
19. Yokoo, M.: Distributed Constraint Satisfaction: Foundations of Cooperation in Multi-agent Systems. Springer, Berlin Heidelberg New York (2001)
20. Yokoo, M., Durfee, E.H., Ishida, T., Kuwabara, K.: Distributed Constraint Satisfaction Problem: Formalization and Algorithms. IEEE Trans. Data Knowl. Engng **10** (1998) 673–685

Examination Timetabling

A Multiobjective Optimisation Technique for Exam Timetabling Based on Trajectories

Sanja Petrovic and Yuri Bykov

School of Computer Science and Information Technology,
University of Nottingham, Jubilee Campus,
Nottingham NG8 1BB, UK
{sxp,yxb}@cs.nott.ac.uk

Abstract. The most common approach to multiobjective examination timetabling is the weighted sum aggregation of all criteria into one cost function and application of some single-objective metaheuristic. However, the translation of user preferences into the weights of criteria is a sophisticated task, which requires experience on the part of the user, especially for problems with a high number of criteria. Moreover, the results produced by this technique are usually substantially scattered. Thus, the outcome of weighted sum algorithms is often far from user expectation.

In this paper we suggest a more transparent method, which enables easier expression of user preferences. This method requires the user to specify a reference solution, which can be either produced manually or chosen among the set of solutions, generated by any automated method. Our aim is to improve the values of the reference objectives, i.e. to produce a solution which dominates the reference one. In order to achieve this, a trajectory is drawn from the origin to the reference point and a Great Deluge local search is conducted through the specified trajectory. During the search the weights of the criteria are dynamically changed.

The proposed technique was experimentally tested on real-world exam timetabling problems on both bi-criteria and nine-criteria cases. All results obtained by the variable weights Great Deluge algorithm outperformed the ones published in the literature by all criteria.

1 Introduction

1.1 Exam Timetabling Problems

University examination timetabling comprises arranging exams in a given number of timeslots. The primary objective of this process is avoiding students' clashes (i.e. a student cannot take two exams simultaneously). This requirement is generally considered as a hard constraint and should be compulsory satisfied in a feasible timetable. However, a number of other restrictions and regulations, which depend on a particular institution, are also to be taken into account when solving exam timetabling problems. Some of them can be also considered as hard constraints, while other constraints are soft: i.e. usually they cannot be

E. Burke and P. De Causmaecker (Eds.): PATAT 2002, LNCS 2740, pp. 181–194, 2003.

completely satisfied, and therefore their violations should be minimised. The soft constraints vary from university to university, as is shown in [2] where the authors analyse responses from over 50 British universities.

The soft constraints usually imply the different importance of the timetable to the timetable officer (decision maker). They are generally incompatible and often conflict with each other. Timetabling problems can be considered to be multiobjective problems where objectives measure the violations of the soft constraints. We believe that multiobjective optimisation methods can bring new insight into timetabling problems by considering simultaneously different criteria during the construction of a timetable.

1.2   Multiobjective Optimisation of Examination Timetabling

The conventional challenge of multiobjective optimisation is assessment of the quality of solutions. Formally, one solution can be considered to be better than another only in the case when the values of all its criteria outperform those of the second ones: i.e. the first solution "dominates" the second one. All solutions which are not dominated by any other one, can be considered to be optimal. However, only one solution from this non-dominated set (often called the "Pareto front") can become the final result. To obtain it, the decision maker must express his/her preferences.

The group of multiobjective methods called "Search-then-Decide" (a posteriori) are designed to produce the set of non-dominated solutions from which the decision maker can select their preferable one. This approach is mostly applicable to small- and middle-sized combinatorial optimisation problems. To our knowledge, there are no publications about the use of these methods for examination timetabling. However, several authors (for example [9, 17]) have applied a posteriori algorithms to class–teacher timetabling, a problem which is similar to examination timetabling.

Traditionally, exam timetabling problems are solved by the "Decide-then-Search" (a priori) approach. In these methods the decision maker specifies his/her preferences regarding the solution before launching the algorithm. The most popular method involves aggregation of the problem's objectives into a cost function in order to apply some single-objective metaheuristic (see survey [8]). Usually, the cost is calculated as a weighted sum of objectives. This method has been applied with simulated annealing [15, 16], tabu search [1], genetic algorithms [10], a memetic algorithm [4], etc. In another method, *lexicographic ordering*, criteria are divided into groups and the search is conducted in several phases by each group. This method has been applied to the examination timetabling by several authors [12, 14]. In [5] the authors investigated a Compromise Programming technique with different distance measures as the means of aggregation of an objective's values while applying the search algorithm, designed as a hybrid of heavy mutation and hill-climbing.

> Set the initial solution s_0
> Calculate the initial cost function $f(s_0)$
> Initial level $B_0 = f(s_0)$
> Specify the input parameter $\Delta B = ?$
> While not stopping condition do
>   Define neighbourhood $N(s)$
>   Randomly select the candidate solution $s^* \in N(s)$
>   If $(f(s^*) \leq f(s))$ or $(f(s^*) \leq B)$
>   Then accept $s^* (s = s^*)$
>   Lower the level $B = B - \Delta B$

Fig. 1. Single-objective Great Deluge algorithm

1.3 Great Deluge

Great Deluge is a local search metaheuristic, introduced by Dueck [11]. In [6] the authors showed its quite promising performance on exam timetabling problems. In this algorithm a new candidate solution (selected from a neighbourhood) is accepted if its objective function is either not worse than a current one or does not exceed the current upper limit B (*level*). The value of the level is reduced gradually throughout the search by some specified decay rate ΔB, which denotes the search speed. Decrease of the level forces the current solution's cost function to correspondingly decrease until convergence. The initial value of B is equal to the cost function of the initial solution, and therefore no additional parameters are required. The pseudocode of the basic variant of the Great Deluge algorithm is given in Figure 1.

Having the search speed as an input parameter leads to the unique property of this algorithm. It provides two options: either the decision maker can estimate beforehand the processing time from the start to the convergence, or alternatively, the value of ΔB can be calculated in order to fit the search into a certain predefined time interval. Experiments with this technique have shown that a longer search usually yields a better final result, and this principle achieves its full strength when applied to large-scale timetabling problems.

With the Great Deluge algorithm the decision maker can obtain higher-quality results at the price of prolongation of the search period (this is viable because processing time is usually not an issue in exam timetabling problems). The algorithm allows him/her to choose a preferable balance between the quality of the solution and the searching time, to fit a solving procedure into his/her personal schedule and to optimise the utilisation of computational resources. In [6] and [7] we presented a comprehensive comparison of the Great Deluge algorithm with a variety of useful metaheuristics on different benchmark problems. In almost all the experiments our approach outperformed the other metaheuristics.

In this paper we present a modified Great Deluge algorithm for multiobjective timetabling problems. The paper is organised in the following way. Section 2 gives the description of the reference points and trajectories in the criteria space. In

Section 3 we introduce the variable weights multiobjective extension of the Great
Deluge algorithm. Results of the experiments are given in Section 4. Finally, in
Section 5 we summarise our contribution and suggest directions for future work.

2   Reference Solution in Criteria Space

The general drawback of the weighted sum approach is the necessity of defining
particular values of weights. This requires experience on the part of the decision maker. The translation of his/her preferences into the form of weights is a
sophisticated task, especially for problems with a high number of criteria. Furthermore, the results produced by the weighted sum are usually substantially
scattered. Thus, the weighted sum technique often produces an outcome that is
far from the decision maker's expectations. Often, the proper setting of weights
can be done only by launching the search procedure several times.

As an alternative to the traditional weighted sum approach we expand the
idea of reference timetable expressed by Paechter et al. [13]. As the reference
they considered a timetable produced either manually or automatically using
another dataset. The authors suggested an algorithm which obtains a solution
genotypically similar to the reference one. They also pointed out that the reference solution may already be located in a local optimum, and therefore it is
worth starting the search for the new solution from scratch.

For the purpose of multiobjective optimisation the reference solution can be
considered in a phenotypic sense: i.e. the decision maker should specify the criteria values of some attainable solution which to a certain degree meet his/her
preferences. This solution can be produced manually or selected from the set
of solutions generated by some automated method. We assume that the decision maker is not satisfied completely with this solution, but this choice gives
information that is helpful for a further search for a better solution.

Having a more or less preferable reference solution, we can consider that all
further solutions which dominate the reference one (where all reference criteria
are outperformed by the new ones) will be even more preferable. In order to find
these solutions, we suggest the following method. We represent the reference
solution as a point in the criteria space and draw a line through this point and
the origin. When the search procedure is launched, an initial solution and all the
following current solutions are also represented as points in the criteria space.
The algorithm should provide the gradual improvement of the current solution
while keeping the corresponding points close to the defined line. The aim is to
approach as close to the origin as possible, driving the search through the defined
trajectory.

In our approach the reference solution is used only for drawing the trajectory,
but does not affect the further search process. We use it only as a benchmark for
assessing the final solution. As the reference solution in most cases already lies in
a local optimum, it cannot be used for the initialisation purpose. Generally, local
search techniques show the best performance when they start from a random
solution. Therefore we suggest keeping this practice for the presented approach.

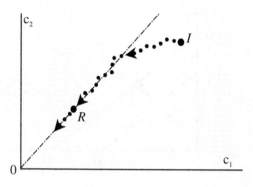

Fig. 2. Following the defined trajectory

For the bi-criteria case the method is illustrated in Figure 2, where the trajectory is depicted as a dash-dotted line.

The search starts from a randomly generated initial solution (point I) and at first approaches the trajectory (generally, the initial solution does not lie on the trajectory). The search then follows the trajectory until it reaches the vicinity of the reference solution (point R) and continues along it until convergence. In the final stage any solution dominates the reference one. The point of convergence is not known in advance, but it will be obviously superior to the reference point.

3   Great Deluge with Variable Weights

In this section we present a technique suitable for driving the search through a predefined trajectory. It operates with a weighted sum cost function, but the weights are varied dynamically during the search. We have developed a special procedure for weight variation to regulate the direction of the search.

The explanation of our method is illustrated with a bi-criteria case (the goal is to minimise criteria c_1 and c_2). We consider a weighted sum aggregation function with weights w_1 and w_2 within the Great Deluge algorithm. The condition of acceptance of a candidate solution $S = (s_1, s_2)$ at any iteration can be expressed by the following inequality:

$$s_1 w_1 + s_2 w_2 \leq B. \tag{1}$$

This formula states that the algorithm accepts any solution in the space bounded by axes c_1 and c_2 (as the criteria values are always positive) and the line

$$c_1 w_1 + c_2 w_2 = B. \tag{2}$$

In Figure 3 this borderline is marked as $G_1 G_2$. The points where it intersects the axes are calculated as $G_1 = B/w_1; G_2 = B/w_2$. The space of acceptance is denoted by the shaded triangle.

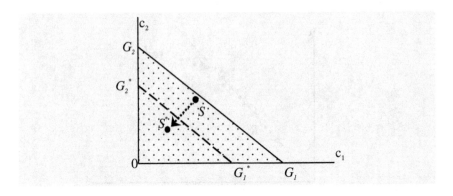

Fig. 3. Borderline in the weighted sum Great Deluge algorithm

The lowering of the *level* value B at each step corresponds to a shift of the borderline towards the origin. The new borderline $G_1^*G_2^*$ is expressed by

$$c_1w_1 + c_2w_2 = B - \Delta B. \tag{3}$$

The new intersection points are calculated as $G_1^* = (B - \Delta B)/w_1$ and $G_2^* = (B - \Delta B)/w_2$. The shifting of the borderline results in obtaining the new current solution (S^*) which is closer to the origin. Let us define $\Delta w = \Delta w/(B - \Delta B)$, and equation (3) can be transformed into the following form:

$$c_1w_1(1 + \Delta w) + c_2w_2(1 + \Delta w) = \Delta B. \tag{4}$$

Due to this formula the decrease of the *level* at any given iteration can be replaced with the appropriate increase of both weights as it causes the same effect (shifting of the borderline).

Hence, finally, each separate increase of a single weight induces a rotation of the borderline such that the new solution improves the corresponding criterion more than the other one. Thus, equation (5) corresponds to the line $G_1^*G_2$ in Figure 4 and equation (6) corresponds to the line $G_1G_2^*$ in Figure 5:

$$c_1w_1(1 + \Delta w) + c_2w_2 = \Delta B, \tag{5}$$
$$c_1w_1 + c_2w_2(1 + \Delta w) = \Delta B. \tag{6}$$

For example, the increase of w_1 in Figure 4 forces solution S to move mostly along the c_1 axis. The increase of w_2 causes the opposite effect (Figure 5).

Thus, instead of reducing a *level* at each step, the proposed algorithm increases a single weight. Although the value of *level* B is invariable, we can consider this technique as a multiobjective extension of the Great Deluge algorithm because it incorporates the same principles.

In order to force the current solutions to follow the given trajectory, the algorithm employs the appropriate rule for selecting the weight to be increased. We suggest the following method (its bi-criteria case is illustrated in Figure 6).

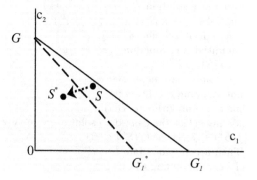

Fig. 4. The increase of w_1

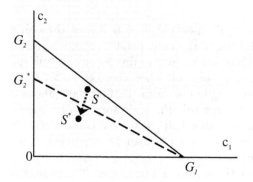

Fig. 5. The increase of w_2

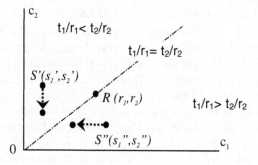

Fig. 6. The selection of increased weight

Set the reference solution $R = (r_1, r_2, \ldots, r_n)$
Set the initial solution $S^0 = (s_1^0, s_2^0, \ldots, s_n^0)$
Specify the initial weights $(w_1^0, w_2^0, \ldots, w_n^0) = ?$
Calculate initial cost function $f(S) = s_1^0 w_1^0 + s_2^0 w_2^0 + \cdots + s_n^0 w_n^0$
level $B = f(S)$
Specify the input parameter $\Delta w = ?$
While not stopping condition do
    Define a neighbourhood $N(S)$
    Randomly select the candidate solution $S^* \in N(S)$
    If $(f(S^*) \leq f(S))$ or $(f(S^*) \leq B)$
        Then accept candidate $S = S^*$
    Find i corresponding to: $\max_{i=1\ldots n}(s_i/r_i)$
    Increase the weight $w_i = w_i(1 + \Delta w)$

Fig. 7. Multiobjective Great Deluge algorithm with variable weights

As the trajectory (dash-dotted line) is drawn through the reference point $R = (r_1, r_2)$ and the origin, then any point (t_1, t_2), where $t_1/r_1 = t_2/r_2$, belongs to the trajectory. Thus, the trajectory divides the criteria space into two halves: one where $t_1/r_1 < t_2/r_2$ and the other where $t_1/r_1 > t_2/r_2$. Obviously, if point S' (the current solution) is placed in the first half (above the trajectory), the directing of the search towards the trajectory can be done while decreasing s_2' (increasing w_2). In the other half (below the trajectory), for point S'' we have to increase w_1. The proposed rule can be expanded into n-criteria space as well. Here, we evaluate the vector $(s_1/r_1, s_2/r_2, \ldots, s_n/r_n)$, choose its maximum element and increase the corresponding weight. The pseudocode of the algorithm is given in Figure 7.

In this algorithm the value of the input parameter Δw affects the computing time in the sense that with higher Δw the search runs faster. However, in contrast to the basic single-objective variant, the search speed is not steady, and therefore we cannot guarantee convergence in the given number of iterations. We expect that this could be arranged with a more advanced mechanism of weight variation.

Additionally to Δw, this algorithm requires the specification of initial weights $(w_1^0, w_2^0, \ldots, w_i^0)$. Their relative proportion defines the angle of an initial border-line, which passes through the initial solution. We have experimentally tested different methods of weight initialisation and found that they affect the duration of the first phase of the search: proper definition of initial weights allows the current solution to reach the trajectory more expeditiously. The best values of initial weights are probably problem dependent. However, in our experiments a fairly good performance was achieved when setting w_1^0 equal to s_1^0/r_i. If the value of some reference criterion r_i is equal to 0 (the constraint in the reference solution is satisfied), we put, instead of r_i, some small value which is less than half of the measurement unit of the criterion. For example, when the criterion has integer value, it is enough to set $r_i = 0.4$.

4   Experiments

The proposed algorithm was experimentally tested on real-world large-scale examination timetabling data. The algorithm was implemented on C++ and launched on PC with AMD Athlon 750 MHz processor under OS Windows98.

4.1   Bi-criteria Case

The first series of experiments was done on a bi-criteria case. We used the following university examination timetabling datasets: the University of Nottingham dataset (Nott-94), placed at ftp://ftp.cs.nott.ac.uk/ttp/Data/Nott94-1, and two datasets (Car-f-92 and Kfu-s-93) from the University of Toronto collection, available at ftp://ftp.mie.utoronto.ca/pub/carter/testprob. The problem formulation was the same as in [4]: the first objective represents the number of conflicts where students have to sit two exams in adjacent periods, and the second objective represents the number of conflicts where students have exams in overnight adjacent periods.

In our first experiment we investigated the ability of the presented technique to follow a defined trajectory. For the Nott-94 problem we specified both reference criteria values to be equal to 300. The trajectory is a line under the 45° angled line (dash-dotted line in Figure 8), and Δw was set to be 10^{-6} (processing time was around 3 min). In order to follow the progress of the search process, we plotted after each 50 000 steps the current solution as a dot and after each 500 000 steps we drew the current borderline as a dotted line. The complete diagram is presented in Figure 8.

The search is first directed towards the trajectory and then follows it producing solutions, which are very close to it. Scatter is relatively high at the beginning of the search and then becomes very low. Looking at the dynamics of the borderline we can notice that at the beginning of the search only w_1 has been increasing until the current solution reaches the trajectory. After that, the two weights increase differently, using the rules described in Section 3.

The second series of experiments were done using published results as reference points. We used the results produced by the Multi-Stage Memetic algorithm presented in [4], while using the same weighted sum cost function. We selected from the best presented results three non-dominated points for every dataset (marked as MSMA in Table 1). The corresponding trajectory was drawn for every reference point. Each launch of our trajectory-based algorithm was started from a random solution and lasted around 30–40 min. Approximately 95% of this time was spent on approaching the reference point and 5% on its improvement. Our final results are shown in Table 1 and marked as TBA.

All our final results dominate the corresponding reference points. This confirms the ability of our algorithm to drive the search through different trajectories, and to produce high-quality solutions, which are better than the reference ones.

Table 1. Reference and our solutions for the bi-criteria case

		Car-f-92 (36 periods)		Kfu-s-93 (21 periods)		Nott-94 (23 periods)	
		MSMA	TBA	MSMA	TBA	MSMA	TBA
1st point	C_1	302	282	222	204	65	53
	C_2	804	799	838	743	324	271
	C_1	313	286	228	218	76	57
	C_2	766	706	704	608	282	187
	C_1	363	327	307	258	100	59
	C_2	576	541	589	562	255	149

4.2 Nine-Criteria Case

We conducted the next series of experiments in order to investigate the effectiveness of the proposed technique when the number of criteria is greater than two. The Nott-94 dataset was considered with nine objectives in the same way as in [5]. Descriptions of the criteria are given in Table 2.

Again we use the solutions presented in [5] as reference points. They were obtained by the multiobjective hybrid of heavy mutation and hill-climbing based on the idea of the Compromise Programming approach. These solutions were produced with different aggregation functions while scheduling exams into 23, 26, 29 and 32 periods. The processing time of each launch was in the range of 5–10 min and any attempt to increase the processing time did not lead to better results. The presented trajectory-based technique was launched for each of these reference points. Our launches lasted approximately 20–25 min and we consider this processing time to be quite acceptable for examination timetabling. The results are compiled in Table 3.

As in the previous experiments, the algorithm produces solutions which dominate the reference ones by all criteria. Thus, the proposed technique provides better satisfaction of user preferences as well as higher overall result quality than conventional weighted sum methods.

5   Conclusions and Future Work

In this paper we have presented a new multiobjective approach, whose main characteristics are as follows:

- instead of reducing the aggregation function, our algorithm aims to improve each criterion separately by changing weights dynamically during the search process;
- the specification of a reference solution may be more transparent to the decision maker than expressing the weights;

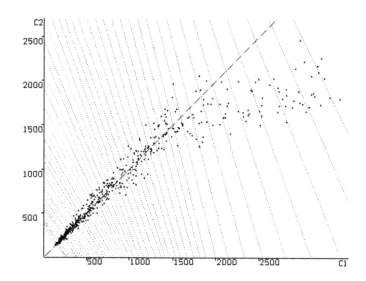

Fig. 8. The progress diagram for the Nott-94 problem

Table 2. Descriptions of criteria

Criterion	Description
C_1	Number of times that room capacities are exceeded
C_2	Number of conflicts, where students have exams in adjacent periods on the same day
C_3	Number of conflicts, where students have two or more exams in the same day
C_4	Number of conflicts, where students have exams in adjacent days
C_5	Number of conflicts, where students have exams in overnight adjacent periods
C_6	Number of times that students have exams that are not scheduled in period of proper duration
C_7	Number of times that students have exams that are not scheduled in the required time period
C_8	Number of times that students have exams that are not scheduled before/after another specified exams
C_9	Number of times that students have exams that are not scheduled immediately before/after another specified exams

Table 3. Reference and our solutions for nine-criteria case

Point		23 periods CPA	TBA	26 periods CPA	TBA	29 periods CPA	TBA	32 periods CPA	TBA
1st	C_1	1038	795	137	0	139	0	25	0
	C_2	1111	651	655	476	513	360	314	184
	C_3	3518	3360	2814	2795	2239	2059	1546	1353
	C_4	4804	4185	2759	2494	2172	1687	1646	1390
	C_5	405	54	265	45	231	43	174	104
	C_6	4	–	–	–	–	–	–	–
	C_7	–	–	–	–	–	–	–	–
	C_8	–	–	–	–	–	–	–	–
	C_9	–	–	–	–	–	–	–	–
2nd	C_1	–	–	–	–	–	–	–	–
	C_2	879	778	604	353	393	292	316	190
	C_3	3623	3524	2544	2174	1957	1482	1332	1104
	C_4	6381	6221	4571	3661	3438	2518	2482	2028
	C_5	264	152	164	38	151	48	53	2
	C_6	–	–	–	–	–	–	–	–
	C_7	–	–	–	–	–	–	–	–
	C_8	–	–	–	–	–	–	–	–
	C_9	–	–	–	–	–	–	–	–
3rd	C_1	2848	1734	2044	889	1559	670	1243	1
	C_2	2608	1367	1872	802	1435	703	1138	488
	C_3	4886	3760	3507	2127	2688	1481	2132	1210
	C_4	4658	2289	3343	1922	2563	1201	2033	1073
	C_5	807	332	475	190	441	128	334	155
	C_6	170	0	119	–	89	–	74	–
	C_7	40	–	24	–	24	–	18	–
	C_8	–	–	–	–	–	–	–	–
	C_9	–	–	–	–	–	–	–	–

- the final solution produced by the algorithm conforms to the reference one, which satisfies the decision maker's preferences. Therefore, it is expected to be more acceptable to the decision maker;

- the solutions dominate the results of other techniques. Therefore, the presented approach can be considered to be a more powerful one.

This paper opens a wide area for further research. The proposed algorithm should be evaluated in other domains, with different numbers of objectives. Different procedures for weight variation could be developed, to direct the search differently. In particular, the question of weight initialisation needs further investigation. In addition, other approaches to trajectory definition should be explored.

References

1. Boufflet, J.P., Negre, S.: Three Methods Used to Solve an Examination Timetabling Problem. In: Burke, E, Ross, P. (Eds.): Practice and Theory of Automated Timetabling I (PATAT'95, Edinburgh, Aug/Sept, selected papers). Lecture Notes in Computer Science, Vol. 1153. Springer-Verlag, Berlin Heidelberg New York (1996) 327–344

2. Burke, E.K., Elliman, D.G., Ford, P.H., Weare, R.F.: Examination Timetabling in British Universities: a Survey. In: Burke, E, Ross, P. (Eds.): Practice and Theory of Automated Timetabling I (PATAT'95, Edinburgh, Aug/Sept, selected papers). Lecture Notes in Computer Science, Vol. 1153. Springer-Verlag, Berlin Heidelberg New York (1996) 76–90

3. Burke, E.K., Jackson, K., Kingston, J.H., Weare, R.: Automated University Timetabling: The State of the Art. Comput. J. **40** (1997) 565–571

4. Burke, E.K., Newall, J.P.: A Multi-stage Evolutionary Algorithm for the Timetabling Problem. IEEE Trans. Evolut. Comput. **3** (1999) 63–74

5. Burke, E.K., Bykov, Y., Petrovic, S.: A Multicriteria Approach to Examination Timetabling. In: Burke, E, Erben W. (Eds.): Practice and Theory of Automated Timetabling III (PATAT 2000, Konstanz, Germany, August, selected papers). Lecture Notes in Computer Science, Vol. 2079. Springer-Verlag, Berlin Heidelberg New York (2001) 118–131

6. Burke, E.K., Bykov, Y., Newall, J.P., Petrovic, S.: A Time-Predefined Local Search Approach to Exam Timetabling Problems. Computer Science Technical Report NOTTCS-TR-2001-6, University of Nottingham (2001)

7. Burke, E.K., Bykov, Y., Newall, J.P., Petrovic, S.: A New Local Search Approach with Execution Time as an Input Parameter. Computer Science Technical Report NOTTCS-TR-2002-3, University of Nottingham (2002)

8. Carter, M.W., Laporte, G.: Recent Developments in Practical Examination Timetabling. In: Burke, E, Ross, P. (Eds.): Practice and Theory of Automated Timetabling I (PATAT'95, Edinburgh, Aug/Sept, selected papers). Lecture Notes in Computer Science, Vol. 1153. Springer-Verlag, Berlin Heidelberg New York (1996) 3–21

9. Carrasco, M.P., Pato, M.V.: A Multiobjective Genetic Algorithm for the Class/Teacher Timetabling Problem. In: Burke, E, Erben W. (Eds.): Practice and Theory of Automated Timetabling III (PATAT 2000, Konstanz, Germany, August, selected papers). Lecture Notes in Computer Science, Vol. 2079. Springer-Verlag, Berlin Heidelberg New York (2001) 3–17

10. Corne, D., Ross, P., Fang, H.L.: Fast Practical Evolutionary Timetabling. In: Fogarty, T.C.: Proc. AISB Workshop Evolut. Comput. (1994) 250–263

11. Dueck, G.: New Optimization Heuristics. The Great Deluge Algorithm and the Record-to-Record Travel. J. Comput. Phys. **104** (1993) 86–92

12. Lotfi, V., Cerveny , R.: A Final-Exam Scheduling Package. J. Oper. Res. Soc. **42** (1991) , 205–216

13. Paechter, B., Rankin, R.C., Cumming, A., Fogarty, T.C.: Timetabling the Classes of an Entire University with an Evolutionary Algorithm. In: Parallel Problem Solving from Nature (PPSNV) Springer-Verlag, Berlin Heidelberg New York (1998)

14. Thompson, J.M., Dowsland, K.A.: Multi-Objective University Examination Scheduling. EBMS/1993/12, European Business Management School, University of Wales, Swansea (1993)

15. Thompson, J.M., Dowsland, K.A.: Variants of Simulated Annealing for the Examination Timetabling Problem. Ann. Oper. Res. **63** (1996) 105–128
16. Thompson, J.M., Dowsland, K.A.: A Robust Simulated Annealing Based Examination Timetabling System. Comput. Oper. Res. **25** (1998) 637–648
17. Tanaka, M., Adachi, S.: Request-Based Timetabling by Genetic Algorithm with Tabu Search. 3rd Int. Workshop Frontiers Evolut. Algorithms (2000) 999–1002

Enhancing Timetable Solutions with Local Search Methods

E.K. Burke[1] and J.P. Newall[2]

[1] School of Computer Science and Information Technology,
University of Nottingham, Jubilee Campus, Wollaton Road,
Nottingham, NG8 1BB, UK
ekb@cs.nott.ac.uk
[2] eventMAP Ltd,
21 Stranmillis Road Belfast BT9 5AF, Northern Ireland,
Jim.Newall@eventmaponline.com

Abstract. It is well known that domain-specific heuristics can produce good-quality solutions for timetabling problems in a short amount of time. However, they often lack the ability to do any thorough optimisation. In this paper we will study the effects of applying local search techniques to improve good-quality initial solutions generated using a heuristic construction method. While the same rules should apply to any heuristic construction, we use here an adaptive approach to timetabling problems. The focus of the experiments is how parameters to the local search methods affect quality when started on already good solutions. We present experimental results which show that this combined approach produces the best published results on several benchmark problems and we briefly discuss the implications for future work in the area.

1 Introduction

1.1 The Examination Timetabling Problem

The examination timetabling problem consists of allocating a number of exams to a limited number of periods or timeslots subject to certain constraints. These constraints may relate to operational limitations (such as no student having to attend two exams at the same time, limitations on seating, etc.) which cannot be overcome, or they may be regarded as being sufficiently important that they should always be observed. In either case we call these hard constraints. We also call a timetable that obeys all hard constraints a feasible timetable. The remaining constraints are those that are considered to be desirable to satisfy, but not essential, such as allowing students study time between exams. These constraints are termed soft constraints. The level of satisfaction of these soft constraints can be considered to be a measure of the quality of a timetable. More information on the various constraints that can exist for the problem can be found in a survey of UK universities [2]. Surveys of practical applications of examination timetabling can be found in [5,6].

E. Burke and P. De Causmaecker (Eds.): PATAT 2002, LNCS 2740, pp. 195–206, 2003.

1.2 Local Search Methods

It is possible to think of local search techniques as methods which typically function by iteratively applying simple moves to a solution. Before a move is applied the effects it has are evaluated and a decision is made to accept or reject the move based on some criteria. The criteria for accepting moves is generally what distinguishes the various varieties of local search.

The simplest form of local search approaches are Descent-Based methods (or Hill Climbing, depending on your point of view). Here a move is only accepted if it produces a solution at least as good as the current solution. For mainly comparative purposes we consider this simple form of local search in this paper.

A more sophisticated approach, which we also consider in this paper, is Simulated Annealing [11]. This differs from descent methods in that it has a stochastic component whereby worsening moves can be accepted. The probability p of accepting a worsening move is given in (1), where Δ represents the change in quality and T represents the current temperature of the system, this being derived from thermodynamics' Boltzmann's distribution. The temperature of the system is periodically lowered according to a cooling schedule. This results in worsening moves being less likely to be accepted as the process continues. The performance of Simulated Annealing is generally regarded to be highly dependent on the choice of parameters such as starting temperature, terminating temperature and the cooling schedule:

$$p = e^{-\frac{\Delta}{T}} . \tag{1}$$

An alternative that is also investigated in this paper, and that has been shown to work particularly well on the examination timetabling problem, is the Great Deluge algorithm proposed by Dueck [9]. This approach functions in a similar fashion to Simulated Annealing and its variant, Threshold Acceptance [10]. However, worsening moves are accepted or rejected according to whether solution quality falls below or above a specified ceiling, which is reduced linearly throughout the process. The method was successfully investigated for timetabling problems by Burke et al. [1], along with an appropriately configured Simulated Annealing process. The main advantage over Simulated Annealing is the lower number of parameters, and that those required can easily be guessed or worked out from a parameter describing the desired run time.

2 Applying Local Search Techniques to Good Initial Solutions

When applying local search techniques such as the Great Deluge or Simulated Annealing to good-quality initial solutions there are factors we need to take into account. Firstly, these methods usually perform a far-ranging search of the solution space before terminating, which may not be desirable when starting from a good initial point, as we could quickly lose any benefit from the initial quality. Fortunately, we can control how far-ranging a search we wish to carry out

```
sinceLastMove := 0
WHILE sinceLastMove < 1,000,000 DO
choose exam e and period t at random s.t. t ! = period(e)
        IF penalty(e, t) ≤ penalty(e, period(e)) THEN
        move exam e to period t
                sinceLastMove := 0
        ELSE
                sinceLastMove+ = 1
        ENDIF
DONE
```

Fig. 1. Pseudo-code describing the Hill Climbing procedure

through parameters for two of the three local search methods that we investigate. It should be noted that the actual method of generating initial solutions is irrelevant here as we are concerned exclusively with establishing how much, and under what conditions, solutions can be enhanced using local search methods. In all the proposed methods we begin with a feasible solution and never accept moves that would produce an infeasible timetable.

2.1 Adaptive Initial Solution Generation

An adaptive heuristic approach is used to generate initial solutions. This basically functions by repeatedly trying to construct a timetable based on an ordering, while promoting any exams that cannot be scheduled to an acceptable level. This in itself produces very competitive results relatively quickly. As the experiments here are largely concerned with the improvements attainable with local search we will not discuss the generation method at length here. For a detailed explanation of the method and experimental results see [3].

As mentioned above, we are not concerned with the actual construction method used. It is likely that the same techniques could be applied just as effectively to other heuristic algorithms such as those presented by Carter et al. [7]. We will now briefly discuss the three local search approaches, and how, where possible, they can be configured to take account of already good solutions.

2.2 Hill Climbing

As Hill Climbing, by its nature, will not accept worsening moves it can only explore a very limited portion of the search space. It is, however, very fast compared to the other methods and will never produce a solution that is worse than the original. For these reasons Hill Climbing provides a good baseline for comparison with the other more thorough search methods.

For these experiments, a simple randomised Hill Climbing method was used, where moves are generated at random (in the same way as for the other methods) and accepted if they do not lead to a worse solution. The process is aborted after

1000 000 unsuccessful successive attempts at generating a move. Figure 1 gives the Hill Climbing procedure used in the tests, where $period(e)$ gives the current period of exam e, and $penalty(e, t)$ gives the penalty arising from scheduling exam e in period t.

2.3   Simulated Annealing

The inherent controlling factor in Simulated Annealing is the temperature. A higher initial temperature will result in increased acceptance of worsening moves, and therefore the search tends to move further away from the starting point. Alternatively, a lower temperature will search more in the vicinity of the initial solution, at the expense that it may not be able to reach other better, though still relatively near, areas of the search space. Figure 2 shows the process used for Simulated Annealing, where T_0 is the initial temperature, T_{min} is the minimum temperature and α represents the cooling schedule.

As selecting a suitable initial temperature T_0 can be a problem for Simulated Annealing we use an approach where we specify a desired average probability of accepting worsening moves. A temperature is then calculated by generating a number of random moves and evaluating the average probability of acceptance. If a temperature of 1 does not provide at least the desired average probability it is doubled and the process repeated. This continues until the generated average probability matches or exceeds the desired average probability. This allows us to specify the temperature indirectly in a less problem-specific way.

Values for T_{min} and α are somewhat easier to determine. As our penalty function has a granularity of 1, a uniform value of $T_{min} = 0.05$ is used throughout to guarantee a Hill Climbing–like phase at the end of the process. Instead of specifying α as a parameter, we give the desired number of moves N as a parameter and calculate α according to

$$\alpha = 1 - \frac{\ln(T_0) - \ln(T_{min})}{N}. \tag{2}$$

2.4   The Great Deluge

The Great Deluge method provides a much simpler mechanism than Simulated Annealing and is known to be effective on exam timetabling problems [1]. Here the controlling factor is the value of an initial ceiling on solution quality, where with a higher ceiling the search is less restricted. We can easily calculate an initial ceiling by multiplying the penalty of the initial solution by some chosen factor.

Figure 3 shows the Great Deluge procedure where b is the amount we wish to reduce the ceiling by at each iteration, and is calculated based on the user specified desired number of iterations N. As it is likely that the ceiling will fall below the current penalty as the solution nears a local optimum, 1000 000 iterations without improvement are allowed in order to accommodate a Hill Climbing–type phase towards the end of the process.

$T := T_0$

　　　WHILE $T > T_{min}$ DO

　　　choose exam e and period t at random s.t. $t\ ! = period(e)$

　　　　　IF $penalty(e,t) \le penalty(e,period(e))$ THEN

　　　　　　　　move exam e to period t

　　　　　ELSE

　　　　　　　　move exam e with probability

　　　　　　　　$\exp (\ (penalty(e,period(e)) - (penalty(e,t))/T)\ )$

　　　　　ENDIF

　　　　　$T := T$

　　　DONE

Fig. 2. Pseudo-code for the Simulated Annealing process

$P := initial\ penalty\ of\ solution$

　　　$B := P * user\ provided\ factor$

　　　$b := B/N$

　　　$sinceLastMove := 0$

　　　WHILE $(B > 0$ AND $P < B)$ OR $sinceLastMove < 1,000,000$ DO

　　　choose exam e and period t at random s.t. $t! = period(e)$

　　　　　$delta := penalty(e,t) - penalty(e,period(p))$

　　　　　IF $P + delta < B$ OR $delta < 0$ THEN

　　　　　move exam e to period p

　　　　　　　$P := P + delta$

　　　　　　　$sinceLastMove := 0$

　　　　　ELSE

　　　　　　　$sinceLastMove+ = 1$

　　　　　ENDIF

　　　　　$B := B - b$

　　　DONE

Fig. 3. Pseudo-code for the Great Deluge approach

3　Results

3.1　Experimental Setup

Experiments were performed on a range of real-world examination timetabling problems available from ftp://ftp.mie.utoronto.ca/pub/carter/testprob (see Table 1).

　　　The objectives of these problems are to firstly create a conflict-free timetable, and secondly to minimise the number of cases where a student has exams s periods apart. The weight w_s for a student taking two exams s periods apart is given by: $w_1 = 16$, $w_2 = 8$, $w_3 = 4$, $w_4 = 2$, $w_5 = 1$. These are summed up

Table 1. The benchmark problems used

Data	Institution	Periods	Number of Exams	Number of Students	Density of Conflict Matrix
car-f-92	Carleton University, Ottawa	32	543	18 419	0.14
car-s-91	Carleton University, Ottawa	35	682	16 925	0.13
ear-f-83	Earl Haig Collegiate Institute, Toronto	24	190	1 125	0.29
hec-s-93	École des Hautes Études Commerciales, Montreal	18	81	2 823	0.20
kfu-s-93	King Fahd University, Dharan	20	461	5 349	0.06
lse-f-91	London School of Economics	18	381	2 726	0.06
sta-f-83	St Andrew's Junior High School, Toronto	13	139	611	0.14
tre-s-92	Trent University, Peterborough, Ontario	23	261	4 360	0.18
uta-s-93	Faculty of Arts and Sciences, University of Toronto	35	622	21 267	0.13
ute-s-92	Faculty of Engineering, University of Toronto	10	184	2 750	0.08
yor-f-83	York Mills Collegiate Institute, Toronto	21	181	941	0.27

and divided by the number of students in the problem to give a measure of the average conflicts per student.

The Great Deluge was applied to the heuristically generated solutions with various initial ceiling values (given in the form of a function of the initial quality). Similarly, Simulated Annealing was applied to the same problems with varying desired initial average probabilities (as described above). Both algorithms were given a desired number of iterations of 20 000 000 and 200 000 000. Each individual experiment was run five times with varying random number seeds. An index of improvement was calculated by summing the average percentage improvement for each of the problems. For example, if five problems improved by 2% and six problems improved by 5% the improvement index will be $(5 \times 2) + (6 \times 5) = 40$.

Often in local search methods it is prudent to keep a copy of the best solution found so far, as the final resting point is not necessarily the best solution encountered. In this case, however, it is more revealing to take the final resting point of the search, as we can then perceive actual reductions in solution quality.

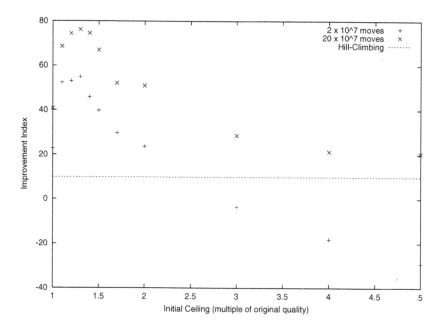

Fig. 4. Results when using Great Deluge to improve solutions

3.2 Results When Using Great Deluge Local Search

The Great Deluge algorithm was run on the generated solution with the initial ceiling set at various values based on the original quality. Figure 4 shows the total improvement index, with the horizontal bar indicating the improvements found by basic stochastic Hill Climbing. It is clear that we need to give the search procedure some extra scope to manoeuvre, though too much scope results in overall degradation of the results. It is even the case that too much scope (a ceiling factor of 5) produces results that are noticeably worse than the original unless given a sufficient amount of time. This said, it seems sensible to set the initial ceiling at the quality of the initial solution, multiplied by a factor of 1.3. Launching the algorithm for 10 times the number of moves results in slightly better improvements. However, the same rules on the initial ceiling still apply. It is worth noting that all reasonable configurations perform considerably better than stochastic Hill Climbing, yielding nearly eight times the improvement of Hill Climbing in the best case.

3.3 Results When Using Simulated Annealing

When experimenting with Simulated Annealing we test a range of temperatures, based on the average probability of accepting worsening moves. The probability is varied between 0.1 and 0.9 and shows the variation in improvement when

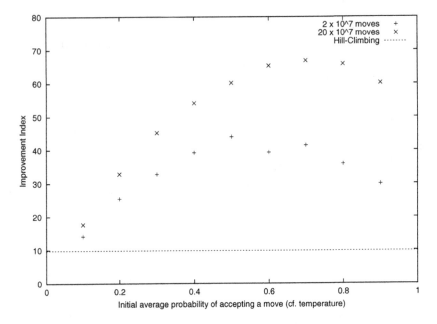

Fig. 5. Results when using Simulated Annealing to improve solutions

using different starting temperatures. The results of this are shown in Figure 5. Simulated Annealing exhibits a similar behaviour to the Great Deluge approach in that there is a "sweet spot" that balances out the need for a far-ranging search against the desire to preserve the quality of the original solution. The differences here though are less pronounced, with none of the tested scenarios producing solutions worse than the originals or, for that matter, than those produced with Hill Climbing. We would, however, expect this to change as the probability approaches one and becomes more like a non-seeded application of Simulated Annealing.

3.4   Analysis

The results indicate a slight advantage in favour of the Great Deluge method when working on improving solutions. The simpler mechanism of the Great Deluge also makes the method more attractive than Simulated Annealing. The results show that for both experiments the value of the parameters can have a dramatic effect on solution quality. Tables 2 and 3 show a comparison between Simulated Annealing (SA) and Great Deluge (GD) when started with and without good solutions, for 20 000 000 and 200 000 000 iterations. It should be noted that the initial

solution for the basic approach was also generated using a similar heuristic method except that soft constraints were not considered. This was necessary as

Table 2. Comparison of both methods with and without heuristic initialisation (20 000 000 iterations)

Data	SA (basic) init prob = 0.999		SA (heuristic) init prob = 0.7		GD (basic) init ceil = 1.0		GD (heuristic) init ceil = 1.3	
	Cost	Time (s)	Cost	Time (s)	Cost	Time (s)	Cost	Time (s)
car-f-92	5.27	31	4.19	61	4.92	90	4.14	81
car-s-91	6.48	35	4.92	81	5.48	108	4.84	109
ear-f-83	41.18	16	37.62	19	39.43	46	37.39	36
hec-s-93	12.43	11	12.02	11	12.34	94	11.94	51
kfu-s-93	16.00	49	15.31	74	15.95	76	15.51	64
lse-f-91	16.16	40	10.77	56	15.80	55	10.75	50
sta-f-83	176.34	14	165.70	16	176.30	32	166.47	42
tre-s-92	9.05	20	8.44	26	8.95	45	8.42	38
uta-s-93	5.10	36	3.29	75	4.79	74	3.26	90
ute-s-92	28.30	23	26.30	26	27.98	52	26.74	37
yor-f-83	40.94	13	39.44	15	39.13	72	39.31	71

Table 3. Comparison of both methods with and without heuristic initialisation (200 000 000 iterations)

Data	SA (basic) init prob = 0.999		SA (heuristic) init prob = 0.7		GD (basic) init ceil = 1.0		GD (heuristic) init ceil = 1.3	
	Cost	Time (s)	Cost	Time (s)	Cost	Time (s)	Cost	Time (s)
car-f-92	4.86	307	4.15	340	4.52	916	4.10	416
car-s-91	5.23	353	4.73	409	4.87	1156	4.65	681
ear-f-83	39.69	163	36.57	167	38.45	451	37.05	377
hec-s-93	11.81	108	11.71	98	11.65	966	11.54	516
kfu-s-93	15.18	486	14.34	506	15.17	749	13.90	449
lse-f-91	13.53	396	10.90	411	12.91	643	10.82	341
sta-f-83	176.10	141	168.37	132	176.44	311	168.73	418
tre-s-92	8.67	196	8.52	192	8.54	473	8.35	304
uta-s-93	4.60	351	3.19	385	4.00	881	3.20	517
ute-s-92	26.46	231	25.88	226	26.06	544	25.83	324
yor-f-83	40.00	125	37.82	121	37.16	786	37.28	695

both Simulated Annealing and the Great Deluge have great difficulty finding feasible solutions for some of these problems. For longer runs of 200 000 000 iterations in all but one case the solution produced by the basic approach is worse that the solution produced using a well configured combination approach. In general, the Great Deluge emerges as the better method, obtaining the best results on seven of the 11 problems. As we would expect though, the benefits of initial solution quality are much more noticeable when dealing with shorter

Table 4. Comparison of results

Data	Tabu Solver (av.) Di Gaspero and Schaerf Cost	Carter et al. Cost	Adaptive with Great Deluge (1.3) (av.) Cost	Time (s)
car-f-92	5.6	6.2–7.6	4.10	416
car-s-91	6.5	7.1–7.9	4.65	681
ear-f-83	46.7	36.4–46.5	37.05	377
hec-s-93	12.6	10.8–15.9	11.54	516
kfu-s-93	19.5	14.0–20.8	13.90	449
lse-f-91	15.9	10.5–13.1	10.82	341
sta-f-83	166.8	161.5–165.7	168.73	418
tre-s-92	10.5	9.6–11.0	8.35	304
uta-s-93	4.5	3.5–4.5	3.20	517
ute-s-92	31.3	25.8–38.3	25.83	324
yor-f-83	42.1	41.7–49.9	37.28	695

20 000 000 iteration runs. Table 4 shows the average individual results for the Great Deluge when given a ceiling factor of 1.3 and allowed 200 000 000 moves, and also compares them with the results of two other methods on the same problem, namely the tabu search method proposed by di Gaspero and Schaerf [8] and the heuristic backtracking approach used by Carter et al. [7]. The times given are in CPU seconds on a 900Mhz Athlon.

On some problems there is substantial improvement on the previously reported results with the others being comparable. The exception is the sta-f-83 problems, on which our proposed method performs the worst (though not by a large margin) of all the techniques. It is also worth noting that the method also outperforms the tabu search approach proposed by White and Xie [12] on the two problems they used for testing.

It would be impractical to display all the produced results here. However, the keen reader can obtain all individual results, together with the produced solutions, from http://www.asap.cs.nott.ac.uk/misc/jpn/patat2002

4   Conclusions

It is clear that under the right circumstances both Simulated Annealing and the Great Deluge method can be used to improve on already good-quality solutions. While obtaining maximal improvement requires a delicate balance of the parameters, using guideline values for these parameters should provide near-maximal improvement. This combination of heuristic construction and a subsequent phase of local search produces very competitive results on the benchmark problems, introducing new best solutions for some of the problems. A possible issue that was not clearly reflected in the experiments is the relationship between the amount of time the search is allowed and how much the search process is limited. It seems logical that if the search process is given more time, then the limitation on the search space can be relaxed a little. There is a suggestion of this in Figure 5, where the best desired average probability is a little higher when run for 10 times the number of moves. Determining this relationship more formally, however, will require substantial experimentation and will be a focus of further work.

Another area in which this work could have a serious impact is in the development of hyper-heuristics (heuristics to choose heuristics). As an aspect of hyper-heuristics deals with what methods are best to use, given a set time frame in which to work, this approach could form part of a larger hyper-heuristic framework, where the framework decides what parameters to use and for how long to perform local searching.

Acknowledgements. The research described in this paper was supported by EPSRC grant no GR/N36837/01. The authors would also like to acknowledge helpful advice given by Yuri Bykov.

References

1. Burke, E.K., Bykov, Y., Newall, J. P., Petrovic, S.: An Investigation of Time Pre-defined Search for the Examination Timetabling Problem. Technical Report NOTTCS-TR-2002-1. School of Computer Science and IT, University of Nottingham (2002)
2. Burke, E.K., Elliman, D.G., Ford, P.H., Weare, R.F.: Examination Timetabling in British Universities – a Survey. In: Burke, E, Ross, P. (Eds.): Practice and Theory of Automated Timetabling I (PATAT 1995, Edinburgh, Aug/Sept, selected papers). Lecture Notes in Computer Science, Vol. 1153. Springer-Verlag, Berlin Heidelberg New York (1996) 76–90
3. Burke, E.K., Newall, J. P.: A New Adaptive Heuristic Framework for Examination Timetabling Problems. Technical Report NOTTCS-TR-2001-5. School of Computer Science and IT, University of Nottingham (2002)(to appear in Ann. Oper. Res.)
4. Burke, E., Petrovic, S.: Recent Research Directions in Automated Timetabling. Eur. J. Oper. Res. **140** (2002) 266–280
5. Carter, M.W.: A Survey of Practical Applications of Examination Timetabling. Oper. Res. **34** (1986) 193–202
6. Carter, M.W., Laporte, G.: Recent Developments in Practical Examination Timetabling. In: Burke, E, Ross, P. (Eds.): Practice and Theory of Automated Timetabling I (PATAT 1995, Edinburgh, Aug/Sept, selected papers). Lecture Notes in Computer Science, Vol. 1153. Springer-Verlag, Berlin Heidelberg New York (1996) 3–21
7. Carter, M.W., Laporte, G., Lee, S.: Examination Timetabling: Algorithmic Strategies and Applications. J. Oper. Res. Soc. **47** (1996) 373–383
8. Di Gaspero, L., Schaerf, A.: Tabu Search Techniques for Examination Timetabling. In: Burke, E, Erben W. (Eds.): Practice and Theory of Automated Timetabling III (PATAT 2000, Konstanz, Germany, August, selected papers). Lecture Notes in Computer Science, Vol. 2079. Springer-Verlag, Berlin Heidelberg New York (2001) 104–117

9. G. Dueck. New optimization Heuristics: the Great Deluge and the Record-to-Record Travel. J. Comput. Phys. **104** (1993)
10. Dueck, G., Scheuer, T.: Threshold Accepting: a General Purpose Algorithm Appearing Superior to Simulated Annealing. J. Comput. Phys. **90** (1990) 161–175
11. Kirkpatrick, S., Gelatt Jr, C.D., Vecchi, M.P.: Optimization by Simulated Annealing. Science 220(4598) (1983) 671–680
12. G. M. White and B. S. Xie. Examination Timetables and Tabu Search with Longer-Term Memory. In: Burke, E, Erben W. (Eds.): Practice and Theory of Automated Timetabling III (PATAT 2000, Konstanz, Germany, August, selected papers). Lecture Notes in Computer Science, Vol. 2079. Springer-Verlag, Berlin Heidelberg New York (2001) 85–103

A Hybrid Algorithm for the Examination Timetabling Problem

Liam T.G. Merlot[1], Natashia Boland[1], Barry D. Hughes[1], and
Peter J. Stuckey[2]

[1] Department of Mathematics and Statistics,
The University of Melbourne, Victoria 3010, Australia
{merlot,natashia,hughes}@ms.unimelb.edu.au
[2] Department of Computer Science and Software Engineering,
The University of Melbourne, Victoria 3010, Australia
pjs@cs.mu.oz.au

Abstract. Examination timetabling is a well-studied combinatorial optimization problem. We present a new hybrid algorithm for examination timetabling, consisting of three phases: a constraint programming phase to develop an initial solution, a simulated annealing phase to improve the quality of solution, and a hill climbing phase for further improvement. The examination timetabling problem at the University of Melbourne is introduced, and the hybrid method is proved to be superior to the current method employed by the University. Finally, the hybrid method is compared to established methods on the publicly available data sets, and found to perform well in comparison.

1 Introduction

The difficulty of developing appropriate examination timetables for tertiary education institutions is increasing. Institutions are enrolling more students into a wider variety of courses including an increasing number of combined degree courses. For example, at the University of Melbourne, approximately 20 000 students have to be fitted into about 650 exams over a two and a half week period. For these 20 000 students, there exist approximately 8000 different individual examination timetables. Consequently, examination timetabling is a difficult combinatorial optimization problem and too complex an issue to be resolved by manual means. Appropriate algorithms are required to provide adequate examination timetables for universities.

The development of an examination timetable requires the institution to schedule a number of examinations ("exams") in a given set of exam sessions ("time slots", or simply "sessions"), so as to satisfy a given set of constraints. A common constraint for universities is that no student may have two exams scheduled at the same time. However, some universities allow a student to have two examinations scheduled at the same time (a "clash"), as long as an appropriate arrangement can be made (such as "quarantining" students between exams). This is the situation at the University of Melbourne, where every semester

E. Burke and P. De Causmaecker (Eds.): PATAT 2002, LNCS 2740, pp. 207–231, 2003.

students are scheduled with two exams in the same session. As quarantining is expensive and inconvenient, we propose a different examination timetabling method for the University of Melbourne that avoids these "clashes".

This paper discusses key features of examination timetabling problems and reviews existing methods for publicly available and other data sets in Section 2. A new hybrid exam scheduling method is presented in Section 3. This hybrid method seeks good-quality schedules, but attempts to avoid unnecessary clashes. The new method is a combination of constraint programming, simulated annealing and hill climbing (local search). Details of the problem for the University of Melbourne are given in Section 4, while benchmark problems in the literature are discussed in Section 5. The hybrid method is demonstrated in Section 4 to be superior to the current timetabling system used by the University of Melbourne, and in Section 5 to be superior or comparable to well-known existing methods, measured against established benchmarks.

2   Previous Work on Examination Timetabling

2.1   Examination Timetabling Problems

The primary form of the exam timetabling problem faced by educational institutions is to allocate a session and a room to every exam, so as to satisfy a given set of constraints. The result is a feasible exam timetable. However, each institution will have some unique combination of constraints, as policies differ from institution to institution. Furthermore, institutions may take different views on what constitutes the *quality* of an exam timetable. In some cases, any feasible timetable will do, while in other cases, timetables exhibiting desirable features are sought. This makes it difficult to give a universal definition of exam timetabling, but, although the exact nature of the constraints and quality measures tends to be unique to individual institutions, they tend to take on only a limited number of forms.

The most common forms of constraint are

1. Clashing: no student may have two exams in the same session.
2. Capacity: the total number of students sitting in all exams in the same session in the same room must be less than the capacity of the room.
3. Total Capacity: the total number of students sitting in all exams in the same session must be less than the total capacity for that session.
4. Exam Capacity: the total number of exams in the same session must be less than some specified number.
5. Exam Availability: some exams are preassigned to specific sessions or can only be held in a limited set of sessions.
6. Room Availability: some rooms are only available in specific sessions.
7. Pairwise Exam Constraints: some pairs of exams must satisfy pairwise scheduling constraints (e.g., one must be held before the other).
8. Exam/Room Compatability: some exams may require specific rooms.

9. Student Restrictions: there may be restrictions on students' individual examination timetables (e.g., no student can have two exams scheduled in three consecutive sessions).

10. Large Exams: large exams should be held earlier in the exam period (e.g., exams with more than 500 students must be held in the first 10 sessions).

Each institution will apply some or all of these constraints. The exact form will be dependent on the institution, and some may be treated as soft constraints (constraints that hold where possible, but can be violated). For example the University of Melbourne treats Clashing and Large Exams as soft constraints. Quality measures (or objectives) of a solution are usually derived from soft constraints, most frequently from Student Restrictions. For example, the number of clashes (instances of a student with two exams scheduled in the same session) is a quality measure for the University of Melbourne, as is the number of instances of a student with an exam scheduled in both the morning and afternoon sessions of the same day. If several different quality measures are used simultaneously, the objective is a linear combination of these measures, with relative weights that reflect their perceived importance.

For some institutions (including the University of Melbourne), the allocation of rooms to the exams in a given session is a secondary problem: exam rooms may be large, or exams easily split between rooms. In these cases, the assignment of sessions to exams has only to respect the total capacity constraint for each session, and the assignment of exams to specific rooms can be done later as a separate activity. Not all institutions are so fortunate.

2.2   Previous Methods

In this section, we review some influential and recent methods for solving exam timetabling problems. Often, different methods have addressed somewhat different versions of the exam timetabling problem, with different constraints and quality measures. Quality measures are often combined to form a mathematical *objective* for the problem, and methods which optimize with respect to that objective developed. We discuss variations of the problem encountered in the literature in more detail in Section 5; here we focus on the methods that have been applied.

Surveys of different methods for exam timetabling by Burke et al. [2] and Carter and Laporte [8] classify the different approaches as cluster methods, sequential construction heuristics, constraint programming, and local search (genetic algorithms, memetic algorithms, simulated annealing and tabu search). In recent years, Carter[1] and Burke[2] have made data sets for exam timetabling publicly available via the internet. Only three different approaches, that we are aware of, have been applied to these publicly available data.

Sequential construction heuristics have been applied to the publicly available data in a variety of forms by Burke et al. [6], Carter et al. [9,10] and Caramia

---

[1] ftp://ftp.mie.utoronto.ca/pub/carter/testprob
[2] ftp://ftp.cs.nott.ac.uk/ttp/Data

et al. [7]. Sequential construction heuristics order the exams in some way (for example, largest exam first), and attempt to allocate each exam to a session in order, while satisfying all the constraints. The different heuristics feature different orders. They also have other differences: Carter et al. [10] allow limited backtracking (deallocation of exams), Burke et al. [6] select exams from a randomly chosen subset of all exams, and Caramia et al. [7] include an optimization step after each exam allocation.

Burke et al. use memetic algorithms for exam timetabling [3,4]. In Burke et al. [4] an initial pool of timetables is generated via a random technique, which attempts to group together exams with similar sets of conflicting exams. Then timetables are randomly selected from the pool, weighted by their objective value, and mutations are applied by rescheduling randomly chosen exams, or all exams in a randomly chosen session. Finally, hill climbing (local search) is applied to the mutated timetable to improve its quality. The process continues with the new pool of timetables. Burke and Newall [3] improve upon their earlier work by applying the memetic algorithm only to the first k exams as defined by a sequential construction method ordering. After the best timetable for the first k exams is found, the exams are fixed in place, and the memetic algorithm applied for the next k exams, until all are fixed.

White and Xie [19] and Di Gaspero and Schaerf [13] use tabu search methods. White and Xie keep two tabu lists, the usual short-term tabu list, and a long-term tabu list which keeps track of the most-moved exams. Di Gaspero and Schaerf [13] use a single tabu list, but when exams are added to this list it is for a randomly determined number of iterations. They also modify the objective function as the algorithm progresses.

There is a considerable body of work on exam and other timetabling problems, which has not been applied to the publicly available data sets. The most closely related to our work appear to be the constraint programming approach used by Boizumault et al. [1] and the simulated annealing approaches explored by Dowsland and Thompson [11,12,16,17,18]. The principal innovation in our work, compared to these others, is the sequential use of these two methods as the first two stages of a total strategy. A similar sequential approach has been taken in work on other problems: White and Zhang [20] use constraint programming to find a starting point for tabu search in solving course timetabling problems, and for high school timetabling Yoshikawa et al. [21] test several combinations of two-stage algorithms, including a greedy algorithm followed by simulated annealing and a constraint programming phase followed by a randomized greedy hill climbing algorithm (which is deemed to be the best combination of those used). In a similar vein, Burke et al. [5] use their work on sequential construction heuristics [6] to generate initial solutions for their memetic algorithm [4].

Our approach is to apply constraint programming, simulated annealing and hill climbing in turn. Although each of these stages is closely related to an existing method, the combination we develop is new, and appears to be particularly effective in practice, as we show later. Here, we motivate the method we use at each stage and comment on the similarities of each to methods described in the published literature.

Firstly, our constraint programming approach is similar to that of Boizumault et al. [1]; in fact our approach can be viewed as a simplification, tailored to finding a feasible solution very quickly rather than solving the entire problem. Other researchers, notably Thompson and Dowsland [16,17,18] have used alternative first-stage methods. Thompson and Dowsland use simulated annealing with a simple neighbourhood to solve a first-stage problem. As in our first stage, overall seating capacity and exam availability (time window) constraints are satisfied. However, we do not permit clashes, but allow some exams to remain unscheduled, if needed, whereas Thompson and Dowsland schedule all exams and seek to minimize the number of clashes. Whilst a direct comparison of these two approaches would be interesting, we have found constraint programming to be very effective. As we discuss in Sections 4 and 5, it is highly successful at providing suitable starting points for our simulated annealing stage and tends to provide compact solutions, using near-minimal number sessions. Running times have never been more than a few seconds on any data set tested.

Secondly, our simulated annealing stage is very similar to the later-stage simulated annealing method of Thompson and Dowsland [16,17,18] which they demonstrated to be very effective on problems similar to our problem class P4, discussed in Section 5.5. Both simulated annealing methods use a Kempe chain neighbourhood and geometric cooling (for more details see Section 3.2). Thompson and Dowsland select a neighbour by choosing two sessions and one exam from the first of these sessions at random and seek to move the selected exam to the second session, but have also experimented with our approach: we choose an exam at random and select a new session from those available for that exam. Thompson and Dowsland observe that when there is little slack in the total seating capacity constraint, the two sampling methods perform similarly. We believe that in cases where the set of sessions available for each exam is quite limited, (as occurs at the University of Melbourne), the latter choice will be more efficient; it will use less time testing infeasible options. We also fix the number of iterations performed at each temperature to a constant. This is one of many cooling schedules tested by Thompson and Dowsland.

Our third stage uses hill climbing to improve the final solution from the simulated annealing stage. The idea of a final hill climbing solution refinement stage is not new. For example, the approach used by Schaerf [15] for high school timetabling combines a metaheuristic with hill climbing: tabu search based on a small neighbourhood employs a (randomized) hill climbing method[3], based on a larger neighbourhood, to improve any local minimum (with respect to the smaller neighbourhood) encountered by the tabu search. Our approach is in some sense the reverse of this: our hill climbing method exhaustively searches a neighbourhood which is in some sense smaller than that used by our simulated annealing stage; the neighbourhood used in simulated annealing is actually very

---

[3] A neighbour is generated at random. It is accepted if it is at least as good as the current solution.

large, but of course simulated annealing only selects a neighbour at random and does not search the entire neighbourhood.

3   A Three-Stage Method

We consider the exam timetabling problem in which a session must be allocated to each exam. We apply three of the constraints defined in Section 2.1: Clashing, Exam Availability and Total Capacity. We also treat the Large Exams constraint as a hard constraint, modelled via the Exam Availability constraint. Room allocation is not critical for the application of most interest to us (see Section 4), so we neglect it here. The approach we take to the exam timetabling problem consists of three stages, yielding a hybrid method:

1. Constraint programming: to obtain a feasible timetable;
2. Simulated annealing: to improve the quality of the timetable;
3. Hill climbing: for further refinement of the timetable.

The first stage of the hybrid method is used primarily to obtain an initial timetable satisfying all the constraints. As we shall see below, our approach aims to achieve this using as few sessions as possible, and although all constraints will be satisfied by the resulting timetable, in some cases, some exams may remain unscheduled.

The second and third stages aim to improve the quality of the timetable, and to schedule as many exams as possible. The methods used in both these stages are optimization methods, which will seek to optimize a given objective function. For this purpose, we formulate an objective function which takes into account both aims. The measures of quality, and hence the objective function formulated, differ according to the particular data set used, but typically the objective function is some combination of penalties applied for proximity of exams (time-wise) in students' timetables. We discuss the objective functions, which we call *objective scores*, used in detail in the context of the data sets considered, in Sections 4 and 5.

It is possible (although rare for data sets we tested) that some exams remain unscheduled after all three stages. In this case, we employ a simple greedy heuristic to schedule the remaining exams. We discuss this heuristic in Section 3.4.

3.1   Constraint Programming

Constraint programming is used to find a first feasible timetable in our hybrid method. Our constraint programming model is defined as follows. We use the notation below:

- $E = \{1, \ldots, n\}$ denotes the given set of n exams,
- s_i is the number of students in exam i, for all $i \in E$,
- $T = \{1, \ldots, v\}$ denotes the set of v sessions,
- $R \subset E \times T$ represents the set of given exam-session restrictions, so that $(a, b) \in R$ indicates that session b cannot be allocated to exam a,

- C_t is the total capacity of session t: that is, the total number of students who can sit exams in session t, for all $t \in T$,
- D_{ij} is the number of students enrolled in both exams i and j, for all $i, j \in E$,
- variable $x_i \in T$ indicates the session allocated to exam i, for all $i \in E$.

Note that the sessions are assumed to be time ordered, so $t_1 < t_2$ for $t_1, t_2 \in T$ if and only if session t_1 occurs before session t_2. We also define the *domain* of each variable x_i to be the set of all sessions that can feasibly be allocated to exam i. Initially, the domain of x_i is the set of sessions $t \in T$ such that $(i, t) \notin R$. During a constraint programming search, the domain of variables may be reduced by the removal of sessions, so as to ensure some form of consistency with respect to the constraints. The initialization of the variable domains ensures that the Exam Availability constraint is satisfied. The Clashing constraint is modelled in our constraint program as follows:

$$x_i \neq x_j \text{ for all exams } i, j \in E \text{ with } i \neq j \text{ and } D_{ij} > 0.$$

The Total Capacity constraint is modelled as

$$\sum_{i \in E} s_i(x_i = t) \leq C_t \text{ for all sessions } t \in T.$$

Here $(x_i = t)$ is a Boolean switch, taking on value 1 if $x_i = t$, and 0 otherwise.

A typical constraint programming method, applied to the above exam timetabling model, will operate as follows. An exam is chosen, and a session in its domain is allocated to it: that is, all sessions except one are removed from the exam's domain. In this case we say the exam has been *scheduled*; otherwise it is *unscheduled*. Consistency of the domains of all variables with respect to the constraints is then checked. For example, any exam which clashes with the chosen exam should have the session allocated to the chosen exam removed from its domain. After consistency is checked, another exam is chosen, and is allocated a session from its domain. This process is repeated until either all exams have been allocated a session, or until infeasibility has been detected, usually as a result of the domain of some variable becoming empty. In this case, the method backtracks and re-allocates sessions to some exams. Clearly, there is a lot of flexibility in this method: which exam is chosen at each step, and which session in its domain is chosen to be allocated to it? The precise form of these choices determines the *search strategy* of the constraint programming method.

Our experiments revealed that the best search strategy for our exam timetabling problem is to choose the unscheduled exam with the smallest current domain size, that is, the exam with the smallest domain size greater than one, and to allocate it the earliest session in its domain. Other search strategies explored were: choosing exams based on the number of students in each exam, choosing exams based on the number of conflicting exams, and choosing sessions based on the total space remaining given the current allocations (most or least)[4].

---

[4] Boizumault et al. [1] found for their data that the best search strategy was to choose the exam with the largest number of students and the session with the most available space.

This constraint programming model and search strategy were implemented using the ILOG package OPL [14]. As the sessions were chosen in order (exams were scheduled as early as possible), the timetables found tended to use a number of sessions close to the minimum number required (and often less than the maximum number allowed). In order to take advantage of this result, and consequently speed up the overall algorithm, additional "dummy" sessions are included (with a capacity equal to the maximum used for all other sessions), which are feasible for all exams, and occur "after" the final session v. Any exams allocated these dummy sessions by OPL are interpreted as unscheduled. A component of the objective in the simulated annealing phase of the algorithm encourages scheduling of unscheduled exams. In Section 5.3, 12 different data sets are used, and for these, two data sets required one dummy session and one required two dummy sessions, but after the simulated annealing phase there were no exams remaining in these dummy sessions.

Our constraint programming approach thus generates timetables satisfying the given constraints, with the proviso that in some cases, some exams may remain unscheduled (although the search strategy we use ensures this is rare).

As the timetables produced by this process are not subject to any quality measures, they are usually of very poor quality. Optimization of exam timetables using constraint programming proved impractical: our experiments showed that the large search space and weak pruning of the minimization constraints made it difficult for a constraint programming method to significantly improve timetable quality. However, as we shall see later, a simulated annealing method *is* able to make substantial improvements.

3.2   Simulated Annealing

The timetable produced by the constraint programming algorithm is used as the starting point for the simulated annealing phase of the hybrid method. This phase is used to improve the quality of the timetable.

In what follows, we refer to a candidate timetable, together with any unscheduled exams (exams scheduled in dummy sessions), as a *solution*. In all cases, all scheduled exams in any solution will satisfy the Clashing, Total Capacity, and Exam Availability constraints, so all exams allocated a given session will not conflict with each other, the total number of students sitting exams allocated the same session will not exceed the capacity of the session, and no exam will be allocated a session for which it is not available.

The number of sessions used for the solutions in the simulated annealing phase (and hill climbing phase) is equal to the number of sessions specified by the problem plus the number of dummy sessions used in the solution generated by the constraint programming phase.

A key component of any simulated annealing method is the *neighbourhood* of a solution. The neighbourhood we use is a slight variant on the Kempe chains neighbourhood used by Thompson and Dowsland [18]. A Kempe chain is determined by an exam i currently allocated session t, and another session $t' \neq t$. Let G be the set of all exams allocated session t, and G' be the set of exams allocated session t'. Note that our definition of a solution ensures that both G and G' are conflict-free sets of exams. The Kempe chain can be thought of as the (unique) minimal pair of sets of exams, $F \subseteq G$ and $F' \subseteq G'$, such that $i \in F$ and both $(G \backslash F) \cup F'$ and $(G' \backslash F') \cup F$ are conflict-free sets of exams. For a given exam i in session t and other session t', the Kempe chain (F and F') can easily be constructed by a simple iterative procedure. We call the timetable obtained by (re-)allocating session t to all exams in F' and (re-)allocating session t' to all exams in F a *neighbour* of the solution. The *neighbourhood* of a solution is defined to be the set of all such neighbours.

In our simulated annealing algorithm, a current solution is maintained, and a neighbour of the current solution chosen at random. We choose a neighbour by selecting an exam i at random from the set of all exams, and selecting a session t' at random from the set of sessions available for i, i.e. with $(i, t') \notin R$, such that $t' \neq t$, where t is the session allocated to exam i in the current solution. Together i, t and t' induce a Kempe chain, and hence a neighbour of the current solution. (Thompson [18] selects two sessions at random, and randomly chooses an exam allocated the first session.)

Our simulated annealing method is quite standard. Once a neighbour has been generated (each generation of a neighbour is defined as an iteration), it is tested for feasibility (of Total Capacity and Exam Availability), and if feasible, the objective function value, or *score*, for this neighbour is calculated. If this neighbour is superior in quality to (has a lower score than) the current solution then the current solution is replaced by the neighbour. Otherwise the current solution may or may not be replaced by the neighbour, depending on the difference in scores and the current *temperature* (as is standard in simulated annealing). Let $o(p)$ denote the objective score for a solution p. Let x be the current solution and let q be the neighbouring solution to x selected at random at the current iteration. Let u be the current temperature. Then if $o(q) > o(x)$, the probability that q will replace x as the current solution is $e^{(o(x) - o(q))/u}$.

The score used varied by problem class (different problem classes have different objectives), so we define the scores used in Sections 4 and 5, where we document each problem class tested.

Thompson and Dowsland [16] discuss different cooling schedules for simulated annealing processes applied to exam scheduling problems. They find that slow cooling schedules are generally more effective, but that no cooling schedule is markedly better than any other. We use a geometric cooling schedule, in which at every a iterations, the temperature, u, is multiplied by α, where a and α are given parameters of the algorithm. Initial experimentation revealed that the parameter settings of a starting temperature of $30\,000$, $\alpha = 0.999$, $a = 10$ and a stopping temperature of 10^{-11}, giving approximately $350\,000$ iterations, were appropriate for all problems discussed in this paper.

Throughout the simulated annealing phase, the algorithm keeps the best solution found so far, and yields as output this best solution, at the conclusion of the algorithm. As with most heuristics, simulated annealing solutions will vary in quality depending on how long the algorithm is permitted to run. It was found that solutions using the standard parameters, generated in under a minute of CPU time on an Alphastation XP900 (466 MHz), were of good quality for the University of Melbourne's examination timetabling problem.

We use simulated annealing to improve the solution determined by the constraint programming phase, and to schedule any remaining unscheduled exams. We could have used simulated annealing starting from a random timetable to satisfy all the constraints. Experiments with a purely simulated annealing approach performed poorly, since finding a feasible solution was in many cases quite difficult, and this overwhelms the search for a good solution. Indeed for one of our data sets there was no solution to the hard constraints as given, and we would never determine this using simulated annealing alone.

3.3   Hill Climbing

The hill climbing algorithm starts with an initial solution, x, called the "current" solution, with objective score $o(x)$. The algorithm processes each exam in some order, determined a priori: we chose the exam subject code order. For each exam, e, in this order, the algorithm considers every neighbour of x (using the Kempe chain definition of neighbour given in Section 3.2) in which e is allocated a different session to the session allocated to it in x. For all such neighbours feasible with respect to the Exam Availability and Total Capacity constraints, the objective score is calculated and the neighbour q with minimum objective score is found. If there are several choices for q, select q to move e to the earliest time. If $o(q) \leq o(x)$, the current solution x is replaced by q: that is, the algorithm sets $x := q$; otherwise x is unchanged. In either case, this completes the processing of e, and the algorithm moves on to the next exam in the given order.

The hill climbing algorithm does not necessarily stop once all exams have been processed: at this point we say that one "iteration" of the hill climbing algorithm has been completed. The hill climbing algorithm may well go on to again process all exams in the given order, several times over; it may perform many iterations. The hill climbing algorithm stops when either all neighbours generated during the last iteration had strictly larger objective scores than the current solution, or after a specified number of iterations. On the data sets we experimented with, there was usually no improvement in the objective score of the current solution after around seven iterations, and consequently 10 hill climbing iterations were deemed to be sufficient; we specified a limit of 10 iterations for the hill climbing algorithm.

Normally the hill climbing stage does not significantly improve the quality of solution produced by the simulated annealing phase. However, in approximately 5–10% of cases, the simulated annealing algorithm had not sufficiently explored the neighbourhood of its best solution. For example, the best solution encountered may have been found early when the temperature was high, and moves that increase the objective are more frequently accepted. The algorithm may move away and settle in an inferior local minimum with a higher objective as the temperature is reduced. When this happened, the hill climbing stage produced a significant improvement on the simulated annealing solution. The overall effect of the hill climbing phase of the algorithm is to make the results more consistent between runs of the three-stage algorithm by improving the less satisfactory simulated annealing solutions.

3.4 A Greedy Heuristic for Scheduling Unscheduled Exams

The constraint programming stage may allocate dummy sessions to some exams (the exams remain unscheduled). Normally the simulated annealing stage will re-allocate normal sessions to these exams. However, it is possible for unscheduled exams to remain after all three stages of the algorithm.

In all tests we have performed, the only cases where this occurred were in the University of Melbourne problem and then only in those data sets where a clash-free solution is impossible. Although the University of Melbourne does allow clashes, when clashes occur, another constraint, known as the "Three-in-a-Day" constraint, must be satisfied; we discuss that constraint in detail below. For the purpose of describing our heuristic, we simply assume that there is some set of *essential constraints* that must be satisfied even when clashes are allowed.

In the event that exams remain unscheduled at the conclusion of the three-stage method, we apply a greedy heuristic to schedule these exams. This heuristic will, of necessity, introduce clashes into the timetable, however it attempts to minimize these. Furthermore, it will ensure the timetable satisfies the essential constraints, and will leave an exam unscheduled rather than violate these constraints.

The heuristic proceeds as follows, where the unscheduled exams are considered in order of subject code. For each exam, e, left in a dummy session (that is, unscheduled), a normal session $t \in \{1, \ldots, v\}$ is chosen so that if exam e was scheduled in session t the essential constraints would hold, and so that the number of clashes introduced by scheduling exam e in session t is minimized over all such sessions. If there are no such sessions, then the exam remains unscheduled (though this never occurred in our tests).

4   The University of Melbourne Problem

At the University of Melbourne, there are 600–700 exams to be scheduled for each of two teaching semesters. The June exam period at the end of semester 1 normally has about 600 exams, to be scheduled in 26–30 sessions (13–15 days with morning and afternoon sessions). In the November exam period at the end of semester 2 there are more exams (at least 650), and these have to be scheduled in 30–34 sessions (15–17 days). There are five rooms in which exams can be held, two of these on campus (with capacities 135 and 405) and three at an off-campus venue (with capacities 540, 774 and 1170).

The constraints that the university imposes can be defined as Capacity, Total Capacity, Exam Availability, Room Availability and Large Exams. It is a matter of university policy that Large Exams is a soft constraint to be respected if possible, but in practice the university's current procedure takes Large Exams as a hard constraint. The Large Exam constraint can thus be treated as an Exam Availability constraint. The university also has some Pairwise Exam Constraints, in that some pairs (or sets) of exams have common content and are required to be held at the same time. We simply combine such exams into a single exam.

As there are only five available rooms, and splitting exams between the off-campus rooms does not present any difficulty, room allocation is not considered a serious issue at the University of Melbourne. Consequently, the University of Melbourne's current software, as well as our algorithm, initially only attempts to allocate a session to every exam, and does not allocate exams to rooms. Thus we only use the Capacity and Room Availability constraints to determine the total capacity for each session, needed for the Total Capacity constraint.

Unlike most educational institutions, the University of Melbourne does not enforce the Clashing constraint. Instead the university enforces a Student Restrictions constraint, called *Three-in-a-Day*, that students cannot have three exams scheduled in the same day (recall there are two sessions per day). By appropriate quarantining, students with two exams scheduled concurrently can be accommodated. This does, however, incur inconvenience and expense. The authors believed that allowing clashes and resolving them by quarantining students was unnecessary, and implemented the tighter Clashing constraint that no student may have two exams scheduled at the same time (as is required by our hybrid algorithm). If, at the end of the three stages of the hybrid algorithm, some exams remain unscheduled, we use the greedy heuristic described in Section 3.4, with essential constraints Total Capacity, Exam Availability and Three-in-a-Day, to allocate sessions to these exams. In all our test cases, the greedy heuristic successfully scheduled all remaining exams.

Data from two different semesters were examined[5]. These were for semesters 1 and 2 of the 2001 academic year at the University of Melbourne (mel01s1 and mel01s2). The semester 1 data set (mel01s1) required 609 exams to be scheduled into 28 sessions (30 sessions, with two sessions lost in the middle to a public holiday), and the semester 2 data set (mel01s2) required 657 exams to be scheduled into 31 sessions. To cope with the Pairwise Exam constraint that some exams had to be held at the same time, such exams were combined into a single exam. This left 521 exams in mel01s1 and 562 exams in mel01s2.

The current University of Melbourne software (which we refer to for brevity as UM) is an unnamed and scantily documented VMS executable code specially written for the university about 10 years ago. Little is known about how the UM software works, but the authors conjecture from running it that it may use some form of simulated annealing algorithm.

The objective score used for University of Melbourne problems in our simulated annealing and hill climbing stages is calculated as follows. Recall that these stages maintain clash-free solutions, and that there are two sessions on each day, so no student can be sitting more than two exams allocated sessions on the same day. Using the notation of Section 3.1, we have normal exam sessions $T = \{1, \ldots, v\}$ ordered chronologically, and introduce dummy sessions

---

[5] The data sets are available at http://www.or.ms.unimelb.edu.au/timetabling

$T' = \{v + 1, \ldots, v'\}$. For a normal session t, we use Boolean $w(t)$ true if t is the last session before a weekend or public holiday. For a given solution x, we calculate the objective score $o(x)$ as follows, where U is the penalty per student for unscheduled exams, w_{sd} is the penalty per student for two exams scheduled on the same day, w_{am} is the penalty per student for an exam scheduled in the afternoon followed by one scheduled in the morning of the next day, w_{g2} is the penalty per student for two exams scheduled with one session between them, and w_{ma} is the penalty per student for an exam scheduled in the morning of one day and an exam scheduled in the afternoon of the next day:

initialize $o(x) := \sum_{\{i \in E \ : \ x_i \in T^*\}} U s_i$
for each exam $i \in E$ with $x_i \in T$ and $w(x_i)$ not true **do**
  for each exam $j \in E$ with $x_j \in T$, $x_i < x_j \le x_i + 3$ and $D_{ij} > 0$ **do**
    if $x_j = x_i + 1$ **then**
      if x_i is a morning session **then** $o(x) := o(x) + w_{sd}D_{ij}$
      else (x_i must be an afternoon session) $o(x) := o(x) + w_{am}D_{ij}$
    else if $w(x_i + 1)$ not true **then**
      if $x_j = x_i + 2$ **then** $o(x) := o(x) + w_{g2}D_{ij}$
      else (x_j must equal $x_i + 3$)
        if $w(x_i + 2)$ not true **then** $o(x) := o(x) + w_{ma}D_{ij}$
      endif
    endif
  endfor
endfor

In all tests on University of Melbourne data reported here[6] we used $U = 10\,000$. After trying a number of different values for the penalty parameters, we found those that gave best results were $w_{sd} = 2$, $w_{am} = 1$, $w_{g2} = 0$, and $w_{ma} = 0$. So in fact we found that a much simpler score would have sufficed for this data.

In addition to the exam and student data sets, the university also provided the data for Exam Availability and Large Exams constraints. As the Large Exams constraint is treated as a hard constraint by the university's timetabling software, this was combined with the Exam Availability constraint in order to restrict the large exams to earlier times. Upon initial examination, it was discovered that due to the Large Exams being treated as a hard constraint, it was impossible to avoid clashes in the timetable for the mel01s2 data set. For example, exams with more than 500 students were required to be held in the first 14 exam sessions, and because of the number of exams in this category, it was not possible to find a clash-free solution. However, if these exams were allowed to be scheduled in the first 15 exam sessions, (one more than previously allowed), it was possible to produce a clash-free solution.

---

[6] If the parameter U is too small, feasible solutions are rarely produced. In experiments conducted for a range of suitably large values of U ($10 \le U \le 10^6$) on both University of Melbourne data and other data, we found that the actual size of this parameter was most likely not a significant factor (probability of significance: 0.17), but to ensure that feasible solutions are found often enough one should select $U > 100$.

Table 1. *Comparison between the University of Melbourne current software (UM) and our hybrid algorithm.* The best timetable was deemed to be the one with the fewest clashes, with ties resolved by choosing the lowest objective score. The corresponding score for the best timetable can exceed the average. Note that it was not possible to compare directly to the UM program for the mel01s1 data set, and consequently the hybrid method was only compared to the timetable produced by the university for that semester. This timetable had been manipulated afterwards to satisfy constraints not known when the timetable was produced, and was not minimizing the same objective

Data set			Hybrid	UM
mel01s1	521 exams	Best clashes	**0**	8
	28 sessions	Corresponding score	**1072**	2085
		Average clashes	0	–
		Average score	1175.0	–
		Average time (s)	56	–
mel01s2a	562 exams	Best clashes	1	1
	31 sessions	Corresponding score	**1448**	2384
		Average clashes	3.2	1.4
		Average score	1634.8	2298
		Average time (s)	74	1008
mel01s2b	562 exams	Best clashes	0	0
	31 sessions	Corresponding score	**1115**	3373
		Average clashes	0	1
		Average score	1300	2632.2
		Average time (s)	73	1008

In order to test the hybrid algorithm under the assumption that clash-free solutions were possible, two different data sets for this semester were used: one used the Exam Availability and Large Exams data sets that the university used for that semester, which forced clashes due to the Large Exams constraint (mel01s2a), and the second modified the Large Exams constraint slightly so that clash-free solutions were possible (mel01s2b). Five different runs were made by the hybrid algorithm on all data sets (with the standard simulated annealing parameters and 10 hill climbing iterations) and five different runs were made by the UM program on the mel01s2a and mel01s2b data sets using the university's parameters. The results are shown in Table 1.

It can be seen that the hybrid method is superior in terms of the objective and time. The algorithms were run on different machines: the hybrid algorithm on an XP900 Alphastation (466 Mhz CPU) and the UM algorithm on a DEC Alphastation 8200 (dual 300 Mhz CPU), but the time difference is still significant. The hybrid method was not designed to cope with clashes, and thus it is not as good at minimizing the number of clashes (on average) compared to the UM program when clashes are unavoidable. Conversely, in the potentially clash-free data sets, the university's program does not always provide a clash-free solution.

Table 2. *Papers on examination timetabling for publicly available data sets.* B1: Burke et al. [4], B2: Burke et al. [6], B3: Burke and Newall [3], Ca: Caramia et al. [7], C: Carter et al. [10], D: Di Gaspero and Schaerf [13] and W: White and Xie [19]. *Note that Burke et al. [6] (B2) use a slightly different version of the KFU-S-93 data set to that of the other authors; they use a different number of sessions

Data Set	Exams	P1	P2	P3	P4
CAR-F-92	543	C,Ca	C,Ca,D,W	B1,D,Ca	B2,B3,D
CAR-S-91	682	C,Ca	C, Ca,D	B1,D,Ca	–
EAR-F-83	190	C,Ca	C,Ca,D	–	–
HEC-S-92	81	C,Ca	C,Ca,D	–	–
KFU-S-93	461	C,Ca	C,Ca,D	B1,D,Ca	B2*,B3,D
LSE-F-91	381	C,Ca	C,Ca,D	–	–
PUR-S-93	2419	C,Ca	C,Ca	–	B3,D
RYE-S-93	486	C,Ca	C,Ca	–	–
STA-F-83	139	C,Ca	C,Ca,D	–	–
TRE-S-92	261	C,Ca	C,Ca,D	B1,D,Ca	–
UTA-S-92	622	C,Ca	C,Ca,D,W	B1,D,Ca	B2
UTE-S-92	184	C,Ca	C,Ca,D	–	–
YOR-F-83	181	C,Ca	C,Ca,D	–	–
NOTT	800	–	–	B1,D,Ca	B3,D

5   Benchmarks

5.1   Publicly Available Data

As stated in Section 2.2, Carter and Burke have made data sets for exam timetabling publicly available. For each of these data sets, attempts have been made to solve up to four different problems. The solutions to these problems provide benchmarks to compare different exam timetabling methods. The problem definitions, and consequently the initial benchmarks for these problems, are presented in the papers of Carter et al. [10] (problems P1 and P2), Burke et al. [4] (problem P3), and Burke and Newall [3] (problem P4). Below we discuss in detail each of these problems and the results of solving them with our method.

Four other papers compared alternative methods to the original benchmarks established by Burke et al. [3,4] and Carter et al. [10]: Burke et al. [6], Di Gaspero and Schaerf [13], Caramia et al. [7], and White and Xie [19]. In Table 2 the publicly available data sets are listed, along with the papers that have tested methods on each problem. It should be noted that the computing resources used differ. We have not been able to convert the reported run times in seconds to equivalent run times for a common standard, so all tabulated times reported below should be taken as indicative only.

5.2   Problem P1: Graph Colouring Benchmarks

Carter et al. [10] look at several of the publicly available data sets, and attempt to find the minimum number of sessions required for a feasible timetable sub-

Table 3. *Problem P1: graph colouring benchmarks.* The constraint programming stage of the hybrid algorithm yields timetables with close to the minimum number of sessions. The time reported for the methods of Carter et al. [10] and Caramia et al. [7] is the time for the run producing the best result. The time for the hybrid algorithm is the average time per run. Reported times have not been converted to account for different computing resources. Data sets are listed in order of number of exams to give an indication of order of difficulty

Data set	No of exams		Hybrid stage 1	Carter et al.	Caramia et al.
HEC-S-92	81	Best	18	**17**	**17**
		Range	–	17–18	17–18
		Time (s)	0.36	0.5	10.6
STA-F-83	139	Best	**13**	**13**	**13**
		Range	–	13–13	13–13
		Time (s)	0.62	2.7	10.2
YOR-F-83	181	Best	23	**19**	**19**
		Range	–	19–21	19–21
		Time (s)	1.05	190.4	226.2
UTE-S-92	184	Best	11	**10**	**10**
		Range	–	10–10	10–10
		Time (s)	0.75	1.6	24.3
EAR-F-83	190	Best	24	**22**	**22**
		Range	–	22–24	22–23
		Time (s)	1.2	8.7	86.3
TRE-S-92	261	Best	21	**20**	**20**
		Range	–	20–23	20–23
		Time (s)	1.03	32.8	214.7
LSE-F-91	381	Best	18	**17**	**17**
		Range	–	17–18	17–18
		Time (s)	2.45	78.0	9.6
KFU-S-93	461	Best	21	**19**	**19**
		Range	–	19–20	19–20
		Time (s)	3.4	97.2	159.6
RYE-S-93	486	Best	22	**21**	**21**
		Range	–	21–23	21–23
		Time (s)	3.95	343.8	225.9
CAR-F-92	543	Best	31	**28**	**28**
		Range	–	28–32	28–32
		Time (s)	5.37	227.2	559.2
UTA-S-92	622	Best	32	32	**30**
		Range	–	32–35	30–34
		Time (s)	6.39	272.3	1023.5
CAR-S-91	682	Best	30	**28**	**28**
		Range	–	28–35	28–32
		Time (s)	7.34	75.1	86.3

ject only to the Clashing constraint. They set no Capacity or Total Capacity constraints, and thus the problem reduces to a graph colouring problem. The sequential construction heuristic of Carter et al. [10] produces a variety of solutions, depending on the order in which the exams were processed. Using a different sequential construction heuristic, Caramia et al. [7] managed to produce equal (and superior in one case) results to Carter et al. [10].

The constraint programming stage of the hybrid method produces solutions that tend to use a relatively small number of sessions, as we demonstrate in Table 3. Here we compare the number of sessions used in the solution produced by the constraint programming stage with the results of Carter et al. (Table 3 in [10]) and Caramia et al. (Table 1 in [7]). The results of the comparison can be seen in Table 3.

The constraint programming stage of the hybrid method only produced one solution with as few sessions as the best of the Carter et al. [10] and Caramia et al. [7] methods. However, in all but one data set, the constraint programming phase produced a solution with only one or two more sessions than the minimum attained by the other methods, and never used more than four additional sessions. These results confirm that the constraint programming stage produces timetables with close to the minimum number of sessions.

5.3   Problem P2: Uncapacitated Benchmarks

In addition to their work on graph colouring benchmarks, Carter et al. [10] developed a second uncapacitated version of the examination timetabling problem. They set a maximum number of sessions, and devise an objective function designed to favour timetables which space out students' exams. The objective function applies a penalty w_t to a timetable whenever a student has to sit two exams scheduled t periods apart, with $w_1 = 16$, $w_2 = 8$, $w_3 = 4$, $w_4 = 2$ and $w_5 = 1$. The total penalty is divided by the number of students to get an average penalty per student; this is the value of the objective function for the given timetable. No account was taken of weekends and there was no differentiation between consecutive exam periods within the same day, versus overnight.

For this problem class, our hybrid algorithm uses an objective score identical to the objective function defined by Carter et al. [10], calculated as follows:

initialize $o(x) := \sum_{\{i \in E \ : \ x_i \in T\}} Us_i$
for each exam $i \in E$ with $x_i \in T$ **do**
    for each exam $j \in E$ with $x_j \in T$, $x_i < x_j \leq x_i + 5$ and $D_{ij} > 0$ **do**
        $o(x) := o(x) + w_{(x_j - x_i)} D_{ij}$
    endfor
endfor

Caramia et al. [7], Di Gaspero and Schaerf [13] and White and Xie [19] compared their algorithms to that used by Carter et al. [10], using the same problem definition, on some or all of the P2 data sets. The method of Caramia et al. [7] proved to be superior on 10 of the 13 data sets, equal to Carter et al. [10]

Table 4. *Problem P2: uncapacitated benchmarks.* The number of sessions is restricted, but there are no capacity constraints. Scores listed under "best" and "average" are per student scores. For the hybrid method, this is obtained by dividing the final score by the number of students. Times reported are average times per run for the hybrid method, and times of the best run for Carter et al. [10] and Caramia et al. [7]. Reported times have not been converted to account for different computing resources. Data sets are listed in order of number of exams to give an indication of order of difficulty

Data set		Hybrid	Carter et al.	Caramia et al.	Di Gaspero and Schaerf	White and Xie
HEC-S-92	Best	10.6	10.8	**9.2**	12.4	–
Exams 81	Average	10.7	15.04	–	12.6	–
Sessions 18	Time (s)	5.4	7.4	11.0	–	–
STA-F-83	Best	**157.3**	161.5	158.2	160.8	–
Exams 139	Average	157.4	167.14	–	166.8	–
Sessions 13	Time (s)	5.1	5.7	6.5	–	–
YOR-F-83	Best	37.4	41.7	**36.2**	41.0	–
Exams 181	Average	37.9	45.6	–	42.1	–
Sessions 21	Time (s)	30	271.4	125.4	–	–
UTE-S-92	Best	25.1	25.8	**24.4**	29.0	–
Exams 184	Average	25.2	30.78	–	31.3	–
Sessions 10	Time (s)	8.6	9.1	5.0	–	–
EAR-F-83	Best	35.1	36.4	**29.3**	45.7	–
Exams 190	Average	35.4	40.92	–	46.7	–
Sessions 24	Time (s)	26	24.7	29.3	–	–
TRE-S-92	Best	**8.4**	9.6	9.4	10.0	–
Exams 261	Average	8.6	10.78	–	10.5	–
Sessions 18	Time (s)	39	107.4	102.8	–	–
LSE-F-91	Best	10.5	10.5	**9.6**	15.5	–
Exams 381	Average	11.0	12.36	–	15.9	–
Sessions 18	Time (s)	35	48.0	92.8	–	–
KFU-S-93	Best	**13.5**	14.0	13.8	18.0	–
Exams 461	Average	14.0	18.76	–	19.5	–
Sessions 20	Time (s)	40	120.2	112.8	–	–
RYE-S-93	Best	8.4	7.3	**6.8**	–	–
Exams 486	Average	8.7	8.68	–	–	–
Sessions 23	Time (s)	70	507.2	89.4	–	–
CAR-F-92	Best	**4.3**	6.2	6.0	5.2	–
Exams 543	Average	4.4	7.04	–	5.6	4.7
Sessions 32	Time (s)	171	47	142.7	–	–
UTA-S-92	Best	**3.5**	**3.5**	**3.5**	4.2	–
Exams 622	Average	3.6	4.8	–	4.5	4.0
Sessions 35	Time (s)	233	664.3	589.4	–	–
CAR-S-91	Best	**5.1**	7.1	6.6	6.2	–
Exams 682	Average	5.2	8.38	–	6.5	–
Sessions 35	Time (s)	296	20.7	34.7	–	–

on the UTA data set, with Di Gaspero and Schaerf [13] superior on the other two data sets.

The hybrid method described in this paper was run on 12 of the data sets, and the results are summarized in Table 4 (compared to results from Table 5 in Carter et al. [10], Table 3 in Caramia et al. [7], Table 1 in Di Gaspero and Schaerf [13] and Table 9 in White and Xie [19]). Note that for all these data sets, the hybrid algorithm produced a clash-free timetable within the maximum allowed number of sessions, without recourse to the fourth greedy heuristic stage. The hybrid method is superior to that of Di Gaspero and Schaerf [13] and to that of White and Xie [19], and better than Carter et al. on nine of 12 data sets (with two ties). However, the method of Caramia et al. [7] produces the best results with superior results to the hybrid method in six of the data sets (again with a tie on the UTA-S-92 data set).

5.4   Problem P3: Capacitated Benchmarks (Set 1)

Burke et al. [4] created a new class of capacitated problem for the publicly available data sets, with three sessions per weekday and a Saturday morning session. It was assumed that the exam period starts on a Monday. They set a maximum number of exam sessions and imposed Clashing and Total Capacity constraints. For the Nottingham data sets (NOTT), an Exam Availability constraint was also applied: exams over two hours in length had to be held in the first session of the day. The objective was to minimize the number of instances of a student having two exams in a row on the same day. For this problem class, our hybrid algorithm uses an identical objective score, calculated as follows:

initialize $o(x) := \sum_{\{i \in E \ : \ x_i \in T^{\cdot}\}} Us_i$
for each exam $i \in E$ with $x_i \in T$ and x_i not the last session of the day **do**
  for each exam $j \in E$ with $x_j = x_i + 1$ and $D_{ij} > 0$ **do**
    $o(x) := o(x) + D_{ij}$
  endfor
endfor

Di Gaspero and Schaerf [13] and Caramia et al. [7] applied their methods to this problem class. The results of Burke et al. [4] were bested by either Di Gaspero and Schaerf [13] or Caramia et al. [7] in every case, each with approximately half the best results on the data sets examined. Our hybrid method, with the objective score calculated as above, was run five different times for each data set. Note that for all these data sets, the hybrid method produced a clash-free timetable within the maximum allowed number of sessions, without recourse to the fourth greedy heuristic stage. The results can be seen in Table 5 (compared to results in Tables 1 and 5 in Burke et al. [4], Table 4 in Caramia et al. [7] and Table 2 in Di Gaspero and Schaerf [13]).

Clearly, the hybrid method is the superior method when applied to these capacitated versions of the data sets, providing the best solution in all instances. The method consistently provides lower scores than Di Gaspero and Schaerf [13],

Table 5. *Problem P3: capacitated benchmarks, set 1.* The objective to be minimized is the number of students with two consecutive exams on the same day. Times reported are average times per run for the hybrid method, times of the best run for Caramia et al. [7] and approximate run times for Burke et al. [4]. Reported times have not been converted to account for different computing resources. Data sets are listed in order of number of exams to give an indication of order of difficulty

Data Set			Hybrid	Burke et al.	Caramia et al.	Di Gaspero and Schaerf
TRE-S-92						
Exams	261	Best	**0**	3	2	4
Sessions	35	Average	0.4	–	–	5
Capacity	655	Time (s)	16	10800	222.4	–
KFU-S-93						
Exams	461	Best	**247**	974	912	512
Sessions	20	Average	282.8	–	–	597
Capacity	1995	Time (s)	45	24240	118.2	–
CAR-F-92						
Exams	543	Best	**158**	331	268	424
Sessions	31	Average	212.8	–	–	443
Capacity	2000	Time (s)	96	24240	80.4	–
UTA-S-92						
Exams	622	Best	**334**	772	680	554
Sessions	38	Average	393.4	–	–	625
Capacity	2800	Time (s)	173	24000	265.1	–
CAR-S-91						
Exams	682	Best	**31**	81	74	88
Sessions	51	Average	47	–	–	98
Capacity	1550	Time (s)	125	21120	31.4	–
NOTT						
Exams	800	Best	**2**	53	44	11
Sessions	26	Average	15.6	–	–	13
Capacity	1550	Time (s)	44	24240	359.1	–
NOTT						
Exams	800	Best	**88**	269	–	123
Sessions	23	Average	104.8	–	–	134
Capacity	1550	Time (s)	42	18000	–	–

which we believe is due to the choice of Kempe chain neighbourhood. The results provided by Burke et al. [4] are no longer competitive. However, it is important to note that on the Nottingham data sets, Burke et al. [4] set a Capacity constraint and allocated rooms to the exams. The hybrid method did not do this, whilst Di Gaspero and Schaerf [13] and Caramia et al. [7] do not state whether they do this or not. The extra constraints applied by Burke et al. [4] may be part of the reason for their relatively poor results on these data sets.

The major difference between the quality of the results produced in the un-capacitated version and the capacitated versions of these problems is between our hybrid method and the algorithm of Caramia et al. [7]. In the uncapacitated version, the algorithm of Caramia et al. [7] is superior in seven of the data sets, and the hybrid method in four, but with the capacitated data sets, the hybrid method is superior in all six instances (Caramia et al. [7] do not attempt the Nottingham data set with 23 sessions). Of these six instances, three were data sets that the hybrid method was superior for the uncapacitated version and two had almost equal results (the Nottingham data set was not attempted as uncapacitated). Therefore, one would expect that the hybrid method would fare better with this comparison. However, the improvement cannot be explained by simply favourable data sets alone, as there is a significant difference in this problem for data sets that were comparable for problem P2.

5.5   Problem P4: Capacitated Benchmarks (Set 2)

Burke and Newall [3] build on previous work undertaken in Burke et al. [4], using a modified version of their memetic algorithm. They looked at some of the publicly available data sets, and optimized these with a different objective function. They consider a situation with three exam sessions per day on weekdays (morning, lunchtime, afternoon) and one on Saturday morning. The objective function considers only students with two exams in two consecutive sessions. They give a penalty of three per student for two exams in a row in the same day, and one per student for two exams in a row overnight. For this problem class, our hybrid algorithm uses an identical objective score, calculated as follows:

initialize $o(x) := \sum_{\{i \in E \ : \ x_i \in T^*\}} Us_i$
for each exam $i \in E$ with $x_i \in T$ and x_i not on a Saturday **do**
  for each exam $j \in E$ with $x_j = x_i + 1$ and $D_{ij} > 0$ **do**
    if x_i is the last session of the weekday **then**
      $o(x) := o(x) + D_{ij}$
    else (x_i must be a morning or lunchtime session of a weekday)
      $o(x) := o(x) + 3D_{ij}$
    endif
  endfor
endfor

The Nottingham data set has the Exam Availability constraint that any exam over 2 hours in length must be held in a morning session.

Di Gaspero and Schaerf [13] compared their algorithm to the results produced by Burke and Newall [3]. Burke et al. [6], independent of their work on memetic algorithms, wrote another paper on sequential construction heuristics using the publicly available data. This method was not compared directly to any other work, though, as it was run on the CAR-F-92 data set with the same problem definition, it can be compared for this problem instance. Unfortunately, the other data this paper used were the KFU-S-93 set with a different number of sessions, and the UTA-S-92 data set, which have not been tested by other authors.

The hybrid method was run on the data sets used in Burke and Newall [3]. The results, together with the results of Burke and Newall [3], Burke et al. [6]

Table 6. *Problem P4: capacitated benchmarks, set 2.* The objective function is based on students with exams in consecutive sessions. Times reported are average times per run. Reported times have not been converted to account for different computing resources. Data sets are listed in order of number of exams to indicate order of difficulty

Data Set			Hybrid	Burke and Newall	Burke et al.	Carter et al.	Di Gaspero and Schaerf
KFU-S-93							
Exams	461	Best	**1337**	1388	–	2700	1733
Sessions	21	Average	1487.8	1608	–	–	1845
Capacity	1995	Time (s)	39	105	–	–	–
CAR-F-92							
Exams	543	Best	2188	**1665**	2555	2915	3048
Sessions	36	Average	2267.6	1765	–	–	3377
Capacity	2000	Time (s)	106	186	–	–	–
NOTT							
Exams	800	Best	731	**498**	–	918	751
Sessions	23	Average	841.2	544	–	–	820
Capacity	1550	Time (s)	44	467	–	–	–

(for the CAR-F-92 problem only), Carter et al. [10] and Di Gaspero and Schaerf [13], can be seen in Table 6 (using data from Tables 2–4 in Burke and Newall [3] and Table 3 in Di Gaspero and Schaerf [13]).

Clearly, the algorithm of Burke and Newall [3] is superior: for two of the data sets their results are clearly the best, and for the third they are only just behind our hybrid algorithm. It should be noted that the values for the memetic algorithm for Burke and Newall [3] in this table have been chosen as the best from five different runs with 18 different parameter settings. This is a total of 90 different runs for each data set, compared to five for the hybrid method. However, if the hybrid method is run many times, the scores do not improve significantly, as the method is tied to the solution produced deterministically by the constraint programming stage, which does not change with multiple runs. Instead, the hybrid method requires more simulated annealing and hill climbing iterations, to allow it to move further away from the initial solution, to improve the solution. To test this theory a series of longer runs were undertaken on these data sets. In addition to the standard run ($\alpha = 0.999$, $a = 10$: approximately 350 000 simulated annealing iterations and 10 hill climbing iterations), a medium-length run ($\alpha = 0.999$, $a = 100$: approximately 3 500 000 simulated annealing iterations and 30 hill climbing iterations) and a long run ($\alpha = 0.999$, $a = 1000$: approximately 35 000 000 simulated annealing iterations and 100 hill climbing iterations) were performed on the data sets. The results can be seen in Table 7.

It is quite clear that the longer the hybrid algorithm is allowed to run, the better the quality of the solution. However, with the CAR-F-92 data set, even though the hybrid algorithm is allowed to run about 50 times longer than the memetic algorithm of Burke and Newall [3], the solution remains slightly inferior

Table 7. *Problem P4: longer hybrid runs.* Our experiments and those of Burke and Newall [3] are on similar but not identical machines

Data set			Hybrid standard	Hybrid medium	Hybrid long	Burke and Newall
KFU-S-93						
Exams	461	Best	1337	1182	**1082**	1388
Sessions	21	Average	1487.8	1255.6	1214.4	1608
Capacity	1995	Time (s)	39	348	3202	105
CAR-F-92						
Exams	543	Best	2188	1809	1744	**1665**
Sessions	36	Average	2267.6	1970.6	1801.4	1765
Capacity	2000	Time (s)	106	828	6671	186
NOTT						
Exams	800	Best	731	481	**401**	498
Sessions	23	Average	841.2	558.8	431.6	544
Capacity	1550	Time (s)	44	317	2818	467

(less than 5% worse). While it looks plausible that if the hybrid algorithm was allowed to perform an even longer run it would provide a superior solution, it is clearly an inferior method for this data set. However, for the other two data sets, the hybrid method does provide superior solutions in less time (a short run for KFU-S-93, and a medium run for NOTT). Unfortunately, with only three different comparisons between the methods, it is difficult to conclude definitively which method is superior.

6 Conclusions

The hybrid method for examination timetabling described in this paper is superior to the method currently used by the University of Melbourne, and performs well in comparison to other, well known methods that have been applied on the publicly available data sets. However, too few of these data sets have had benchmarks established on them (problem P4 has only been developed for five of the 14 data sets), and only three comparisons have been made between the hybrid method and the Burke and Newall [3] memetic algorithm with sequential construction. It is therefore not possible to make a definitive assessment of the relative quality of solutions found by the hybrid method and by existing methodologies. In addition, it is desirable to perform comparisons of the algorithms using the same computer resources.

In spite of the shortcomings of the comparisons, the hybrid method is still proven as a worthwhile algorithm, among the best currently in use for examination timetabling. The constraint programming stage provides a fast route to a first feasible solution. This solution is improved by the simulated annealing stage, with the Kempe chain neighbourhoods proving effective at diversifying solutions (though occasionally more time is needed). The hill climbing stage further improves the solutions, and reduces the effect of unfavourable fluctuations in the simulated annealing stage.

We suggest that the dominant methods of the future for the examination timetabling problem will combine solution construction with local search. The stages of the hybrid method may be integrated more fully, to produce a still more powerful algorithm.

Acknowledgements. The authors would like to thank Aiden Tran and Gerry Barretto for providing the University of Melbourne data, and Michael Carter, Luca Di Gaspero and James Newall for help they gave in understanding their previous work.

References

1. Boizumault, P., Delon, Y., Peridy, L.: Constraint Logic Programming for Examination Timetabling. J. Logic Program. **26** (1996) 217–233
2. Burke, E.K., Jackson, K., Kingston, J., Weare, R.F.: Automated University Timetabling: the State of the Art. Comput. J. **40** (1997) 565–571
3. Burke, E.K., Newall, J.: A Multistage Evolutionary Algorithm for the Timetable Problem. IEEE Trans. Evolut. Comput. **3** (1999) 63–74
4. Burke, E.K., Newall, J., Weare, R.F.: A Memetic Algorithm for University Exam Timetabling. In: Burke, E, Ross, P. (Eds.): Practice and Theory of Automated Timetabling I (PATAT 1995, Edinburgh, Aug/Sept, selected papers). Lecture Notes in Computer Science, Vol. 1153. Springer-Verlag, Berlin Heidelberg New York (1996) 241–250
5. Burke, E.K., Newall, J., Weare, R.F.: Initialisation Strategies and Diversity in Evolutionary Timetabling. Evolut. Comput. J. (Special Issue on Scheduling) **6** (1998) 81–103
6. Burke, E.K., Newall, J., Weare, R.F.: A Simple Heuristically Guided Search for the Timetable Problem. Proc. Int. ICSC Symp. Eng. Intell. Syst. (1998) 574–579
7. Caramia, M., Dell'Olmo, P., Italiano, G.F.: New Algorithms for Examination Timetabling. In: Näher, S., Wagner, D. (Eds.): Algorithm Engineering 4th Int. Workshop, Proc. WAE 2000 (Saarbrücken, Germany, September) Lecture Notes in Computer Science, Vol. 1982. Springer-Verlag, Berlin Heidelberg New York (2001) 230–241
8. Carter, M., Laporte, G.: Recent Developments in Practical Examination Timetabling. In: Burke, E, Ross, P. (Eds.): Practice and Theory of Automated Timetabling I (PATAT 1995, Edinburgh, Aug/Sept, selected papers). Lecture Notes in Computer Science, Vol. 1153. Springer-Verlag, Berlin Heidelberg New York (1996) 373–383
9. Carter, M., Laporte, G., Chinneck, J.: A General Examination Scheduling System. Interfaces **24** (1994) 109–120
10. Carter, M., Laporte, G., Lee, S.T.: Examination Timetabling: Algorithmic Strategies and Applications. J. Oper. Res. Soc. **47** (1996) 373–383
11. Dowsland, K.: Using Simulated Annealing for Efficient Allocation of Students to Practical Classes. In: Vidal, R.V.V. (Ed.): Applied Simulated Annealing. Lecture Notes in Economics and Mathematical Systems, Vol. 396. Springer-Verlag, Berlin Heidelberg New York (1993) 125–150

12. Dowsland, K.: Off-the-Peg or Made-to-Measure? Timetabling and Scheduling with SA and TS. In: Burke, E, Carter, M. (Eds.): Practice and Theory of Automated Timetabling II (PATAT 1997, Toronto, Canada, August, selected papers). Lecture Notes in Computer Science, Vol. 1408. Springer-Verlag, Berlin Heidelberg New York (1998) 37–52

13. Di Gaspero, L., Schaerf, A.: Tabu Search Techniques for Examination Timetabling. In: Burke, E, Erben W. (Eds.): Practice and Theory of Automated Timetabling III (PATAT 2000, Konstanz, Germany, August, selected papers). Lecture Notes in Computer Science, Vol. 2079. Springer-Verlag, Berlin Heidelberg New York (2001) 104–117

14. Van Hentenryck, P.: The OPL Optimization Programming Language. MIT Press, Cambridge, MA (1999)

15. Schaerf, A.: Tabu Search Techniques for Large High-School Timetabling Problems. Proc. Natl Conf. Am. Assoc. Artif. Intell. AAAI Press, Menlo Park, CA (1996) 363–368

16. Thompson, J., Dowsland, K.: General Cooling Schedules for a Simulated Annealing Timetabling System. In: Burke, E, Ross, P. (Eds.): Practice and Theory of Automated Timetabling I (PATAT 1995, Edinburgh, Aug/Sept, selected papers). Lecture Notes in Computer Science, Vol. 1153. Springer-Verlag, Berlin Heidelberg New York (1996) 345–363

17. Thompson, J., Dowsland, K.: Variants of Simulated Annealing for the Examination Timetabling Problem. Ann. Oper. Res. **63** (1996) 105–128

18. Thompson, J., Dowsland, K.: A Robust Simulated Annealing Based Examination Timetabling System. Comput. Oper. Res. **25** (1998) 637–648

19. White, G.M., Xie, B.S.: Examination Timetables and Tabu Search with Longer-Term Memory. In: Burke, E, Erben W. (Eds.): Practice and Theory of Automated Timetabling III (PATAT 2000, Konstanz, Germany, August, selected papers). Lecture Notes in Computer Science, Vol. 2079. Springer-Verlag, Berlin Heidelberg New York (2001) 85–103

20. White, G.M., Zhang, J.: Generating Complete University Timetables by Combining Tabu Search with Constraint Logic. In: Burke, E, Carter, M. (Eds.): Practice and Theory of Automated Timetabling II (PATAT 1997, Toronto, Canada, August, selected papers). Lecture Notes in Computer Science, Vol. 1408. Springer-Verlag, Berlin Heidelberg New York (1998) 187–198.

21. Yoshikawa, M., Kaneko, K., Nomura, Y., Watanabe, M.: A Constraint-Based Approach to High-School Timetabling Problems: a Case Study. Proc. Natl Conf. Am. Assoc. Artif. Intell. AAAI Press, Menlo Park, CA (1994) 1111–1116

GRASPing the Examination Scheduling Problem

Stephen Casey and Jonathan Thompson

School of Mathematics
Cardiff University, UK
ThompsonJM1@cardiff.ac.uk

Abstract. This paper presents a Greedy Randomised Adaptive Search Procedure for solving the examination scheduling problem. GRASP is a two-phased multi-start or iterative method consisting of a construction phase and an improvement phase. Each iteration builds a feasible solution using a probabilistic selection procedure, and then optimises the solution using a local search technique.

1 Introduction

GRASP has proved successful on a variety of problems including the maximum independent set problem [11], p-hub location problems [14] and scheduling problems including airline flight scheduling [10] and parallel machine scheduling [16]. GRASP has also been successfully applied to the graph colouring problem [17], which is frequently used as a model for examination scheduling. Further encouragement is provided by the success of greedy construction heuristics in the solution of examination time-tabling problems. Carter et al. [6] use a system based on greedy construction with some backtracking as the basis for a robust examination scheduling package, while Corne et al. [8] use a genetic algorithm to control the order of examinations into a greedy heuristic.

This paper demonstrates that GRASP can successfully solve the examination scheduling problem. Various modifications and variations, based on features of the underlying problem, are evaluated experimentally. The results are used to identify an efficient GRASP implementation, which is compared to previous works on standard datasets in the public domain. The rest of the paper is organised as follows. Section 2 describes the background problem and includes a brief review of existing solution methods. The following section outlines GRASP and modifications that have been proposed elsewhere. Section 4 presents our solution algorithm and the following section describes a set of experiments designed to optimise our implementation, including a comparison with existing solution methods. Section 6 concludes.

2 The Examination Scheduling Problem

Given a set of exams E and a fixed number of timeslots P, the objective of the examination scheduling problem is to allocate each exam e to a timeslot p so that

E. Burke and P. De Causmaecker (Eds.): PATAT 2002, LNCS 2740, pp. 232–244, 2003.

various hard constraints are satisfied and various soft constraints are satisfied where possible. Typical examples of hard constraints are

1. No student to be scheduled to take two exams simultaneously.
2. A restriction on the number of seats available.
3. Time-windows. These define a subset of the timeslots during which a given exam cannot be scheduled.
4. Pre-assignments. A given exam has to be assigned to a certain slot.

A timetable that satisfies all hard constraints is said to be a feasible timetable. There are also soft constraints, which it is desirable to satisfy but not essential. These usually vary significantly from institute to institute, and include (but are not limited to)

1. Minimise the number of students sitting exams in consecutive timeslots.
2. Schedule larger exams early, to give more time for them to be marked.
3. Minimise the number of exams of different length that are simultaneously scheduled in the same room.
4. Ensure that no student has more than x exams in p consecutive time periods.

The importance of these objectives also varies between different institutions, making it difficult to identify one algorithm as being optimal across all datasets. Examination scheduling is a well-researched problem and surveys have been produced by Carter et al. [5] and Burke et al. [2]. Most relevant here are applications of meta-heuristics and algorithms which have been applied to the datasets in the public domain.

Laporte and Descroches [15] employ steepest descent to improve a greedily produced starting solution. Having assigned exams to non-clashing groups, Johnson [13] models the problem of assigning groups to timeslots so as to minimise second-order conflict as a TSP and solve it using simulated annealing. A two-phased approach was proposed by Dowsland and Thompson [20], who used simulated annealing to produce a feasible schedule, and then used the same method to minimise second-order conflict, while maintaining feasibility. A neighbourhood based on Kempe chains was shown to improve solution quality. A similar method was employed by Merlot et al. [18] who produced high-quality results by using a slower cooling schedule. Tabu search has been used by, amongst others, Di Gaspero and Schaerf [9] who employ a variable sized tabu list and dynamic neighbourhood selection. Genetic algorithms have been applied to examination scheduling by Corne et al. [8] and Burke et al. [3].

Other methods have been applied to standard test problems in the public domain including Carter et al. [7] who use graph colouring heuristics with backtracking, and Caramia et al. [4] who use a greedy construction heuristic and optimise after each exam allocation.

3   GRASP and Known Variants

The basic method of GRASP consists of two phases. The first is a probabilistic greedy algorithm that constructs a solution and the second optimises using some form of local descent. The former phase generally consists of ordering elements according to some criteria, and then probabilistically choosing from the top n (candidate list) one element to be assigned. This assignment is made according to some criteria, and the process continues. GRASP can be considered in pseudocode as

$S* = \{\}$ The best solution so far
$F(S*) = \inf$ $F(S*)$ is the cost of solution $S*$.
Construction phase – construct a feasible solution S'
Local descent phase – produces locally optimal solution S''
If $F(S'') < F(S*)$ then
    $S* = S('')$ Endif
For $i = 1, 2, \ldots,$ *niter niter* is the number of iterations.
Construction phase – construct a feasible solution S' which is
different to previous solutions in some way
    Local descent phase – produces locally optimal solution S''
    If $F(S'') < F(S*)$ then
       $S* = S('')$ Endif
Endfor
Return final solution $S*$

The success of GRASP has led to several improvements being proposed. Some are pertinent to our GRASP implementation, including

1. Memory function. It is logical to utilise information gathered in previous iterations in the construction of further solutions. One example is provided by Fleurent and Glover [12] who maintain a set of the best S solutions found so far, known as the elite. Subsequent construction phases attempt to create solutions which are similar in some way to the elite.

2. Reactive GRASP. Defining the optimal size of the candidate list can be difficult, in particular it will vary between applications and datasets. Prais and Ribeiro [19] determine a discrete set of values $n_1, n_2, \ldots, n_m$. During an initial warm-up period, a number of GRASP iterations are conducted in which a value of n is chosen randomly and the quality of the solutions recorded. Subsequently, A_i is calculated as the average result of the runs with n_i. Further runs select a value for n probabilistically according to the quality of these previous results.

3. Hybrids. GRASP has been used in conjunction with other meta-heuristics. Tabu search has been used to improve the local search element (Laguna and Gonzalez-Velarde [16]), and Ahuja et al. [1] used GRASP solutions as the initial population in a genetic algorithm implementation.

4   GRASP Implementation

We consider the problem of minimising the proximity of exams in students' timetables subject to the timetable being free of clashes. Further constraints are not considered here, though we will explain how other constraints can be incorporated into our GRASP implementation. Given the two-phase nature of GRASP, and the nature of the examination timetabling problem, it seems appropriate to use the first greedy phase to identify a feasible, clash-free timetable and the second descent phase to optimise the spacing of the exams while maintaining feasibility. Where possible, it may be preferable to make some effort to optimise the secondary objectives in the first phase as this could reduce the computational time required in the second phase. However, placing too much emphasis on the objective of minimising second-order conflict will increase the probability of producing infeasible timetables.

In the construction phase, the exams are ordered according to some criteria. The criteria considered were proposed by Carter et al. [7], namely

- Largest Enrolment (LE)
- Largest Degree (LD)
- Largest Weighted Degree (LWD)
- Saturation Degree (SD)
- Random Ordering (RO).

The degree of an exam is the number of clashing exams, weighted degree is the sum of the students taking the clashing exams and saturation degree is the number of clashing exams that have not yet been scheduled and which cannot be allocated to the current timeslot. Thus saturation degree is the only criterion which entails updating, and is more computationally expensive. Random ordering negates the need for a probabilistic choice.

At each construction phase, the exams are ordered according to the chosen criterion and an exam is chosen from the top n, where n is a parameter. The choice is not completely random, but is according to roulette wheel selection. Each exam is then allocated to the first available period. In our implementation, this means the first period that does not contain any clashing exams but this can easily be extended to the first period that satisfies all primary constraints. If no feasible allocation is possible, then a new procedure is entered, similar to the backtracking method used by Carter et al. [7]. This finds the period with the least number of clashing exams in, these are removed and given a higher weighting than any others to ensure that they will be re-scheduled next. The previously unassignable exam is scheduled to the period that is now conflict free. To ensure that this process is not repeated indefinitely, a tabu list is kept and a maximum cycle limit is also used to ensure that the construction phase will stop even if no feasible solution can be found.

To aid the second phase in reducing second-order conflict, phase one may allocate exams to the best rather than the first available period, where best is defined as the feasible period which minimises the increase in the secondary objectives. However, this may decrease the probability of finding a feasible solution. Once a feasible solution is found, the optimisation process is entered. After evaluating random descent and steepest descent, a compromise proved most successful, in which exams are ordered according to their contribution to the cost, and considered in turn. All periods are considered and the best move is chosen if it causes a decrease in cost. The next exam is then considered. This was continued until no improvements were found within a set cycle limit. The process then returns to phase 1 with a blank timetable, and continues for a set number of iterations. Clearly, this process can be extended to include other hard and soft constraints. In the descent phase, any moves that would force the solution into infeasibility are rejected and the quality of all feasible moves is measured according to the soft constraints.

Various improvements are considered, as outlined in Section 3. These focus on the concept of maximising the algorithm performance in a limited time. The improvements are as follows:

1. Improved Descent. Thompson and Dowsland [20] demonstrated that improved results are achieved with a neighbourhood based on Kempe chains. These stem from the graph colouring model of examination scheduling and are formed by considering the subgraph created from the vertices in two colour classes. The set of Kempe chains, which can be produced from this sub-graph, consists of the sets of linked vertices. Given that the two colour classes are conflict free, the vertices in the Kempe chain can swap colour classes without introducing any conflict into the solution. The original neighbourhood is a subset of the Kempe chain neighbourhood, as if vertex v can be coloured c without producing a clash, the Kempe chain will consist of a single vertex.

2. Reactive GRASP is implemented as follows. As it is felt desirable to limit the time used for our GRASP implementation, a warm-up period of evaluating different values of n is too expensive. As an alternative, a series of experiments may be conducted to identify the optimal value of n.

3. A limited form of simulated annealing is implemented to widen the search in phase two. This hybrid method is designed to ensure that a thorough investigation of the solution space is conducted within the local vicinity. However simulated annealing should not spend excessive time travelling across the solution space, as the iterations of GRASP will perform this function.

4. Memory Function. In contrast to the Kempe chain neighbourhood and limited simulated annealing which are designed to maximise the efficiency of the search, the purpose of the memory function is to guide the search to different, more profitable parts of the solution space. This is similar to the diversification function in tabu search, and the diversification functions used by Thompson and Dowsland [20] to widen their simulated annealing search. Three memory functions were evaluated. These are
   a) Place exams into new timeslots.
   b) Split pairs of exams which have previously been scheduled together.
   c) Prioritise exams according to their contribution to the cost function.
   The first two memory functions are intended to explicitly guide the search to new parts of the solution space and the third is intended to improve the solution quality. Placing exams into new periods may not cause a large change in cost, as cost is dependent on which sets of exams are in each group and which exams are in consecutive slots. Thus, splitting pairs of

Table 1. Test data characteristics

Name	University	Exams	Students	Timeslots	Density
CAR-F	Carleton, Canada	543	18 419	32	0.14
CAR-S	Carleton, Canada	682	16 925	35	0.13
EAR-F	Earl Haig, Canada	190	1 125	24	0.29
HEC-S	Ecole des Hautes Etudes Commerciales, Canada	80	2 823	18	0.20
KFU-S	King Fahd, Saudi Arabia	461	5 349	20	0.06
LSE-F	London School of Economics, UK	381	2 726	18	0.06
STA-F	St Andrews, Canada	139	611	13	0.14
TRE-S	Trent, Canada	261	4 360	23	0.18
UTE-S	Toronto, Canada	184	2 750	10	0.08
YOR-F	York Mills, Canada	181	941	21	0.27

exams which have previously been in the same period may have a greater probability of guiding the search to markedly different solutions.

The memory functions are used throughout the construction phase, and the extent to which they are used can be controlled, as even placing a single exam into a new timeslot means that a new part of the solution space has been reached. Experiments have shown that it is best not to use the memory function in the improvement phase, as this prevented the search from attaining high-quality solutions.

5 Results

As stated previously, there are standard examination timetabling datasets in the public domain. This section describes our preliminary experiments which are intended to optimise our GRASP implementation and then compares the results with existing methods. The publicly available datasets are obtainable from ftp://ftp.mie.utoronto.ca/pub/carter/testprob/

These are real-world timetabling problems, originating from different universities across the world. Table 1 shows the characteristics of the datasets used in our experiments. The density column refers to the number of edges in the underlying graph model as a percentage of the maximum number of edges possible.

The cost function used was proposed by Carter et al. [7] and is $\sum_{i=1}^{5} w_i$, where w_i is the penalty applied when a student has two exams within i periods. The penalties used are: $w_1 = 16$, $w_2 = 8$, $w_3 = 4$, $w_4 = 2$, $w_5 = 1$. There is no penalty for students with two exams within six or more periods. All experiments have used 10 GRASP iterations and have been run on a 1000MHz Pentium computer. Five random seeds were used.

The purpose of our experiments were first to determine a suitable value of n and the best ordering criterion for the greedy phase, and then to examine the

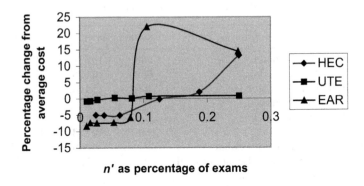

Fig. 1. Comparison of different candidate list sizes

improvements that could be obtained from a suitable choice of neighbourhood, a limited form of simulated annealing and a suitable memory-led diversification strategy. These will be dealt with in turn in the remainder of this section.

Our initial experiments were intended to identify the optimal strategy for determining the value of n, the candidate list size. It is likely that an absolute value of n will not be suitable for all datasets, given the wide differences in their size. However, it may be possible to determine a suitable value as a percentage of the total number of exams in each dataset (denoted as n'). Initial experiments compared the results for different percentages for three datasets. Figure 1 shows the results with the y-axis representing the percentage change from the mean result, with positive values being above average results.

Although there is not a similar trend for all datasets, a low value of n', in the range 0.01 to 0.1 appears to be optimal. Further experiments found a range of 0.02 to 0.07 gave the best results. We then implemented the idea of reactive GRASP, where the value of n is chosen probabilistically according to previous results. However, the run times required were excessive in comparison to existing solution methods. The results did confirm the findings above, and for all datasets here, values of n between 2 and 6 were deemed appropriate, depending on the size of the dataset.

Subsequently, we compared the efficiency of the five methods of ordering the exams in phase one. These were largest enrolment (LE), largest weighted degree (LWD), largest degree (LD), saturation degree (SD) and random ordering (RO). However, RO proved so unsuccessful that it was quickly discarded. In our experiments, we performed five runs for values of n from 2 to 6 inclusive, giving 25 runs in total. Figure 2 shows, for three datasets, the frequency of feasible solutions obtained in phase one.

Clearly, an ordering based on saturation degree is the most successful across all datasets and this confirms the findings of Carter et al. The drawback of this method is an increase in run time, but this is not significant in the context of the run time for the entire GRASP algorithm.

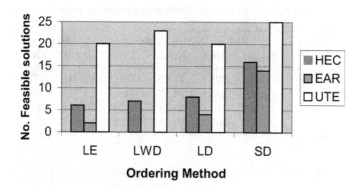

Fig. 2. Comparison of different ordering strategies

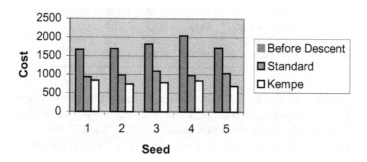

Fig. 3. Effect of different descent strategies

The following experiments compare a neighbourhood based on Kempe chains and a neighbourhood based on moving single exams (standard neighbourhood). Figure 3 compares the results of five descents with the two neighbourhoods from the same starting solution using the HEC dataset. The advantage of using the larger neighbourhood is shown and, despite the Kempe chain neighbourhood being computationally expensive, it has proved impossible for the standard neighbourhood to match the Kempe chain results, even when it is given additional run time.

We then wished to determine the merits of including some form of simulated annealing in the descent phase. Clearly it is nonsensical to use simulated annealing to at-tempt to search the entire solution space; the iterative nature of GRASP will do this. Rather it is envisaged that a limited form of simulated annealing will ensure that the local vicinity of the constructed solution will be thoroughly searched. Using the HEC dataset, a variety of simulated annealing parameter settings were evaluated. Initial experiments showed that higher starting temperatures with rapid cooling worked best. The mean results are shown in Table 2, with the temperature being multiplied by the cooling ratio every iteration. Table 2 also gives the computational time for each method, in seconds.

Table 2. Results of limited simulated annealing

Temp. cooling	1000 Result	Time	2000 Result	Time	5000 Result	Time
0.75	10.82	323.4	11.65	256.0	11.39	300.7
0.9	11.15	235.8	10.87	358.0	11.23	392.3
0.95	10.75	488.3	10.86	322.9	10.83	617.4
0.99	10.85	3050	10.58	2628	10.69	1858

Table 3. Results using different diversification strategies

Strategy	HEC Mean	SD	STA Mean	SD
None	11.2	0.22	152.5	0.014
Same period	11.2	0.16	152.6	0.017
Split pairs	11	0.05	152.5	0.010
Prioritise cost	11.3	0.24	152.6	0.077

The comparable results without simulated annealing were a mean value of 11.12 with a computational time of 188.5 s. It appears that cooling at a rate of 0.95 is not overly expensive, but produces significantly better results than straightforward descent. Cooling at a rate of 0.99 is too computationally expensive. As the purpose of examination scheduling is to find the single best solution, it seems worthwhile to use additional time improving the quality of the descent phase. A starting temperature of 1000 was selected and simulated annealing was employed in our GRASP algorithm.

The final preliminary experiments compared the different memory strategies in terms of their effect on the costs and their success in guiding the search to new parts of the solution space. Table 3 shows the mean and standard deviation of the cost function over five runs with the HEC and STA datasets.

The memory function based on splitting pairs of exams which have previously been in the same period proved the most successful on the HEC dataset and this result was confirmed on other datasets (results not shown here). The STA results are given to show that even when other strategies can match the mean cost function, splitting pairs of exams proves the most consistent in producing high-quality results. The methods were also compared in terms of their effect at guiding the search to different parts of the solution space. Given that each run produces 10 pre-descent solutions and 10 post-descent solutions, each set of solutions were compared and the average number of times that exams occur in the same period in each pair of solutions was com-piled. Figure 4 shows the average for each memory function using the HEC dataset and demonstrates the value of a diversification method (memory function) for locating new parts of the solution space. The x-axis labels are Pairs (splitting pairs of exams which have previously been together), None (no memory function), Examcost (order-

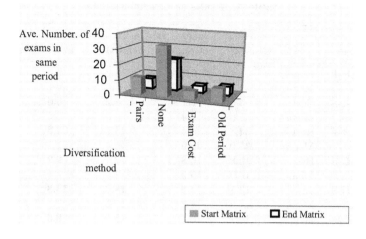

Fig. 4. Similarity of solutions using different diversification strategies

ing exams according to their contribution to the cost function) and OldPeriod (placing exams into new periods). The memory functions are reasonably similar in their results, and it cannot be concluded that one is more efficient at locating new parts of the solution space than any other.

Our experiments have shown that the most efficient GRASP implementation is one in which the candidate list size should be approximately 0.05 of the number of exams in the dataset, the vertices should be ordered according to saturation degree, the descent phase should use a Kempe chain neighbourhood, a limited form of simulated annealing should be employed and a memory function based on splitting pairs of exams which have previously been in the same period should be used. Table 4 compares the results of this implementation with previous results given by Carter et al. [7], Di Gaspero and Schaerf [9], Caramia et al. [4] and Merlot et al. [18]. The best and mean results are given for each method except for Caramia et al. who give no average figures. The mean result for GRASP is the mean of the best solutions in each of five runs.

Comparing the four methods across all datasets gives inconsistent results. GRASP compares well with the methods of Carter et al. and Di Gaspero and Schaerf on the majority of datasets and produces the best results of all on the STA-F-83 dataset. The algorithms of Caramia et al. and Merlot et al. narrowly outperform GRASP on the remaining datasets. It is clear that GRASP is a very consistent method, as the mean cost function is similar to the best cost function for most datasets. This is important because in practice, an examination timetabler can have confidence in the result of a single run.

It is difficult to compare timings for different algorithms, given the differences in computing platforms. The run times for the GRASP algorithm are reasonable and the number of iterations can be reduced if a smaller run time is required. The longest run time was 1513 s, which is practical, given that the examination scheduling problem is one which only needs to be solved a small number of times each year.

Table 4. Results on benchmark data

Name		GRASP	Carter et al.	Di Gaspero and Schaerf	Caramia et al.	Merlot et al.
CAR-F	Best	4.4	6.2	5.2	6.0	4.3
	Mean	4.7	7.0	5.6		4.4
CAR-S	Best	5.4	7.1	6.2	6.6	5.1
	Mean	5.6	8.4	6.5		5.2
EAR-F	Best	34.8	36.4	45.7	29.3	35.1
	Mean	35.0	40.9	46.7		35.4
HEC-S	Best	10.8	10.8	12.4	9.2	10.6
	Mean	10.9	15.0	12.6		10.7
KFU-S	Best	14.1	14.0	18.0	13.8	13.5
	Mean	14.3	18.8	19.5		14.0
LSE-F	Best	14.7	10.5	15.5	9.6	10.5
	Mean	15.0	12.4	15.9		11.0
STA-F	Best	134.9	161.5	160.8	158.2	157.3
	Mean	135.1	167.1	166.8		157.4
TRE-S	Best	8.7	9.6	10.0	9.4	8.4
	Mean	8.8	10.8	10.5		8.6
UTE-S	Best	25.4	25.8	29.0	24.4	25.1
	Mean	25.5	30.8	31.3		25.2
YOR-F	Best	37.5	41.7	41.0	36.2	37.4
	Mean	38.1	45.6	42.1		37.9

6   Conclusions

GRASP has proved to be a powerful solution method for the examination scheduling problem. An efficient construction phase is required to ensure that feasible timetables are produced. The descent phase must then maintain feasibility while optimising the secondary objectives. We have used a limited form of tabu search and efficient orderings of exams in order to maximise the probability of locating the feasible region. The descent phase has been enhanced with a neighbourhood based on Kempe chains and a limited form of simulated annealing. Memory functions have been used to guide the search to different parts of the solution space.

Given a limited run time, it is important to consider whether an optimal implementation of GRASP places the emphasis on performing more iterations, or on getting the most out of each of a limited number of iterations. Our results tend towards the latter, as it is advantageous to find a few deep local minima rather than many shallow minima. GRASP has the advantage of being simple to understand and apply, and performs robustly across datasets with different characteristics.

References

1. Ahura, R.K., Orlin, J.B., Tiwari, A.: A Greedy Genetic Algorithm for the Quadratic Assignment Problem. Technical Report. Sloan School of Management, Massachusetts Institute of Technology, Cambridge, MA (1997)
2. Burke, E.K., Elliman, D.G., Ford, P.H., Weare, R.F.: Examination Timetabling in British Universities – a Survey. In: Burke, E, Ross, P. (Eds.): Practice and Theory of Automated Timetabling I (PATAT 1995, Edinburgh, Aug/Sept, selected papers). Lecture Notes in Computer Science, Vol. 1153. Springer-Verlag, Berlin Heidelberg New York (1996) 76–90
3. Burke, E.K., Newall, J., Weare, R.F.: A Memetic Algorithm for University Examination Timetabling. In: Burke, E, Ross, P. (Eds.): Practice and Theory of Automated Timetabling I (PATAT 1995, Edinburgh, Aug/Sept, selected papers). Lecture Notes in Computer Science, Vol. 1153. Springer-Verlag, Berlin Heidelberg New York (1996) 241–250
4. Caramia, M., Dell'Olmo, P., Italiano, G.F.: New Algorithms for Examination Timetabling. In: Nher, S., Wagner D. (Eds.): Algorithm Engng 4th Int. Workshop (WAE 2000, Saarbrucken, Germany). Lecture Notes in Computer Science, Vol. 1982. Springer-Verlag, Berlin Heidelberg New York (2001) 230–241
5. Carter, M.W.: A Survey of Practical Applications of Examination Timetabling. Oper. Res. **34** (1986) 193–202
6. Carter, M.W., Laporte, G., Chinneck, J.W.: A General Examination Scheduling System, Interfaces **24** (1994), 109–120
7. Carter, M.W., Laporte, G., Lee, S.T.: Examination Timetabling: Algorithmic Strategies and Applications. J. Oper. Res. Soc. **47** (1996) 373–383
8. Corne, D., Ross, P., Fang, H.L.: Fast Practical Evolutionary Timetabling. In: Fogarty, T. (Ed.): Lecture Notes in Computer Science, Vol. 865. Springer-Verlag, Berlin Heidelberg New York (1994)
9. Di Gaspero, L., Schaerf, A.: Tabu Search Techniques for Examination Timetabling. In: Burke, E, Erben W. (Eds.): Practice and Theory of Automated Timetabling III (PATAT 2000, Konstanz, Germany, August, selected papers). Lecture Notes in Computer Science, Vol. 2079. Springer-Verlag, Berlin Heidelberg New York (2001) 104–117
10. Feo, T.A., Bard, J.: Flight Scheduling and Maintenance Base Planning. Manage. Sci. **35** (1989) 1415–1432
11. Feo, T.A., Resende, M.G.C., Smith, S.H.: A Greedy Randomized Adaptive Search Procedure for Maximum Independent Set. Oper. Res. **42** (1994) 860–878
12. Fleurent, C., Glover, F.: Improved Constructive Multistart Strategies for the Quadratic Assignment Problem Using Adaptive Memory. INFORMS J. Comput. **11** (1999) 198–204
13. Johnson, D.: Timetabling University Examinations. J. Oper. Res. Soc. **41** (1990) 39–47
14. Klincewicz, J.: Avoiding Local Optima in the p-hub Location Problem Using Tabu Search and GRASP. Technical Report. AT&T Laboratories, Holmdel, NJ (1989)
15. Laporte, G., Descroches, S.: Examination Timetabling by Computer, Comput. Oper. Res. **11** (1984) 351–360
16. Laguna, M., Gonzalez-Velarde, J.: A Search Heuristic for Just-in-Time Scheduling in Parallel Machines. J. Intell. Manufact. **2** (1991) 253–260
17. Laguna, M., Marti, R.: A GRASP for Coloring Sparse Graphs. Technical Report. Graduate School of Business, University of Colorado, Boulder, CO (1998)

18. Merlot, L.T.G., Boland, N., Hughes, B.D., Stuckey, P.J.: A Hybrid Algorithm for the Examination Timetabling problem. In: Proc. 4th Int. Conf. Pract. Theory Automat. Timetabling (2002) 348–371
19. Prais, M., Ribeiro, C.C.: Reactive GRASP: An Application to a Matrix Decomposition Problem in TDMA Traffic Assignment. Technical Report. Department of Computer Science, Catholic University of Rio de Janeiro, Brazil (1998)
20. Thompson, J.M., Dowsland, K.A.: Variants of Simulated Annealing for the Examination Timetabling Problem. Ann. Oper. Res. **63** (1996) 105–128

University Course and
School Timetabling

Search Strategy for Constraint-Based Class–Teacher Timetabling

Wojciech Legierski

Institute of Automatic Control, Silesian Technical University,
Akademicka 16, 44-100 Gliwice, Poland
wlegierski@ia.polsl.gliwice.pl
http://www.ia.polsl.gliwice.pl/~wlegiers

Abstract. The paper deals with a scheduling problem: the computation of class–teacher timetables. Two cases are taken into consideration: high school problems and university department problems. The timetable was constructed using constraint programming techniques. The timetabling needs to take into account a variety of complex constraints and use special-purpose search strategies. The concurrent constraint language Mozart/Oz was used, which provides high-level abstraction, and allows the expression of complex constraints and the creation of a complicated, custom-tailored distribution strategy. This strategy, consisting of six stages, was crucial for finding a feasible solution. The space-based search allows the incorporation of local search into constraint programming; this is very useful for timetable optimization. Technical details and results of the implementation are presented.

1 Introduction

A timetable construction is an NP-complete scheduling problem. It is not a standard job–shop problem because of the additional classroom allocation. It is large and highly constrained, but above all the problem differs greatly for different schools and educational institutions. It is difficult to write a universal program, suitable for all imaginable timetabling problems. Although manual construction of timetables is time-consuming, it is still widespread, because of the lack of appropriate computer programs.

In recent years two main approaches seem to have been successful. The first approach is based on local search procedures such as simulated annealing [2], tabu search [1] and genetic algorithms [9]. These methods express constraints as some cost functions, which are minimized by a heuristic search of better solutions in a neighbourhood of some initial feasible solution. Their greatest disadvantages are (1) the difficulty of taking into account hard constraints and (2) the need to determine their parameters through experimentation. Although they are good for optimizing the initial feasible solution, they have problems with finding it.

The second approach, presented in this paper, is based on constraint programming (CP) [3,4,6,7]. Its main advantage is declarativity: a straightforward

E. Burke and P. De Causmaecker (Eds.): PATAT 2002, LNCS 2740, pp. 247–261, 2003.

statement of the constraints serves as part of the program. This makes the program easy to modify, which is crucial in timetabling problems. The constraints are handled through a system of constraint propagation, which reduces domains of variables, coupled with backtracking search. In modern CP languages, both features do not need to be programmed explicitly. The main disadvantages of this approach are (1) difficulties with expressing soft constraints and (2) possible problems with improving the initial feasible solution, which – as a rule – may be determined without difficulties. An attempt to overcome the drawbacks with soft constraints was discussed by Rudová [11]. White and Zhang [15] successfully combined local search with constraint satisfaction to reduce their drawbacks. They determined an initial solution using constraint logic programming and then optimized it using tabu search.

The rationale of this paper is as follows:

- the ability to express complex constraints in a simple, declarative way is crucial for introducing the requirements of the high school and university timetabling problem into the program and is decisive for their successful solution,
- a custom-tailored distribution (labelling) strategy is able to introduce soft constraints during a search, leading quickly to a "good" timetable,
- incorporation of local search into CP gives the ability to optimize effectively the timetable.

Therefore it is desirable to have a programming language with the ability to formulate custom-tailored distribution and search strategies. Such features are not often found in CP languages, but they are included in the Mozart system, which is an implementation of the multiparadigm Oz language. This language has been already used by Henz and Würzt [4] for a class–teacher timetable with 91 courses, 34 teachers and 7 rooms. Rooms were used only as a constraint (the number of courses scheduled in one timeslot cannot exceed that limit). The system used standard techniques: a first-fail distribution strategy and a branch-and-bound method for optimization, which gives some improvement in this small problem. However, they did not take full advantage of the language's flexibility. The work and program described below was derived from their approach. But when a much more complicated timetable with additional classroom allocation is treated, then standard techniques are insufficient. The contribution of this paper consists in the development of special methods which are necessary for larger timetables.

The paper is organized as follows. First, the timetable problem is formulated (Section 2). Then the architecture of Mozart/Oz is briefly described (Section 3). After presenting constraints for the timetable (Section 4), a custom-tailored distribution strategy is developed (Section 5) as well as the method for finding a better solution (Section 6). Implementation results are presented in Section 7 and some conclusions are drawn in Section 8.

2   Class–Teacher Timetabling

Most school timetabling problems are formulated for classes of students that have a common learning program consisting of courses, and group of teachers assigned to courses. The problem consists in allocating start times and allocating rooms for all classes, courses and teachers without violating constraints. In Poland this problem is typical for most universities, where students attend courses common to the entire class: individual student preferences are not taken into account.

The presented program was designed to deal efficiently with two timetabling problems. The first problem is to find a weekly timetable for a high school running 253 courses for 10 classes, 43 teachers and 24 rooms. There are $8 \times 5 = 40$ possible timeslots for each course. Some courses are given for a half of a class (called under-class courses). There are no common courses attended by more than two classes at the same time. There are also no courses run in an odd–even weekly cycle. The main difficulties are as follows:

- the existing regulations do not allow high school students to have time gaps between courses,
- any of the rooms may be chosen for many courses,
- each class of students has to take 30 courses on average.

The second problem was a timetable for the Institute of Automation at the Silesian Technical University (STU) for 23 classes of students. The problem involves 223 courses, 21 of them being common for five classes, 54 teachers, 42 rooms and 40 timeslots. There are some courses run in an odd–even weekly cycle, but the courses are not divided into two under-classes. The main difficulties of this timetable are as follows:

- many courses must be placed in specific timeslots,
- there are some teachers with very constrained time available,
- for some courses (e.g. laboratories) participation of more than one teacher is necessary.

3   Space-Based Search in Mozart/Oz

Some information about Mozart/Oz [8] is necessary to understand the presented concepts. This is based mainly on the contributions by Schulte [13,14]. Mozart implements a concurrent constraint language Oz providing functional and object-oriented programming. It differs in structure from CLP languages based on Prolog (CHIP, SICStus Prolog, Eclips) or from CP C++ libraries (ILOG Solver). The constraint propagation is not imposed by arc-consistency algorithms, but by concurrent computational agents (called propagators). They make contributions to the "constraint store" that stores basic constraints of variables and their domains. Propagators connected to a constraint store form together a computation space. The search is done from computational spaces that are either copied or recomputed during the distribution process. The search leads to a search tree where each node in turn corresponds to a computation space. Mozart/Oz search efficiency is due to two major factors:

- the distribution strategy, which determines what variables are selected for instantiation and which value they are instantiated with.
- the search method, which determines how the search tree is explored.

The distribution strategy and the search method will be referred to as the search strategy.

Space-based search is competitive with trailing-based search implemented in other CP languages, see Schulte [12,14]. It gives a framework within which a custom-tailored distribution and search method can be easily implemented. This option is usually not available in other CP languages.

4   Constraints

The ability to express complex constraints is crucial for CP. Mozart/Oz does this in an efficient way. The timetables of the two case studies have to fulfil the following hard and soft constraints:

C1.   Basic constraints that can be introduced during problem formulation:
- courses have to be scheduled in certain timeslots,
- some teachers are sometimes unavailable,
- some courses must be held simultaneously.

C2.   Many-to-one constraints:
- there are common courses (lectures) for several classes,
- there are courses (e.g. laboratories) conducted by many teachers simultaneously.

C3.   Joint constraints for a number of courses and classes of teachers:
- some courses must be held on different days and often should have a special order during a week,
- some teachers do not want to teach on all days in a week and more than several hours during a day.

C4.   No-overlapping constraints and exceptions:
- a teacher can teach a group and a group can have one course at a time
- the courses of classes and teachers cannot overlap,
- two courses of a class can overlap if they are split into odd and even weeks or in under-classes,
- two courses of a teacher can overlap if they are split into odd and even weeks.

C5.   "One course in a room" constraints:
- in one room there must be only one course at a time.

C6.   Soft constraints that are not obligatory, but should be preserved to some extent, if permitted by other hard constraints:
- no-gaps-for-class constraint; gaps between courses for classes should be as few as possible (high school classes must not have gaps, this is treated as a soft constraint),
- no-gaps-for-teachers constraint; gaps between courses for teachers should be as few as possible (mainly for high schools),

- preferable-rooms constraint; courses should be scheduled first in preferable rooms,
- equal-number-of-courses constraint; the number of courses each day should be roughly equal,
- early-rise constraint; courses should start in the early morning hours.

4.1   Reified Constraints (C3)

Constraints C1 and C2 can be easily implemented, but C3 constraints relate to other courses so they are taken into account after problem formulation, as is defined below. The main difficulty is that they connect the start time of one course with start times of other courses that are unknown, in a complicated and indirect way. To cope with them reified constraints are necessary and they are provided by the Mozart/Oz language. Generally, reified constraints are used to reflect the fact that some constraint is fulfilled by attaching to it a proper value of some 0/1-value variable. Through this variable the instantiation of a course on a specific day can be detected and courses connected with the instantiated course can be constrained. It is crucial that reified constraints not only check the validity of constraints and transpose them into the 0/1-value variable, but that they impose either the validity or the negation of constraints if their 0/1-value variables are instantiated. They impose negation of the constraints that some courses must not be on specific days. This reduces the start domains of the remaining courses by removing from them all values not corresponding to the day on which the course has already been scheduled. For example, if a teacher wants to teach only for two days in a week and two of his courses have been already instantiated for two other days, then the remaining courses of this teacher are constrained to be scheduled only on these days.

4.2   No-Overlapping Constraints (C4)

Timetabling problems can be considered as job–shop problems, where resources are rooms, classes and teachers and tasks are courses run using these resources. The Mozart/Oz language has strong propagators to implement capacity constraints, such as `Schedule.serialize`. It employs edge-finding, an algorithm which takes into consideration all tasks using any resource [16]. This propagator is very effective for job–shop problems. However, in the studied cases this propagator is not suitable, because most tasks frequently have duration of one or two timeslots and the computational effort is too large in comparison with the results. The experiments show that using too complicated a propagator can double time and memory consumption. The constraints C4 are introduced with the help of a built-in propagator `FD.disjoint` which, although may cut holes into domains, must be applied for any two courses that cannot overlap. Those constraints enable the handling of some special situations such as different schedules for even and odd weeks or the existence of an under-class with partially different schedules. Results of experiments with these approaches are presented in Table 1.

Table 1. Comparison of two propagators

Timetable for	Parameters	Propagators	
		Serialize	Disjoint
High school	time (s)	73	36
(253 courses)	memory (MB)	723	337
University	time (s)	32	12.5
(223 courses)	memory (MB)	269	156

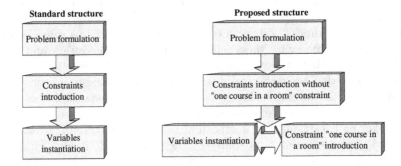

Fig. 1. The standard and proposed structure of program.

4.3 "One Course in a Room" Constraint (C5)

Room allocation was performed simultaneously with search for start times. The difficulty in timetabling is often due to the rather small number of rooms available; performing room allocation after scheduling courses would lead to many unfeasible solutions. In the literature this constraint is incorporated by two global constraints: alldifferent and the global cardinality constraint [10] (in Mozart/Oz terminology, distinct and reified sum constraint [4]). These approaches divide rooms into some categories (e.g. for lectures, for classes), and then constrain in each timeslot the number of courses that requires some category to the number of rooms of the given category. Unfortunately, this does not suit the case where most rooms have special features different from other rooms. Additional difficulties arise when courses have more than one timeslot duration.

Therefore the "one course in a room" constraint for the described problem is the most difficult constraint to incorporate efficiently. The essence of the difficulty is that it depends on two undetermined variables (start time and room) of any course and the corresponding variables of all the other courses. It can be written as rather complicated reified constraints, which check every room, in every timeslot, for every course. But this creates a lot of propagators that have to be checked during the entire search. It is computationally very expensive.

This problem is solved by a change of the standard CP program structure. The idea of the standard structure relies on constraint propagation, common in

Table 2. Comparison of standard and proposed approaches to the "one course in a room" constraint

Timetable for	Parameters	Approaches	
		Standard	Proposed
High school	time (s)	132	11
(145 courses)	memory (MB)	641	79
University	time (s)	106	17
(184 courses)	memory (MB)	504	110

all CP languages, whereby all constraints should be introduced before searching for solutions by variable instantiation. The idea of introducing this constraint only after some variables have already been instantiated (Figure 1.) proved to be fruitful for reducing the computational effort inherent in these constraints. Constraint C5 is introduced after instantiation of room and start time of a specific course and refers to all not-labelled courses. The program does not need to search at each step every possible room and start time to find an inconsistency. Put simply, after instantiating some course, it considers the corresponding room at the corresponding time as tentatively off-limits for all other courses. This approach (called constraining while labelling) gave very good results for the two case studies. Results of the experiments are presented in Table 2 for some smaller problems. All soft constraints were introduced by the distribution strategy and are described in the next section.

5 Distribution Strategy

The timetabling problem is not a standard scheduling problem, for which we can effectively use the standard distribution strategies. Rather than simplifying the problem to fit the standard, distribution strategies should be developed to solve the problem with all its complexities. This approach to the problem suggests a special distribution strategy. For timetabling, the distribution strategy tries to answers the question, Which courses should be scheduled first and in which timeslots?.

The main assumptions for a good distribution strategy are (1) reduction of backtracks (decreases the probability of constraint violation) and (2) searching for a "good" timetable right away (fulfilling as many soft constraints as possible). To put into effect these assumptions the distribution strategy was partitioned into six stages to instantiate courses in the proper order. An outline of the distribution strategy is shown in Figure 2. At each stage a special heuristic is implemented to choose a suitable course and start time for this course. Because this heuristic is similar for each stage it is described in detail when discussing the fifth stage, where most courses are scheduled. The program in the first stage labels courses with already assigned start times and looks for a proper room from the list of available rooms. The second stage is based on

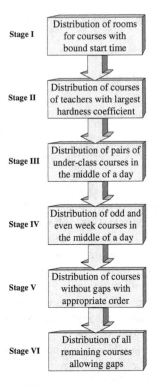

Fig. 2. The main structure of the distribution strategy

a hardness coefficient defined for each teacher giving courses. It is defined as follows: $hardness\_coefficient = \frac{duration\_of\_all\_courses\_run\_by\_teacher}{availability\_times\_of\_teacher}$.

The coefficient range is $(0, 1)$ and in the second stage only courses with coefficient greater then 0.8 are labelled. This stage brings particular improvement in the university case, because there are a number of teachers from other departments, with a heavily constrained set of free timeslots at their disposal.

Stages III and IV are very similar. The first refers to high school timetabling, where under-classes appear, and the second refers to university timetabling, where even and odd weeks are distinguished. The idea of that stage consists of connecting courses in pairs for each class and labelling two courses in the same timeslot. This timeslot is chosen in the middle of the day, because these kind of courses create gaps if they are between standard courses and are not paired with some other courses. The remaining courses, which cannot be connected in pairs, are left to be scheduled either at the beginning or at the end of a day.

The fifth stage is the most important and will be described in detail. Its outline, as presented in Figure 3, is partitioned into six steps. In the first step an appropriate course has to be chosen. Most often the *first-fail* heuristic is used, where a variable with the smallest domain is selected. A much more complicated heuristic was introduced to fulfil the distribution assumptions. It is based on a special cost function, which depends on

- domain of a start time (as in *first-fail*),
- duration of course,
- hardness coefficient for teacher,
- number of ways to schedule course without gaps,
- type of course; the cost function is smaller if the course is common for a few classes; also, standard courses are more preferable than those run for under-classes or even, odd weeks, with the exception of the first hour in a day, where it is preferable to schedule those courses.

In general this function is rather complex and its parameters (weights of each element) are determined experimentally. At the second step a proper value of start time for the course is chosen. This heuristic is based on two substeps:

- choose the start time in a less loaded day,
- choose the start time avoiding gaps for classes.

The first substep has already been used in experiments presented by Guéret et al. [3]. A timetable in which all days are rather equally loaded is desirable. At this substep a "week load" list is introduced, being a list of all days of the week with number of courses already placed in each day. The initial assignment of these lists can be adjusted to specific preferences: for example, courses should not be on Friday. At the second substep an attempt is made to fulfil the main soft constraint about no gaps for courses attended by the same class. The chosen value of the start time can only be before or after courses already placed in a specific day (if no courses are scheduled, in the middle of the day). It is also possible that there are gaps between courses resulting from previous stages. In that case a course is preferably scheduled in gaps. It may be justified as follows:

- this constraint can be written explicitly (e.g. it is required for a high school timetable), but as with the "one course in a room" constraint, it is very computationally demanding.
- a violation of this constraint can be detected at the end of the search, when all courses for a class are scheduled.

These drawbacks suggest introducing the constraint during search and at once schedule courses without gaps.

At the third step the first value from the list of rooms assigned to each course is chosen. The room list determines in which room the course can take place and in what order they should be assigned due to preferences.

At steps 4, 5 and 6 distribution is performed. The Mozart/Oz language allows programmers to design the most suitable distribution, the way that choice points are created, their order and number. Choice points can be seen as alternatives which can be chosen during a search. Every choice point when chosen becomes a new computational space and a new node in the search tree. At the fourth step the first choice point is created by assigning for two variables (start time

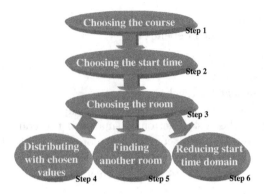

Fig. 3. Diagram of fifth stage.

and room) describing the course's two chosen values. Then constraint C5 ("one course in a room") is introduced. If the partial schedule is feasible, the program tries to label the next course. If this partial schedule is not feasible, at the fifth step the second choice point is checked: another value from the room list is chosen for the room variable. As many choice points as there are possible rooms may be checked. If all of them fail, the program goes to the sixth step, where it removes from the start time domain the chosen value and backtracks to the beginning.

At the sixth (and final) stage of the distribution strategy the no-gaps constraint is relaxed. All remaining courses are labelled in their entire remaining domains, and steps 1 and 2 from the fifth stage are transformed to a first-fail strategy by choosing a value in a proper day. Generally, at this stage only a few courses have to be scheduled.

This distribution strategy was crucial for finding timetables in reasonable time. All of the stages and steps are important and contribute to the efficiency of the program.

6   Search Method and Optimization

There are several search methods for a single feasible solution, but for timetabling problems depth-first search (DFS) seems to be best. This obvious way to explore a search tree from left to right is justified by the distribution strategy, which attempts to make the start with leftmost leaves most preferable. Limited discrepancy search (LDS) was tested, but it did not give a feasible solution, the reason being the systematicity of this method, which is similar to the ineffectiveness of branch-and-bound, described in Section 6.1.

All search methods in Mozart/Oz make recomputation available as an essential technique for search engines. Mozart/Oz language architecture is based on copying rather than trailing. Normally the search engine creates copies of choice nodes in exploration steps in order not to compute them in the case of

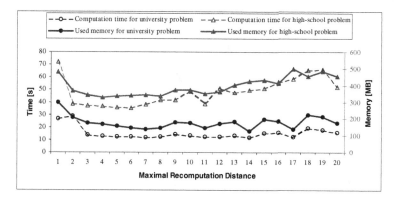

Fig. 4. Computation time and memory consumption due to maximal recomputation distance

backtracking. It is possible to reduce the number of copies, and thereby memory consumption, by using a recomputation. Instead of creating a copy in each distribution step, at every mth (called MRD, the maximal recomputation distance) exploration step a copy of the current node is memorized. Recomputation saves space, possibly at the expense of increased runtime, but can also save time, due to the optimistic attitude to the search. The relationship between recomputation distance, computation time and memory consumption is shown in Figure 4. It is seen that memory consumption and computation time cannot be easily interpreted, but there is already improvement of program efficiency for $m = 3$, and then with a small local minimum efficiency slowly decreases. More details of the advantages of this Mozart/Oz feature are presented in Schulte [12].

6.1   Optimization with BAB

Finding a feasible solution to a timetabling problem is often not sufficient. The timetable may need to satisfy requirements specific to an institution. Finding a "good" timetable right away was described in the previous section, but there is still a need for better solutions. The most popular method for finding a better solution, branch-and-bound (BAB) search, was checked. The chosen criteria depend mainly on the overall number of gaps for all classes, which is the most important soft constraint and on starting courses in the morning hours. This method performed very poorly, for three reasons. The distribution strategy as described in Section 5 already tries to minimize the gaps; therefore branch-and-bound has little chance to improve the already good situation. The second reason is that the main advantage of BAB – the pruning of the branches with poorest cost function – is not effective, because it can be fully computed after instantiation of all courses. Performing a complete branch-and-bound search is impossible because of the problem size and virtual memory restrictions. In any event, BAB is a systematic method. It maintains the ordering of variables from the distribution strategy and tries to change their values beginning from the

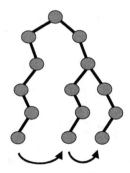

Fig. 5. Searching for the next solution

bottom of a search tree, and makes redundant the work of checking for different configuration of the last variables. For this reasons another optimization method was developed.

6.2   Local Search for Constraint Programming

Local search techniques are based on searching for a better solution in the neighbourhood of an already found one. For the timetabling problem this neighbourhood can be, for example, a timetable with one course placed differently or with two courses exchanged. It seems easy to make this kind of operation and compute a cost function, but when the problem is described using CP some restrictions are necessary. The constraint propagation needs to know explicitly the state of variables (either their values or domains) and the search must be done by instantiating variables one by one. From a timetable point of view it is possible to replace during the search only the latest scheduled course.

Several works have studied co-operation between local search and CP. Jussien and Lhomme [5] distinguish three categories of hybrid approaches that can be found in the literature:

- performing a local search before or after a CP approach with systematic search [15],
- performing a systematic search improved with a local search at some point of the search: at each leaf of the tree (i.e. over complete assignments) but also at nodes in the search tree (i.e. on partial assignments),
- performing an overall local search, and using systematic search and constraint propagation either to select a candidate neighbour or to prune the search space.

The first two approaches are techniques of switching between CP and local search at some point in the search. The third one tries to combine local search and CP more tightly. The presented approach falls into the first and third category. The main idea of the presented incorporation of local search into CP is based on memorizing the initial space, where all constraints are incorporated. Then, after finding the first solution only with pure CP, the search engine makes following steps:

Table 3. Results for the two case studies

	University			High-school		
	Time (s)	Mem. (MB)	Cost	Time (s)	Mem. (MB)	Cost
FF	19	258	7185.5	7.5	119	29 909
FF+BAB	no improvement			28	450	29 908
CTDS	29.5	340	429.0	10.5	177	18 657
CTDS+BAB	no improvement			no improvement		
CTDS+LS	36	340	180.5	17	177	17 158

1. Finds a course which makes gaps between courses (takes course schedule at the last, sixth stage).
2. Finds a second course to exchange with the first one.
3. Creates a new search of the original problem from memorized initial space.
4. Instantiates to the values from solution all courses besides the two previously chosen (this can be done in one step because they surely do not violate constraints).
5. Schedules the first course in place of the second one.
6. Finds the best start time for the second course.
7. Computes the cost function; if it is better, the solution is memorized, else another exchange is made.

The search engine stops after finding a given number of better solutions. The main idea of this searching is illustrated in Figure 5. The next solution is looked for not in neighbourhood branches, but in branches in different parts of the search tree. The crucial point to put this idea into effect was the ability to programm the search engine. It has to store a copy of the initial space that a computed solution is injected into. Finding a next solution does not need exploration of an entire branch because, as described in the 4th step of this search, most courses are instantiated very quickly.

7   Implementation Results

The two case studies, a high school and a university problem, were tested on a Pentium III/850 MHz, 256 MB station. The results with MRD $= 3$ are presented in Table 3.

As described in Section 6.1, minimized cost depends mainly on the overall number of gaps for all classes and on starting courses from the morning hours. The first-fail (FF) distribution strategy finds a feasible solution very fast, but the cost shows how undesirable this solution is. Using the BAB method for improving the solution has no effect. When the custom-tailored distributed strategy (CTDS) is used, the first solution is already very good one. If this solution is optimized with the proposed local search techniques, improvement is very effective. This connection of the distribution strategy and the special search for a better result allows one reduce to zero the number of gaps in the high school timetable and by 10% the cost function of the university timetable. Its higher cost can be explained by the characteristic of the university timetable, which has a large number of courses with fixed start time and teachers with hardness coefficient above 0.8. However, the number of gaps for one group averaged 2 hours a week, which is a good result.

The program written in Mozart/Oz language uses text files for input and output data. MS Excel and Visual Basic for Applications were used to create the user interface. MS Excel gives a simple way to save the data, make tables, print, etc. Visual Basic was convenient for generating the input text file and interpreting the output file.

8   Conclusions

The approach presented in the paper allows the generation of feasible solutions for real high school and university department timetabling data. Features of the concurrent CP Mozart/Oz language allow the formulation of complex timetabling constraints. Although various CP languages have been already used for timetabling (CHIP [7,3], Coastool/C++ [6], ILOG Solver [17] and others), none of them, to my knowledge, provides the functionality to go as deep into the search strategy as presented here. The strategy is custom-tailored specially for timetabling problems. Using the standard distribution strategy and search method seems to be insufficient; the custom-tailored search strategy for the timetabling problem yields a quick solution. The idea of incorporating local search into CP was presented. It has already given good results. Optimization in CP is numerically very demanding. The commonly used BAB is often too slow to produce results in a reasonable time for large problems. The development of new methods of finding a better solution is crucial and they should be adapted to the nature of real problems.

Acknowledgements. The author is grateful to Professor Antoni Niederliński for his support and comments on a draft of the paper.

References

1. Boufflet J.P., Negre S.: Three Methods Used to Solve an Examination Timetable Problem. In: Burke, E, Ross, P. (Eds.): Practice and Theory of Automated Timetabling I (PATAT 1995, Edinburgh, Aug/Sept, selected papers). Lecture Notes in Computer Science, Vol. 1153. Springer-Verlag, Berlin Heidelberg New York (1996) 327–344
2. Elmohamed S., Coddington P., Fox G.: A Comparison of Annealing Techniques for Academic Course Scheduling. In: Burke, E, Carter, M. (Eds.): Practice and Theory of Automated Timetabling II (PATAT 1997, Toronto, Canada, August, selected papers). Lecture Notes in Computer Science, Vol. 1408. Springer-Verlag, Berlin Heidelberg New York (1998) 92–112

3. Guéret C., Jussien N., Boizumault P., Prins C.: Building University Modular Timetabling Using Constraint Logic Programming. In: Burke, E, Ross, P. (Eds.): Practice and Theory of Automated Timetabling I (PATAT 1995, Edinburgh, Aug/Sept, selected papers). Lecture Notes in Computer Science, Vol. 1153. Springer-Verlag, Berlin Heidelberg New York (1996) 130–145

4. Henz M., Würtz J.: Using Oz for College Timetabling, In: Burke, E, Ross, P. (Eds.): Practice and Theory of Automated Timetabling I (PATAT 1995, Edinburgh, Aug/Sept, selected papers). Lecture Notes in Computer Science, Vol. 1153. Springer-Verlag, Berlin Heidelberg New York (1996) 162–177

5. Jussien N., Lhomme O.: Local Search with Constraint Propagation and Conflict-Based Heuristic, Artif. Intell. **139** (2002) 21–45

6. Kaneko K., Yoshikawa M., Nakakuki Y.: Improving a Heuristic Repair Method for Large-Scale School Timetabling Problems. In: Principles and Practice of Constraint Programming (Proc. CP'99). Springer-Verlag, Berlin Heidelberg New York (2000) 275–288

7. Lajos G.: Complete University Modular Timetabling Using Constraint Logic, Practice and Theory of Automated Timetabling. In: Burke, E, Ross, P. (Eds.): Practice and Theory of Automated Timetabling I (PATAT 1995, Edinburgh, Aug/Sept, selected papers). Lecture Notes in Computer Science, Vol. 1153. Springer-Verlag, Berlin Heidelberg New York (1996) 146–161

8. Mozart Consortium. The Mozart Programming System. Documentation and system available at http://www.mozart-oz.org

9. Rich D.C.: A Smart Genetic Algorithm for University Timetabling. In: Burke, E, Ross, P. (Eds.): Practice and Theory of Automated Timetabling I (PATAT 1995, Edinburgh, Aug/Sept, selected papers). Lecture Notes in Computer Science, Vol. 1153. Springer-Verlag, Berlin Heidelberg New York (1996) 181–197

10. Régin J.C.: Generalized Arc Consistency for Global Cardinality Constraints. In: Proc. AAAI'96 (Portland, OR) (1996) 209–215

11. Rudová H.: Constraint Satisfaction with Preferences, Ph.D. Thesis, Brno (2001)

12. Schulte C.: Comparing Trailing and Copying for Constraint Programming. In: Proc. 16th Int. Conf. Logic Program. (1999)

13. Schulte C.: Programming Constraint Inference Engines. In: Proc. 3rd Int. Conf. on Principles and Practice of Constraint Programming (Schloss Hagenberg, Linz, Austria). Lecture Notes in Computer Science (1997)

14. Schulte C.: Programming Constraint Services. PhD Thesis, Saarbrócken (2000)

15. White G.M., Zhang J.: Generating Complete University Timetables by Combining Tabu Search and Constraint Logic. In: Burke, E, Carter, M. (Eds.): Practice and Theory of Automated Timetabling II (PATAT 1997, Toronto, Canada, August, selected papers). Lecture Notes in Computer Science, Vol. 1408. Springer-Verlag, Berlin Heidelberg New York (1998) 187–198

16. Würtz J.: Oz Scheduler: A Workbench for Scheduling Problems. In: Proc. 8th Int. Conf. on Tools with Artificial Intelligence (1996) 149–156

17. Zervoudakis K., Stamatopoulos P.: A Generic Object-Oriented Constraint-Based Model for University Course Timetabling. In: Burke, E, Erben W. (Eds.): Practice and Theory of Automated Timetabling III (PATAT 2000, Konstanz, Germany, August, selected papers). Lecture Notes in Computer Science, Vol. 2079. Springer-Verlag, Berlin Heidelberg New York (2001) 28–47

Multi-neighbourhood Local Search with Application to Course Timetabling

Luca Di Gaspero[1] and Andrea Schaerf[2]

[1] Dipartimento di Matematica e Informatica,
Università di Udine,
via delle Scienze 206, I-33100, Udine, Italy
`digasper@dimi.uniud.it`
[2] Dipartimento di Ingegneria Elettrica, Gestionale e Meccanica,
Università di Udine,
via delle Scienze 208, I-33100, Udine, Italy
`schaerf@uniud.it`

Abstract. A recent trend in local search concerns the exploitation of several different neighbourhood functions so as to increase the ability of the algorithm to navigate the search space.

In this paper we investigate the use of local search techniques based on various combinations of neighbourhood functions, and we apply this to a timetabling problem. In particular, we propose a set of generic operators that automatically compose neighbourhood functions, giving rise to more complex ones. In the exploration of large neighbourhoods, we rely on constraint techniques to prune the list of candidates. In this way, we are able to select the most effective search technique through a systematic analysis of all possible combinations built upon a set of basic, human-defined, neighbourhood functions.

The proposed ideas are applied to a practical problem, namely the Course Timetabling problem. Our algorithms are systematically tested and compared on real-world instances. The experimental analysis shows that neighbourhood composition leads to much better results than traditional local search techniques.

1   Introduction

Local search is a successful meta-heuristic paradigm for the solution of constraint satisfaction and optimization problems. Main local search strategies, such as hill climbing, simulated annealing and tabu search (see, e.g., [1]), have proved to be very effective in a large number of problems.

One of the most critical features of local search is the definition of the neighbourhood structure. In fact, for most popular problems, many different neighbourhood structures have been considered and experimented with. For example, for Job–Shop Scheduling, at least ten different ones have appeared in the literature (see [19]). Moreover, for most common problems, there is more than one neighbourhood structure that is sufficiently natural and intuitive to deserve systematic investigation.

E. Burke and P. De Causmaecker (Eds.): PATAT 2002, LNCS 2740, pp. 262–275, 2003.
© Springer-Verlag Berlin Heidelberg 2003

One of the attractive properties of the local search paradigm is its flexibility, in the sense that different techniques and neighbourhoods can be combined and alternated to give rise to complex algorithms. The main motivation for considering the combination of diverse neighbourhoods is related to the diversification of search needed to escape from local minima. In fact a solution that is a local minimum for a given definition is not necessarily a local minimum for another one, and thus an algorithm that uses both has more chances to move toward better solutions.

There are actually many ways to combine different neighbourhoods and different algorithms. In this work, we formally define and investigate the following three:

Neighbourhood union. We consider as neighbourhood the union of many neighbourhoods. The algorithm at each iteration selects a move belonging to any of the components.

Neighbourhood composition. We consider as atomic moves, chains of moves belonging to different neighbourhoods.

Token-ring search. Given an initial state and a set of algorithms based on different neighbourhood functions, the token-ring search makes circularly a run of each algorithm, always starting from the best solution found by the previous one.

These three notions are not completely new, and they have been proposed in the literature in similar forms (under various names). For example, the effectiveness of token-ring search for two neighbourhoods has been stressed by several authors (e.g. [7]). In particular, when one of the two algorithms is not used with the aim of improving the cost function, but exclusively for diversifying the search region, this idea falls under the name of *iterated local search* [11]. As an example, in [15] we employ the alternation of tabu search using a small neighbourhood with hill climbing using a larger neighbourhood for the solution of the high-school timetabling problem.

The alternation of simple search and move chains is also the basis of the so-called *Variable Neighbourhood Search* strategy proposed by Hansen and Mladenović [8], which has been used in many applications (see, e.g., [3]).

Our contribution consists in the attempt to systematize the different ideas in a general multi-neighbourhood framework, and to perform a comprehensive experimental analysis on a real application. In addition, we want to exploit constraint propagation techniques in local search, in the spirit of [13], so as to speed up the exploration of large neighbourhoods.

Our case study is the Course Timetabling (CTT) problem [2,16]. An application built around the algorithms presented here is actually used to make the working timetable at the Faculty of Engineering of the University of Udine. However, the version of the problem we consider in this paper is simplified, by eliminating very specific constraints, so as to reduce it to a more general form. Experiments are performed on real instances (simplified accordingly), and the data are available at http://www.diegm.uniud.it/schaerf/projects/coursett/.

The experimental results confirm that algorithms based on combinations of neighbourhoods performs much better than basic ones.

2   Local Search

Local search is a family of general-purpose search techniques, which was first introduced more than 35 years ago [10]. It has become quite popular in AI after the seminal papers by Minton et al. [12] and Selman et al. [18]. Local search techniques are *non-exhaustive* in the sense that they do not guarantee to find a feasible (or optimal) solution, but they search non-systematically until a specific stop criterion is satisfied.

2.1   Local Search Basics

Given an instance p of a search or optimization problem P, we associate a *search space* S with it. Each element $s \in S$ corresponds to a potential solution of p, and is called a *state* of p. Local search relies on a function N which assigns to each $s \in S$ its *neighbourhood* $N(s) \subseteq S$. Each $s' \in N(s)$ is called a *neighbour* of s.

A local search algorithm t starts from an initial state s_0, which can be obtained with some other technique or generated randomly, and enters a loop that *navigates* the search space, stepping from one state s_i to one of its neighbours s_{i+1}.

The neighbourhood of a state s can be described in terms of changes (called *moves*) that are applied to transform s in the members of $N(s)$. A move is typically composed by a limited set of attributes (or variables) that describes the changes to the state. Given a state s and a move m, we denote by $s \circ m$ the state obtained from s applying the move m. Therefore a neighbourhood can be seen as a set of moves, even though not all moves can be applied in any state s, because some moves might be *infeasible*, i.e. they lead to a state outside the search space.

Local search techniques differ from each other according to the strategy they use both to select the *move* in each state and to stop the search. In all techniques, the search is driven by a *cost function* f that estimates the quality of the state. For optimization problems, f generally accounts for the number of violated constraints and for the objective function of the problem.

Two of the most common local search techniques are *hill climbing* (HC) and *tabu search* (TS). We describe them here; however, a full description of HC and TS is beyond the scope of this paper (see, e.g., [7]). We only present the formulations and the concepts which are used in this work.

2.2   Hill Climbing

HC is actually not a single local search technique, but rather a family of techniques based on the idea of performing only moves that improve or leave unchanged (i.e. *sideways* moves) the value of the cost function f.

We employ the so-called *randomized non-ascending* strategy which selects a random move m_i at each iteration i, and if $f(s_i \circ m_i) \leq f(s_i)$ then let $s_{i+1} = s_i \circ m_i$, otherwise let $s_{i+1} = s_i$.

HC does not stop when it reaches a local minimum. In fact, the search might loop infinitely by cycling among two or more states at equal cost. To provide against this situation, the stop criterion is based on the number of iterations elapsed from the last strict improvement. Specifically, given a fixed value n the algorithm stops after n iterations that do not improve the value of the cost function, i.e. it stops at iteration j such that $f(s_j) = f(s_{j-1}) = \cdots = f(s_{j-n})$.

2.3   Tabu Search

At each state s_i, TS explores exhaustively the current neighbourhood $N(s_i)$. Among the elements in $N(s_i)$, the one that gives the minimum value of the cost function becomes the new current state s_{i+1}, independently of whether $f(s_{i+1})$ is less or greater than $f(s_i)$.

Such a choice allows the algorithm to *escape* from local minima, but creates the risk of cycling among a set of states. In order to prevent cycling, the so-called *tabu list* is used, which determines the forbidden moves. This list stores the most recently accepted moves. The *inverses* of the moves in the list are forbidden.

The simplest way to run the tabu list is as a queue of fixed size k. That is, when a new move is added to the list, the oldest one is discarded. We employ a more general mechanism which assigns to each move that enters the list a random *tenure*, i.e. each move remains in the list for a random number of steps varying between two values k_{min} and k_{max}. When its tabu period is expired, a move is removed from the list. In this way the size on the list is not fixed, but varies dynamically between k_{min} and k_{max}.

There is also a so-called *aspiration* mechanism that overrides the tabu status: If a move m leads to a state whose cost function value is better than the current best, then its tabu status is dropped and the resulting state is acceptable as the new current one.

Also in this case, like HC, the search is stopped when no improvement of the cost function is found after n iterations.

As a final remark, we must mention that we use just one of the simplest forms of TS: more involved ones include sophisticated prohibition strategies and mechanisms for long-term memory. However, as a matter of fact, the algorithm described here is the most employed in the TS literature.

3   Multi-neighbourhood Search

Consider a problem, a search space S for it and a set k of neighbourhood functions $N_1, \ldots, N_k$ defined on S. Given also a set of n local search techniques (in this work $n = 2$, namely HC and TS), we can define $k \times n$ different search algorithms, called *runners*, by combining any technique with any neighbourhood function.

The functions N_i are obviously problem dependent, and they are defined by the person who investigates the problem. In this section, we show that, given a set of human-defined neighbourhood functions, we can automatically create new runners, using some composition operators.

3.1 Neighbourhood Union

Given k neighbourhood functions $N_1, \ldots, N_k$, we call a *union*, written as $N_1 \oplus \cdots \oplus N_k$, the neighbourhood function such that, for each state s, the set $N_1 \oplus \cdots \oplus N_k(s)$ is equal to $N_1(s) \cup \cdots \cup N_k(s)$.

According to the above definition, a HC runner that uses the neighbourhood $N_1 \oplus \cdots \oplus N_k$ selects at each iteration a random move from any N_i, whereas a TS runner explores all N_i exhaustively and selects the overall best solution.

The random distribution for selecting a move in $N_1 \oplus \cdots \oplus N_k$ from s is the following: we first select a random i (with $1 \leq i \leq k$) and then a random state $s' \in N_i(s)$. The selection thus is not uniform, because it is not weighted based on the cardinality of the sets $N_i(s)$.

3.2 Neighbourhood Composition

Given k neighbourhood functions $N_1, \ldots, N_k$, we call *composition*, denoted by $N_1 \otimes \cdots \otimes N_k$, the neighbourhood function defined as follows. Given two states s_a and s_b, then s_b belongs to $N_1 \otimes \cdots \otimes N_k(s_a)$ if there exist $k-1$ states $s_1, \ldots s_{k-1}$ such that $s_1 \in N_1(s_a)$, $s_2 \in N_2(s_1)$, ..., and $s_b \in N_1(s_{k-1})$.

Intuitively, a composite move is an ordered sequence of moves belonging to the component neighbourhoods, i.e. $m = m_1 m_2 \ldots m_k$ with $m_i \in N_i$. Differently from the union operator, for composition the order of the N_i is relevant, and it is meaningful to repeat the same N_i in the composition.

Given the k neighbourhood functions and an integer h, we call *total composition* of step h the union of all possible compositions (also with repetitions) of all k neighbourhoods. We denote a total composition by $\odot_h N_1, \ldots, N_k$. A move in this neighborhood is an ordered sequence of h moves $m_1 m_2 \ldots m_h$ such that $m_i \in N_1 \oplus \cdots \oplus N_k$. In other words, each move m_i ($1 \leq i \leq h$) can be chosen in any neighborhood N_j ($1 \leq j \leq k$).

3.3 Token-Ring Search

Given an initial state s_0, and a set of q *runners* $t_1, \ldots, t_q$, the token-ring search, denoted by $t_1 \triangleright \cdots \triangleright t_q$, makes circularly a run of all t_i. Each t_i always starts from the final solution of the previous runner t_{i-1} (or t_q if $i = 1$).

The token-ring search keeps track of the global best state, and stops when it performs a fixed number of rounds without an improvement of this global best. The component runners t_i stop according to their own specific criteria.

3.4   Local Search Kickers

As noticed by several authors (see, e.g., [11]), local search can benefit from alternating regular runs with some perturbations that allow the search to escape from the attraction area of a local minimum.

In our settings, we define a form of perturbation, that we call *kick*, in terms of neighbourhood compositions. A *kicker* is a runner that makes just one single move, and uses a neighbourhood composition (total or simple) of a relatively long length. A kicker can perform either a random kick, i.e. a random sequence of moves, or a best kick, which means an exhaustive exploration of the composite neighbourhood searching for the best sequence.

Random kicks roughly correspond to the notion of random walk used in [17]. The notion of best kicks is based on the idea of ejection chains (see, e.g., [14]), and generalizes it to generic chains of moves (from different neighbourhoods). Experiments with kickers as part of a token-ring search, called Run & Kick, are shown in our case study, and, as highlighted in Section 5, the use of best kicks turned out to be very effective in our test instances.

Notice that the cardinality of a composition is the product of the cardinalities of all the base neighbourhoods, therefore if the base neighbourhoods have some few thousand members, the computation of the best kick for a composition of length 3 or more is normally intractable. In order to reduce this complexity, we introduce the problem-dependent notion of *synergic moves*. For every pair of neighbourhood functions N_1 and N_2, the user might define a set of constraints that specifies whether two moves m_1 and m_2, in N_1 and N_2 respectively, are synergic or not. This relationship is typically based on equality constraints of some variables that represent the move features. If no constraint is added, the kicker assumes that all moves are synergic.

A move sequence belonging to the neighbourhood composition is evaluated only if all pairs of adjacent moves are synergic. The intuition behind the idea of synergic moves is that a combination of moves that are not all focused on the same features of the current state s have little chance to produce improvements. In that case, in fact, the improvements would have been found by one of the runners that make one step at the time. Conversely, a good sequence of "coordinated" moves can be easily overlooked by a runner based on a simple neighbourhood function.

In order to build kicks, i.e. chains of synergic moves, the kicker makes use of a constraint-based backtracking algorithm that builds it starting from the current state s, along the lines of [13]. Differently from [13], all variables describing a move are instantiated simultaneously, and backtracking takes place only at "move granularity" rather than at the level of each individual variable. That is, the algorithm backtracks at level i if the current move m_i has no synergic move in the neighbourhood N_{i+1} that is feasible if applied in the state reached from s executing the moves of the partial sequence built up to level i.

Notice that the use of a backtracking algorithm for the exploration of the composite neighbourhood does not mean that this process is exponential in nature. In fact, the size of the compound neighborhood for a kick of step n is bound by the product of the size of the component neighborhoods N_i $(i = 1, \ldots, n)$. More correctly, the size of the compound neighborhood grows exponentially w.r.t. the number of neighborhoods involved, but in our experimentation we always choose a constant value for n that is small enough to ensure an efficient computation of kicks.

Different definitions of synergy are possible for a given problem. In general, there is a trade-off between the time necessary to explore the neighbourhood and the probability to find good moves. In our case study, we experiment with two different definitions of synergy and compare their results.

4   A Case Study: Course Timetabling

The CTT problem consists in the weekly scheduling of lectures for a set of courses. There are various formulations of the CTT problem (see, e.g., [16]), which mostly differ from each other in the hard and soft constraints (or objectives) they consider. For the sake of generality, we consider in this work a basic version of the problem.

4.1   Problem Definition

There are q courses $c_1, \ldots, c_q$, p periods $1, \ldots, p$, and m rooms $r_1, \ldots, r_m$. Each course c_i consists of l_i lectures to be scheduled in distinct time periods, and it is attended by s_i students. Each room r_j has a capacity cap_j, in terms of number of seats. There are also g groups of courses, called *curricula*, such that any two courses of a curriculum have students in common.

The output of the problem is an integer-valued $q \times p$ matrix T, such that $T_{ik} = j$ (with $1 \leq j \leq m$) means that course c_i has a lecture in room r_j at period k, and $T_{ik} = 0$ means that course c_i has no class in period k. We search for the matrix T such that the following *hard* constraints are satisfied, and the violations of the *soft* ones are minimized. Hard constraints must be always satisfied in the final solution of the problem, whereas soft constraints can be violated, but at the price of deteriorating the solution quality.

(1) Lectures (hard): The number of lectures of course c_i must be exactly l_i.
(2) Room occupancy (hard): Two distinct lectures cannot take place in the same room in the same period.
(3) Conflicts (hard): Lectures of courses in the same curriculum must be all scheduled at different times.
    We define a conflict matrix CM of size $q \times q$, such that $cm_{ij} = 1$ if there is a curriculum that includes both c_i and c_j, $cm_{ij} = 0$ otherwise.
(4) Availabilities (hard): Teachers might be not available for some periods. We define an availability matrix A of size $q \times p$, such that $a_{ik} = 1$ if lectures of course c_i can be scheduled at period k, $a_{ik} = 0$ otherwise.
(5) Room capacity (soft): The number of students that attend a course must be less than or equal to the number of seats of all the rooms that host its lectures.

(6) Minimum working days (soft): The set of periods p is split into wd days of p/wd periods each (assuming p divisible by wd). Each period therefore belongs to a specific week day. The lectures of each course c_i must be spread into a minimum number of days d_i (with $d_i \leq k_i$ and $d_i \leq wd$).

(7) Curriculum compactness (soft): The daily schedule of a curriculum should be as compact as possible, avoiding gaps between courses. A gap is a free period between two lectures scheduled in the same day and that belong to the same curriculum.

4.2 Search Space, Cost Function, and Initial State

In order to solve CTT by local search, first we have to define the search space. Our search space is composed of all the assignment matrices T_{ik} for which the constraints (1) and (4) hold. States for which the hard constraints (2) and (3) do not hold are allowed, but are considerably penalized within the cost function.

The cost function is thus a weighted sum of the violations of the aforementioned hard constraints plus the violations of the soft constraints (5)–(7).

The weight of constraint type (5) is the number of students without a seat, whereas the weight of constraint types (6) and (7) is fixed to 5 and 2, respectively. Hard constraints are assigned the weight 1000.

The initial solution is selected at random. That is, we create a random matrix T that satisfies constraints (1) and (4).

4.3 Neighbourhood Functions

In the CTT problem, we are dealing with the assignment of a lecture to two kinds of resources: the time periods and the rooms. Therefore, one can very intuitively define two basic neighbourhood structures which deal separately with each one of these components. We call these neighbourhoods Time and Room (or simply T and R for short) respectively.

The first neighbourhood is defined by simply changing the period assigned to a lecture of a given course to a new one which satisfies the constraints (4). A move of the Time type is identified by a triple of variables $\langle C, P, Q \rangle$, where C represents a course, and P and Q are the old and the new periods of the lecture, respectively.

The Room neighbourhood, instead, is defined by changing the room assigned to a lecture in a given period. A move of this type is identified by a triple of variables $\langle C, P, R \rangle$, where C is a course, P is a period and R is the new room assigned to the lecture.

Obviously, there are some constraints (part of the so-called *interface* constraints in [13]) for a given move m to be applicable. In detail, a Time move $\langle C = c_i, P = k_1, Q = k_2 \rangle$ is feasible in a given state only if in that state the course c_i has a lecture at time k_1, it has no lecture at time k_2, and the teacher of c_i is available at k_2. Instead, we consider a Room move $\langle C = c_i, P = k, R = r_j \rangle$ as applicable in a state if the course c_i has a lecture at time k which is assigned to a room $r_{j'}$ with $j \neq j'$.

Table 1. Features of the instances used in the experiments

Instance	q	p	$\sum_i l_i$	m	Conflicts	Occupancy
1	46	20	207	12	4.63%	86.25%
2	52	20	223	12	4.75%	92.91%
3	56	20	252	13	4.61%	96.92%
4	55	25	250	10	4.61%	100.00%

Given these two basic neighbourhoods we define the neighbourhood union Time⊕Room whose moves are either a Time or a Room. Conversely, the neighbourhood composition Time⊗Room involves both the resources at once. For the composite neighbourhood, we define a move $\langle C_1, P_1, Q_1 \rangle$ of type Time and a move $\langle C_2, P_2, R_2 \rangle$ of type Room as synergic under the constraints $C_1 = C_2 \wedge Q_1 = P_2$.

4.4 Runners and Kickers

We define eight runners, obtained equipping HC and TS with the four neighbourhoods: Time, Room, Time⊕Room and Time⊗Room.

We also define two kickers both based on the total composition $\odot_h$ Time,Room of the basic neighbourhoods. The two kickers differ from each other in the definition of the synergic moves for the four combinations. The first one is more strict and requires that the moves "insist" on the same period and on the same room. The second one is more relaxed and also allows combination of moves on different rooms.

All the above runners and kickers are combined in various token-ring strategies, as described in the next section.

5   Experimental Results

To our knowledge, no benchmark instance for the CTT problem has been made available in the scientific community. For this reason we decided to test our algorithms with four real-world instances from the School of Engineering of our university, which will be made available through the web. Real data have been simplified to adapt to the problem version of this work, but the overall structure of the instances is not affected by the simplification.

The main features of these instances are reported in Table 1. All of them have to be scheduled in 5 days of 4 or 5 periods each.

The column denoted by $\sum_i l_i$ reports the overall number of lectures, while the columns "Conflicts" and "Occupancy" show the density of the conflict matrix, and the percentage of occupancy of the rooms $(\sum_i l_i/(m \cdot p))$, respectively. The first feature is a measure of instance size, whereas the other two are the main indicators of instance constrainedness.

Table 2. Results for the plain multi-neighbourhood HC and TS algorithms

Instance	HC(T⊕R)	HC(T⊗R)	HC(T)▷HC(R)
1	288	**285**	295
2	**18**	22	101
3	**72**	169	157
4	**140**	159	255

Instance	TS(T⊕R)	TS(T⊗R)	TS(T)▷TS(R)
1	**238**	277	434
2	**35**	175	262
3	**98**	137	488
4	**150**	150	2095

The proposed algorithms are coded in C++ and have been tested on a PC running Linux equipped with an AMD Athlon 1.5 GHz processor and 384 MB of central memory. In order to obtain a fair comparison among all algorithms, we fix an upper bound on the overall computational time (600 s per instance) of each solver during multiple trials, and we record the best value found up to that time. In this way, each algorithm can take advantage of a multi-start strategy proportionally with its speed, thus having increased chances to reach a good local minimum.

5.1   Multi-neighbourhood Search

We run the HC and TS multi-neighbourhood algorithms on the three instances with the best parameter settings found in a preliminary test phase. Namely, the tabu list is a dynamic one and the tabu tenure varies in the range 20–30. Concerning the number of idle iterations allowed, it is one million for HC and 1000 for TS.

All algorithms found a feasible solution for all trials. Concerning the objective function, the best costs found by the algorithms are summarized in Table 2, where the neighbourhood is in parentheses. The best results found by each technique are displayed in bold face.

From the results, it turns out that the HC algorithms are superior to the TS ones for three out of four instances. Concerning the comparison of neighbourhood operators, the best results are obtained by the Time⊕Room neighbourhood for both HC and TS.

Notice that the thorough exploration of Time⊗Room performed by TS does not give good results. This highlights the trade-off between the steepness of search and the computational cost.

5.2   Multi-neighbourhood Run & Kick

In this section, we evaluate the effect of $\odot_h$Time,Room kickers in joint action (i.e. token-ring) with the proposed local search algorithms.

We take into account three types of kicks. The first two are the best kicks with the strict and the more relaxed definition of move synergy (denoted in Tables 3 and 4 by b and b^*, respectively). In the aim of maintaining the computation time below a certain level we experiment with these kickers only with steps $h = 2$ and $h = 3$.

We compare these kicks with random kicks of length $h = 10$ and $h = 20$ (denoted in Tables 3 and 4 by r). In preliminary experiments, we have found that shorter random walks are almost always undone by the local search algorithms in token-ring alternation with the kicker. In contrast, longer walks perturb the solution too much, leading to a waste of computation time.

The results of the multi-neighbourhood Run & Kick are reported in Tables 3 and 4. In the column "Kick" is reported the length of the kick and the selection mechanism employed.

For each technique we list the best state found and the percentage of improvement obtained by Run & Kick w.r.t. the corresponding plain algorithm presented in the previous section. As before, the best results for each instance are displayed in bold face.

Comparing these results with those of the previous table, we see that the use of kickers can provide a remarkable improvement on the algorithms. In particular, kickers implementing the best kick strategy of length 2 increase the ability of the local search algorithms independently of the search technique employed. Unfortunately, the same conclusion does not hold for the best kicks of length 3. In fact, the time limit granted to the algorithms makes it possible only to perform a single best kick of this length at early stages in the search. Therefore, for instances of this size the improvement in the search made by these kicks is hidden because of their high computational cost.

Furthermore, it is possible to see that for TS the random kick strategy obtains moderate improvements in joint action with T$\oplus$R and T$\otimes$R neighbourhoods, favouring a diversification of the search. Conversely, the behaviour of the HC algorithms with this kind of kick is not uniform, and it deserves further investigation.

Concerning the influence of different synergy definitions, it is possible to see that the stricter one has a positive effect in joint action with TS, while it seems to have little or no impact with HC. In our opinion this is related to the thoroughness of neighbourhood exploration performed by TS.

Another effect of the Run & Kick strategy, which is not shown in the tables, is the improvement of algorithm robustness measured in terms of standard deviations of the results.

Table 3. Results for the HC & Kick algorithms

Instance	Kick	HC(T⊕R)		HC(T⊗R)		HC(T)▷HC(R)	
1	b_2	207	−28.1%	212	−25.6%	**200**	−32.2%
1	b_2^*	206	−28.5%	217	−23.9%	203	−31.2%
1	b_3	271	−5.9%	518	81.8%	439	48.8%
1	b_3^*	341	18.4%	515	116%	773	171%
1	r_{10}	271	−5.9%	275	−3.5%	414	30.3%
1	r_{20}	284	−1.4%	294	3.2%	440	49.2%
2	b_2	18	0.0%	21	−4.6%	27	−73.3%
2	b_2^*	18	0.0%	**17**	−22.7%	23	−77.2%
2	b_3	71	294%	67	205%	239	137%
2	$b_3.$	79	339%	92	318%	481	376%
2	r_{10}	19	5.6%	21	−4.6%	156	54.5%
2	r_{20}	24	33.3%	19	−13.6%	182	80.2%
3	b_2	64	−11.1%	94	−44.4%	78	−50.3%
3	b_2^*	**55**	−23.6%	87	−48.5%	79	−49.7%
3	b_3	182	153%	329	94.7%	853	443%
3	b_3^*	235	226%	436	158%	1632	940%
3	r_{10}	94	30.6%	202	19.5%	206	31.2%
3	r_{20}	95	31.9%	113	−33.1%	181	15.3%
4	b_2	132	−5.71%	146	−8.18%	**113**	−55.69%
4	b_2^*	139	−0.71%	151	−5.03%	142	−44.31%
4	b_3	250	78.57%	565	255.35%	1242	387.06%
4	b_3^*	180	28.57%	3417	2049.06%	19267	7455.69%
4	r_{10}	115	−17.86%	250	57.23%	3292	1190.98%
4	r_{20}	130	−7.14%	172	8.18%	4344	1603.53%

6 Discussion and Conclusions

We have proposed a set of multi-neighbourhood search strategies to improve local search capabilities. This is only a step toward a full understanding of the capabilities of multi-neighbourhood techniques.

Our neighbourhood operators are completely general, in the sense that, given the basic neighbourhood functions, the synthesis of the proposed algorithms requires only the definition of the synergy constraint, but no further domain knowledge.

With respect to other multi-neighbourhood meta-heuristics, such as Variable Neighbourhood Search [8] and Iterated Local Search [11], we have tried to give a more general picture in which these previous (successful) proposals fit naturally.

Our software tool [4,5,6] generates automatically the code for exploration of a composite neighbourhood starting from the code for the basic ones. This is very important, from the practical point of view, in order that the test for composite techniques be very inexpensive not only in terms of design efforts, but also in terms of human programming resources.

Table 4. Results for the TS and Kick algorithms

Instance	Kick	TS(T⊕R)		TS(T⊗R)		TS(T)▷TS(R)	
1	b_2	**208**	−12.6%	214	−22.7%	210	−57.0%
1	b_2^*	208	−12.6%	210	−24.2%	226	−53.7%
1	b_3	287	20.6%	424	53.1%	347	−20.0%
1	b_3^*	273	14.7%	464	67.5%	399	−8.1%
1	r_{10}	265	11.3%	314	13.4%	546	11.9%
1	r_{20}	220	−7.6%	274	−1.1%	569	16.6%
2	b_2	**13**	−62.9%	40	−77.1%	27	−89.7%
2	b_2^*	18	−48.6%	34	−80.6%	47	−82.1%
2	3_b	82	134%	445	154%	491	87.4%
2	b_3^*	97	177%	798	356%	1703	550%
2	r_{10}	17	−51.4%	40	−77.1%	544	108%
2	r_{20}	20	−42.9%	32	−81.7%	726	177%
3	b_2	**76**	−22.5%	83	−50.9%	101	−79.3%
3	b_2^*	**78**	−20.4%	97	−42.6%	145	−70.3%
3	b_3	227	132%	312	127%	1019	109%
3	b_3^*	259	164%	476	248%	1348	176%
3	r_{10}	71	−27.6%	147	−13.0%	832	70.5%
3	r_{20}	72	−26.5%	139	−17.8%	966	98.0%
4	b_2	78	−48.00%	99	−34.00%	105	−94.99%
4	b_2^*	87	−42.00%	126	−16.00%	88	−95.80%
4	b_3	103	−31.33%	201	34.00%	1356	−35.27%
4	b_3^*	177	18.00%	2189	1359.33%	12020	473.75%
4	r_{10}	134	−10.67%	123	−18.00%	4105	95.94%
4	r_{20}	101	−32.67%	159	6.00 %	4324	106.40%

The typical way to solve CTT is by a decomposition: first schedule lectures neglecting the rooms, then assigns the rooms (see, e.g., [9]). In our framework, this would correspond to a token-ring A(Time)▷A(Room) (where A is any technique) with one single round, with the initial solution in which all lectures are in the same room. Experiments show that this choice gives much worse results than those shown in this paper.

It is worth noticing that for CTT, it is natural to compose the neighbourhoods because they are complementary, as they work on different features of the current state (the search space is not connected under them). However, preliminary results with other problems show that multi-neighbourhood search also helps for problems that have completely unrelated neighbourhoods, and thus could be solved also relying on a single neighbourhood function.

References

1. Aarts, E., Lenstra, Jan Karel.: Local Search in Combinatorial Optimization. Wiley, Chichester (1997)
2. Burke, E., Erben, W. (Eds.): Practice and Theory of Automated Timetabling III (PATAT 2000, Konstanz, Germany, August, selected papers). Lecture Notes in Computer Science, Vol. 2079. Springer-Verlag, Berlin Heidelberg New York (2001)
3. den Besten, M., Stützle, T.: Neighborhoods Revisited: An Experimental Investigation into the Effectiveness of Variable Neighborhood Descent for Scheduling. In: Pinho de Sousa, J. (Ed.): Proc. 4th Metaheuristics Int. Conf. (MIC-01) (2001) 545–550
4. Di Gaspero, L., Schaerf. A.: EASYLOCAL++: An Object-Oriented Framework for Flexible Design of Local Search Algorithms. Technical Report UDMI/13/2000/RR. Dipartimento di Matematica e Informatica, Università di Udine (2000). Available at http://www.diegm.uniud.it/schaerf/projects/local++
5. Di Gaspero, L., Schaerf. A.: A Case-Study for EASYLOCAL++: the Course Timetabling Problem. Technical Report UDMI/13/2001/RR. Dipartimento di Matematica e Informatica, Università di Udine (2001). Available at http://www.diegm.uniud.it/schaerf/projects/local++
6. Di Gaspero, L., Schaerf. A.: EASYLOCAL++: An Object-Oriented Framework for Flexible Design of Local Search Algorithms. Softw. – Pract. Exper. (to appear)
7. Glover, F., Laguna, M.: Tabu search. Kluwer, Dordrecht (1997)
8. Hansen, P., Mladenović, N.: An Introduction to Variable Neighbourhood Search. In: Voß, S., Martello, S., Osman, I.H., Roucairol, C. (Eds.): Meta-Heuristics: Advances and Trends in Local Search Paradigms for Optimization. Kluwer, Dordrecht (1999) 433–458
9. Laporte, G., Desroches, S.: Examination Timetabling by Computer. Comput. Oper. Res. **11** (1984) 351–360
10. Lin, S.: Computer Solutions of the Traveling Salesman Problem. Bell Syst. Tech. J. **44** (1965) 2245–2269
11. Lourenço, H.R., Martin, O., Stützle, T.: Applying Iterated Local Search to the Permutation Flow Shop Problem. In: Glover, F., Kochenberger, G. (Eds.): Handbook of Metaheuristics. Kluwer, Dordrecht (2001) (to appear)
12. Minton, S., Johnston, M.D., Philips, A.B., Laird, P.: Solving Large-Scale Constraint Satisfaction and Scheduling Problems Using a Heuristic Repair Method. In: Proc. 8th Natl Conf. Artif. Intell. (AAAI'90) AAAI Press/MIT Press, Boston, MA (1990) 17–24
13. Pesant, G., Gendreau, M.: A Constraint Programming Framework for Local Search Methods. J. Heuristics **5** (1999) 255–279
14. Pesch, E., Glover, F.: TSP Ejection Chains. Discr. Appl. Math. **76** (1997) 175–181
15. Schaerf, A.: Local Search Techniques for Large High-School Timetabling Problems. IEEE Trans. Syst. Man Cybern. **29** (1999) 368–377
16. Schaerf, A.: A Survey of Automated Timetabling. Artif. Intell. Rev. **13** (1999) 87–127
17. Selman, B., Kautz, H.A., Cohen, B.: Noise Strategies for Improving Local Search. In: Proc. 12th Natl Conf. Artif. Intell. (AAAI'94) (1994) 337–343
18. Selman, B., Levesque, H., Mitchell, D.: A New Method for Solving Hard Satisfiability Problems. In: Proc. 10th Natl Conf. Artif. Intell. (AAAI'92) (1992) 440–446
19. Vaessens, R., Aarts, E., Lenstra, J.K.: Job Shop Scheduling by Local Search. INFORMS J. Comput. **8** (1996) 302–317

Knowledge Discovery in a Hyper-heuristic for Course Timetabling Using Case-Based Reasoning

E.K. Burke[1], B.L. MacCarthy[2], S. Petrovic[1], and R. Qu[1]

[1] School of Computer Science and Information Technology,
Jubilee Campus, University of Nottingham, Nottingham NG8 1BB, UK
{ekb,sxp,rxq}@cs.nott.ac.uk
[2] School of Mechanical, Materials, Manufacturing Engineering and Management,
University of Nottingham, University Park, Nottingham NG7 2RD, UK
bart.maccarthy@nottingham.ac.uk

Abstract. This paper presents a new hyper-heuristic method using Case-Based Reasoning (CBR) for solving course timetabling problems. The term *hyper-heuristics* has recently been employed to refer to "heuristics that choose heuristics" rather than heuristics that operate directly on given problems. One of the overriding motivations of hyper-heuristic methods is the attempt to develop techniques that can operate with greater generality than is currently possible. The basic idea behind this is that we maintain a case base of information about the most successful heuristics for a range of previous timetabling problems to predict the best heuristic for the new problem in hand using the previous knowledge. Knowledge discovery techniques are used to carry out the training on the CBR system to improve the system performance on the prediction. Initial results presented in this paper are good and we conclude by discussing the considerable promise for future work in this area.

1 Introduction

1.1 Case-Based Reasoning

What is Case-Based Reasoning? Many techniques from Artificial Intelligence (AI) and Operational Research (OR) solve timetabling problems directly by employing heuristics, meta-heuristics and hybrids on the problem in hand [15, 17,33]. Case-Based Reasoning (CBR) [27] is a Knowledge-Based technique that solves problems by employing the knowledge and experience from previous similar cases. Solutions or problem solving strategies that were used in solving earlier problems (cases) are maintained in a store (case base) for reuse. Adaptation usually needs to be carried out for the new problem employs domain knowledge of some kind. The solved new problems may be retained and the case base is thus updated. Leake [29] described CBR as follows:

> In CBR, new solutions are generated not by chaining, but by retrieving the most relevant cases from memory and adapting them to fit new situations.

E. Burke and P. De Causmaecker (Eds.): PATAT 2002, LNCS 2740, pp. 276–287, 2003.

A case usually consists of two major parts: the problem itself represented in a certain form to describe the conditions under which it should be retrieved; and the solution of the problem or the lessons it will teach. Throughout this paper, the term *source case* is used to denote the cases in the case base and the term *target case* is used to denote the new problem to be solved.

A similarity measure is usually defined by a formula to calculate the similarity between source cases and the target case. The most similar source cases are retrieved for the target case. The development of this similarity measure for large real-world problems such as those encountered in course timetabling presents one of the major research challenges in this area.

Case-Based Reasoning in Scheduling and Optimization Problems. Timetabling has been studied extensively over the years [15,17,33,6,12,4,5]. It can be thought of as a special type of scheduling problem. The potential for CBR has been discussed for different scheduling problems [30,34]. A brief survey of CBR in scheduling was presented in [10] where three Case-Based Scheduling systems, SMARTplan [28], CBR-1 [3] and CABINS [31], were reviewed. The authors claimed that CBR is a very good approach in expert scheduling systems and emphasized potential research in dynamic scheduling. Other studies in case-based scheduling concerned a variety of scheduling problems and issues, i.e. optimization [21], nurse rostering [35] and educational timetabling problems [9, 10].

1.2   Course Timetabling and Hyper-heuristics in Scheduling

Timetabling Problems. A general timetabling problem involves assigning a set of events (meetings, matches, exams, courses, etc.) into a limited number of timeslots subject to a set of constraints. Constraints are usually classified into two particular types: *hard constraints* and *soft constraints*. Hard constraints should under no circumstances be violated. Soft constraints are desirable but can be relaxed if necessary.

Over the last 40 years, there has been a considerable amount of research on timetabling problems [15,17,33,7]. In the early days of educational timetabling research, graph colouring [39] and integer linear programming techniques were widely used [14]. Some of the latest approaches can be seen in [6,12].

This paper concentrates on educational course timetabling problems. Modern heuristic techniques have been successfully applied to course timetabling [6, 4,5]. Tabu Search (e.g. [19]) and Simulated Annealing (e.g. [1]) have been successfully applied. Evolutionary Algorithms/Genetic Algorithms (GAs) (e.g. [18]) and Memetic Algorithms (that hybridizes GAs with local search techniques) (e.g. [13]) have also been extensively studied. Constraint-Based techniques have also been widely employed (e.g. [40]).

Hyper-heuristics in Scheduling. Hyper-heuristics can be defined to be "heuristics that choose heuristics" or as "algorithms to pick the right algorithm for the right situation" [8]. The main reason for using the term hyper-heuristics rather than the widely used term meta-heuristics is that hyper-heuristics represent a method of selecting from a variety of different heuristics (that may include meta-heuristics).

Some research in scheduling has investigated this approach although it does not always use the term "hyper-heuristics". An approach was presented in [22] on open shop scheduling problems using GAs to search a space of abstractions of solutions to "evolve the heuristic choice". GAs have been employed to construct a schedule builder that chooses the optimal combinations of heuristics [26]. Another approach in [37] used a GA to select the heuristic to order the exam in a sequential approach for exam timetabling problems. A hybrid GA investigated for vehicle routing problems has demonstrated promising results [36,2].

Some other research on hyper-heuristics has also been carried out on a variety of scheduling problems. Guided local search was used to select appropriate choices from a set of heuristics for the travelling salesman problem [38]. In [8] a hyper-heuristic approach was used to select from a set of lower-level heuristics according to the characteristics of the current search space in a sales summit scheduling problem.

2   Case-Based Heuristic Selection for Course Timetabling

2.1   Knowledge Discovery for Course Timetabling

The overall goal of our approach is to investigate CBR as a selector to choose (predict) the best (or a reasonably good) heuristic for the problem in hand according to the knowledge obtained from solving previous similar problems. The goal is to avoid a large amount of computation time and effort on the comparison and choosing of different heuristics. A large number of approaches and techniques in AI and OR have been studied to solve a wide range of timetabling problems successfully over the years. Comparisons have been carried out in some papers on using different approaches in solving a specific range of problems. Thus, the development of heuristics for timetabling is very well established and a reasonable amount of knowledge exists on which specific heuristic works well on what specific range of timetabling problems. This provides a large number of cases that can be collected, studied and stored in the case base, providing a good starting point for solving new course timetabling problems.

In knowledge engineering, techniques in knowledge discovery and machine learning have been employed with success in a number of ill-structured domains. Knowledge discovery is the process of studying and investigating a collection of datasets to discover information such as rules, regularities, or structures in the problem domain. It was defined in [23] as a "non-trivial process of identifying valid, novel, potentially useful, and ultimately understandable patterns in data". A key step in the knowledge discovery process is data mining that may employ a wide range of techniques from AI, machine learning, knowledge acquisition and statistics. Knowledge discovery is usually carried out on databases and the application areas include medicine, finance, law and engineering [32].

In our CBR system the previous most similar cases provide information that facilitates the prediction of the best heuristic for the target case. The retrieval in CBR is a similarity-driven process that is carried out on cases described in specific forms. Thus, the key issues are the case representation (that should be in a proper form to describe the relevant context within the timetabling problem) and how it influences the similarity between cases, which is what drives the retrieval to provide an accurate prediction on heuristic selection.

Knowledge discovery techniques are employed to extract knowledge of meaningful relationships within the case-based heuristic selector via iterative training processes on cases of course timetabling problems. There are two iterative training stages used in the process. The first stage tries to discover the representation of cases with a proper set of features and weights. The second stage trains the case base so that it contains the proper collection of source cases. Both of the processes are carried out iteratively. The overall objective is to obtain the highest accuracy on retrievals for predictions of heuristics for target cases.

2.2   Knowledge Discovery Process on Case-Based Heuristic Selection

Getting Started. In most knowledge discovery approaches, the development starts from the data preparation. Cases in the system are represented by a list of feature–value pairs. A set of features is used to describe the relevant characteristics of the timetabling problems, and a value is given for each of these features in each case. The current CBR system examines the source cases and target case that are produced artificially with specific characteristics as their problem part. These include problems with different size, different timeslots, different rooms, etc. Some heuristics will work well on some problems and less well on others. This means that the system has many types of problems that are studied and collected. Appendix A presents a description of the problem specifications. For every source case and target case, five heuristics (described in Appendix B) are used to solve the problem beforehand. By checking the penalties of the timetables produced, these heuristics are stored with each case in an ascending order as its solution part.

The retrieval is a similarity-driven process that searches through the case base to find the most similar source cases. The similarity measure employs a nearest-neighbour method that calculates a weighted sum of the similarities between each pair of individual features between cases. Formula (1) presents the similarity measure between the source case C_s and the target case C_t in the system:

$$S(C_s, C_t) = \frac{1}{\sqrt{\sum_{i=0}^{j} w_i(fs_i - ft_i)^2 + 1}} . \tag{1}$$

The notation is as follows:

- j is the number of features in the case representation,
- w_i is the weight of the ith feature reflecting the relevance on the prediction,

- fs_i, ft_i are the values of the ith feature in source case C_s and target case C_t respectively.

The possible values of the features describing timetabling problems are all integers (see Appendix C). So the higher the value of $S(C_s, C_t)$, the more similar the two cases are.

The performance of the system is tested on different sets of target cases. The training on the system is targeted at a reasonably high accuracy on all of the (quick) retrievals for the target cases. Within each retrieval, the best two heuristics of the retrieved case are compared with the best heuristic of the target case. If the best heuristic of the target case maps onto any of the best two heuristics of the retrieved case, the retrieval is concluded as successful. Actually, in the training processes, we found that sometimes the penalties of the timetables produced by different heuristics are close or equal to each other. We choose the best two heuristics to be stored with each source case so that we have the best heuristic stored and still retain some randomness.

Training on the Case Representation. An initial case base is built up which contains a set of different source cases with artificially selected specific constraints and requirements from Appendix A. An initial list of features is first randomly selected to represent cases. Each of the features is initially assigned with the same normalized weights. There are 11 features (details of which are given in Appendix C) in the initial case representation.

Our knowledge discovery on the case representation to train the features and their weights in the system adopts the iterative methodology presented in [20]. In every iteration, we

(a) analyse the retrieval failures;
(b) propose new features to address retrieval failures;
(c) select a discriminating set of features for the new case representation;
(d) evaluate the competence of this representation.

In our CBR system, the training for case representation is a recursive failure-driven process carried out to refine the initial features and their weights. A schematic diagram of the knowledge discovery on case representation is given in Figure 1. The knowledge discovery process in the system includes the following steps:

> *Adjusting feature weights.* The best two heuristics of the retrieved case are compared with the best one of the target case to see if the retrieval is successful (the best heuristic of the target case mapped onto one of the best two of the retrieved case). Adjustments on feature weights are iterative error-driven processes: the weights of the features that result in the failures of the retrieval are penalized (decreased) and those that can contribute successful retrievals are rewarded (increased) to discriminate the source cases that should be retrieved from the others that should not be retrieved.

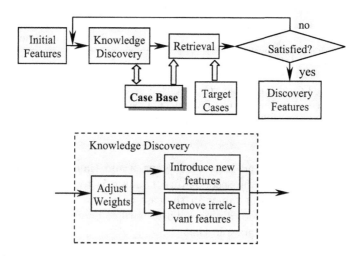

Fig. 1. Schematic diagram of knowledge discovery on features and their weights

Removing irrelevant features. After certain rounds of iterative adjustments, the weights of some of the features may be small enough to be removed from the feature list. This means that these features are either irrelevant or less important and thus are not needed in the case representation. Retaining the irrelevant features may confuse the retrieval process, as the similarities between cases may be too close to each other, thus reducing the number of the successful retrievals and decreasing the system performance.

Introducing new features. When the adjustment of feature weights does not result in a successful retrieval for a target case, new relevant features are added. New features are proposed by studying if they can distinguish the correct source case from the others, if they can give a prediction of success, or if they can express the specific characteristics in a particular case.

Due to the complexity of the problem, at the beginning we do not know what features are relevant to the similarity-driven retrieval and which should be used to represent cases. Also we do not know their weights as we do not know how important they are to properly calculate the similarity that influences the heuristic selection. By using the recursive knowledge discovery process presented above, irrelevant and less important features are removed from the initial feature list. The feature vector that gives the highest accuracy on retrievals for all of the target cases will be employed as the basis for the second stage of knowledge discovery. The trained case representation (with six features left) after the first stage of training is presented in Appendix D.

Training on the Case Base. Case selection is a particularly important issue in building up a case base. Sometimes, keeping irrelevant source cases can decrease

Table 1. Accuracies of system performance on initial and trained case bases

Case base	Retrieval accuracy
"OneSet" (45 cases)	42%
"TwoSet" (90 cases)	60%
Trained Case Base from "OneSet" (15 cases)	70%
Trained Case Base from "TwoSet" (14 cases)	71%

the system performance and increase the space and time requirements of the system. The objective of the second-stage training is to select a collection of relevant cases without redundancy for the case base.

Firstly, we build up two initial case bases with source cases of nine different sizes with $10, 15, \ldots, 50$ courses in them:

(a) "OneSet" – For each size, five source cases are produced, each has one of the five heuristics listed in Appendix A as its best heuristic. We name this case base "OneSet" as it contains one set of the five heuristics for cases with different sizes (thus in "OneSet" there are $9 \times 5 = 45$ source cases).

(b) "TwoSet" – For each size, 10 source cases are produced, each two have one of the five heuristics listed in Appendix A as their best heuristic. It is named "TwoSet" and in total there are $9 \times 5 \times 2 = 90$ source cases in "TwoSet".

The target cases are produced with the size of $10, 20, 30, \ldots, 100$ courses, for each size with 10 instances. Thus, there are $10 \times 10 = 100$ target cases to be tested on the two initial case bases. The best heuristics for each of them is obtained beforehand to evaluate the retrieval.

A database is built up containing these two case bases and the target case set. The training process on these two initial case bases is carried out recursively using the "Leave-One-Out" strategy: Each time a source case is removed from the case base we test to see if the number of successful retrievals on the case base for all of the target cases is increased. If removing a source case decreases the number of successful retrievals, it will be restored back to the case base as it may contribute to successful retrievals for certain types of cases. Otherwise, if the number of successful retrievals increases or does not change, it will be removed from the case base as a redundant case. The process stops when the highest number of retrievals is obtained on all the target cases.

Finally, after the second stage of training, there are 14 and 15 source cases left in the original two case bases, respectively. To test the system performance, an experiment is carried out on both the initial and trained case bases for another set of target cases that are, of course, not the same as those of the training set. The accuracies of the system performance on these case bases are shown in Table 1.

We can observe that the initial "TwoSet" provides better performance than that of "OneSet". By storing more source cases, the system is equipped with more knowledge and thus is capable of providing better performance during the retrieval. We can also see that the second training process removes quite a lot of source cases that are redundant or that are harmful to the performance of the CBR system. With a smaller number of more relevant source cases retained in the case bases, the system performance is improved to provide higher accuracies of predictions of suitable heuristics. To obtain better system performance, a higher number of relevant source cases need to be selected in the case base.

3   Conclusion and Future Work

This paper presents the first step of our work in developing a hyper-heuristic method using CBR for heuristic selection on course timetabling problems. Knowledge discovery techniques employ relatively simple methods and just a few training processes are carried out. The results are good and indicate the possible advantages of employing knowledge discovery techniques in the course timetabling domain. We believe better results may be obtained after further training processes are carried out which employ a range of knowledge discovery techniques.

There are many more complex and elaborate techniques that can be investigated and integrated into the CBR system to improve its performance. For example, for the case representation we currently employ a simple technique that is manually carried out to choose the features and adjust their weights. This can be seen as a feature selection task, which is the problem of selecting a set of features to be used as the data format in the system to achieve high accuracy of prediction. Feature selection is an important issue in machine learning [25] for which a variety of traditional techniques exist. Some recent work employing AI methods, such as evolutionary algorithms [24] to optimize the feature selection, also provide a wide range of possible research directions. For complex timetabling problems, these more efficient algorithms can be employed to carry out the searching on features more effectively when dealing with larger data sets. Our future work will study and compare these different techniques to optimize the case representation to improve the system performance on a wider range of larger timetabling problems. New features are being studied and introduced into the system. For example, some refined features such as the number of rooms with a range of capacities and the number of courses with more than a certain number of constraints can be introduced to give a more specific description of problems. Other issues relating to knowledge discovery in the CBR system may include how to deal with the incomplete data in case bases and how to involve domain knowledge in the system. User interaction in knowledge discovery is also important on tasks like judgment and decision making, in which humans usually perform better than a machine.

The current system uses five simple heuristics to implement the analysis and testing on the case-based heuristic selection. Future work will study more heuristics in the system. Also, the testing cases are artificially produced to give a systematic analysis on as many types of problem as possible. After the initial study of using CBR as a heuristic selector we have increased our understanding of the area. Real-world benchmark timetabling data (such as that presented in [16]) will be collected and stored in the case base for solving real-world problems. Adaptation may also need to be conducted to utilize domain knowledge on some of the heuristics retrieved for the new problem.

References

1. Abramson, D.: Constructing School Timetables Using Simulated Annealing: Sequential and Parallel Algorithms. Manage. Sci. **37** (1991) 98–113
2. Berger, J., Sassi, M., Salois, S.: A Hybrid Genetic Algorithm for the Vehicle Routing Problem with Windows and Itinerary Constraints. Proc. Genet. Evolut. Comput. Conf. 1999 (GECCO'99), Morgan Kaufmann, San Mateo, CA (1999) 44–51
3. Bezirgan, A.: A Case-Based Approach to Scheduling Constraints. In: Dorn, J. and Froeschl, K.A. (Eds.): Scheduling of Production Processes. Ellis Horwood, New York (1993) 48–60
4. Burke, E, Ross, P. (Eds.): Practice and Theory of Automated Timetabling I (PATAT 1995, Edinburgh, Aug/Sept, selected papers). Lecture Notes in Computer Science, Vol. 1153. Springer-Verlag, Berlin Heidelberg New York (1996)
5. Burke, E, Carter, M. (Eds.): Practice and Theory of Automated Timetabling II (PATAT 1997, Toronto, Canada, August, selected papers). Lecture Notes in Computer Science, Vol. 1408. Springer-Verlag, Berlin Heidelberg New York (1998)
6. Burke, E, Erben W. (Eds.): Practice and Theory of Automated Timetabling III (PATAT 2000, Konstanz, Germany, August, selected papers). Lecture Notes in Computer Science, Vol. 2079. Springer-Verlag, Berlin Heidelberg New York (2001)
7. Burke, E.K., Jackson, K.S., Kingston, J.H., Weare, R.F.: Automated Timetabling: the State of the Art. Comput. J. **40** (1997) 565–571
8. Burke E., Kendall, G., Newall, J., Hart, E., Ross, P., Schulenberg, S.: Hyperheuristic: an Emerging Direction in Modern Search Technology. In: Glover, Kochenberg (Eds.): Handbook of Meta-heuristics. Kluwer, Dordrecht (2003) 457–474
9. Burke, E.K., MacCarthy, B. Petrovic, S., Qu, R.: Structured cases in CBR – Reusing and Adapting Cases for Timetabling Problems. J. Knowledge-Based Syst. **13** (2000) 159–165
10. Burke, E.K., MacCarthy, B.L. Petrovic, S., Qu, R.: Case-Based Reasoning in Course Timetabling: an Attribute Graph Approach. In: Aha, D.W., Watson, I. (Eds.): Case-Based Reasoning Research and Development. Proc. 4th Int. Conf. on Case-Based Reasoning (ICCBR2001, Vancouver, Canada). Lecture Notes in Artificial Intelligence, Vol. 2080. Springer-Verlag, Berlin Heidelberg New York (2000) 90–104
11. Burke, E., Newall, J. and Weare, R.: A Simple Heuristically Guided Search for the Timetabling Problem. Proc. Int. ICSC Symp. Engng Intell. Syst. (EIS'98) (1998) 574–579
12. Burke, E., Petrovic, S.: Recent Research Directions in Automated Timetabling. Eur. J. Oper. Res. **140** (2002) 266–280
13. Carrasco, A.P., Pato, M.V.: A Multiobjective Genetic Algorithm for the Class/Teacher Timetabling Problem. In: Burke, E, Erben W. (Eds.): Practice and Theory of Automated Timetabling III (PATAT 2000, Konstanz, Germany, August, selected papers). Lecture Notes in Computer Science, Vol. 2079. Springer-Verlag, Berlin Heidelberg New York (2001) 3–17

14. Carter, M.W.: A Lagrangian Relaxation Approach to the Classroom Assignment Problem. IFOR **27** (1986) 230–246
15. Carter, M.W., Laporte, G.: Recent Developments in Practical Examination Timetabling. In: Burke, E, Ross, P. (Eds.): Practice and Theory of Automated Timetabling I (PATAT 1995, Edinburgh, Aug/Sept, selected papers). Lecture Notes in Computer Science, Vol. 1153. Springer-Verlag, Berlin Heidelberg New York (1996) 3–21
16. Carter, W.M., Laporte, G.: Examination Timetabling: Algorithmic Strategies and Applications, J. Oper. Res. Soc. **74** (1996) 373–383
17. Carter, M.W., Laporte, G.: Recent Developments in Practical Course Timetabling. In: Burke, E, Carter, M. (Eds.): Practice and Theory of Automated Timetabling II (PATAT 1997, Toronto, Canada, August, selected papers). Lecture Notes in Computer Science, Vol. 1408. Springer-Verlag, Berlin Heidelberg New York (1998) 3–19
18. Corne, D., Ross, P.: Peckish Initialisation Strategies for Evolutionary Timetabling. In: Burke, E, Erben W. (Eds.): Practice and Theory of Automated Timetabling III (PATAT 2000, Konstanz, Germany, August, selected papers). Lecture Notes in Computer Science, Vol. 2079. Springer-Verlag, Berlin Heidelberg New York (2001) 227–240
19. Costa, D.: A Tabu Search Algorithm for Computing an Operational Timetable. EJOR **76** (1994) 98–110
20. Cunningham, P., Bonzano, A.: Knowledge Engineering Issues in Developing a Case-Based Reasoning Application. Knowledge-Based Syst. **12** (1999) 371–379
21. Cunningham, P., Smyth, B.: Case-Based Reasoning in Scheduling: Reusing Solution Components. Int. J. Prod. Res. **35** (1997) 2947–2961
22. Fang, H.L., Ross, P., Corne, D.: A Promising Hybrid GA/Heuristic Approach for Open-Shop Scheduling Problems. 11th Eur. Conf. Artif. Intell (ECAI'94). Wiley, New York (1994)
23. Fayyad, U., Piatetsky-Shapiro, G., Smyth, P.: From Data Mining to Knowledge Discovery in Databases. In: Fayyad, U., Piatetsky-Shapiro, G., Smyth, P., Uthurusamy, R. (Eds.): Advances in Knowledge Discovery and Data Mining. AAAI Press, Menlo Park, CA (1996) 1–34
24. Freitas, A: A Survey of Evolutionary Algorithms for Data Mining and Knowledge Discovery. In: Ghosh, A., Tsutsui, S. (Eds.): Advances in Evolutionary Computation. Springer-Verlag, Berlin Heidelberg New York (2002)
25. Hall, M.A., Smith, L.: Practical Feature Subset Selection Machine Learning. Proc. Austral. Comput. Sci. Conf. (1996)
26. Hart, E., Ross, P., Nelson, J.: Solving a Real-world Problem Using an Evolving Heuristically Driven Schedule. Evolut. Comput. **6** (1998) 61–80
27. Kolodner, J.L. Case-Based Reasoning. Morgan Kaufmann, San Mateo, CA (1993)
28. Koton, P.: SMARTlan: A Case-Based Resource Allocation and Scheduling System. Proc. Workshop on Case-Based Reasoning (DARPA). (1989) 285–289
29. Leake, D. (Ed.): Case-Based Reasoning: Experiences, Lessons and Future Directions. AAAI Press, Menlo Park, CA (1996)
30. MacCarthy, B.L., Jou, P.: Case-Based Reasoning in Scheduling. In: Khan, M.K., Wright, C.S. (Eds.): Proc. Symp. Adv. Manu. Process., Syst. Techniques (AMPST'96). MEP Publications (1996) 211–218
31. Miyashita, K., Sycara, K.: CABINS: A Framework of Knowledge Acquisition and Iterative Revision for Schedule Improvement and Reactive Repair. Artif. Intell. **76** (1995) 377–426

32. Piatetsky-Shapiro, G.: Knowledge Discovery in Databases. AAAI Press, Menlo Park, CA (1991)
33. Schaef, A.: A Survey of Automated Timetabling. Artif. Intell. Rev. **13** (1999) 87–127
34. Schmidt, G.: Case-Based Reasoning for Production Scheduling. Int. J. Prod. Econ. **56/57** (1998) 537–546
35. Scott, S., Simpson, R., Ward, R.: Combining Case-Based Reasoning and Constraint Logic Programming Techniques for Packaged Nurse Rostering Systems. Proc. 3rd UK Case-Based Reasoning Workshop. (1997)
36. Shaw, P.: Using Constraint Programming and Local Search Methods to Solve Vehicle Routing Problems. Proc. CP'98. (1998) 417–431
37. Terashima-Marin, H., Ross, P., Valenzuela-Rendon, M.: Evolution of Constraint Satisfaction Strategies in Examination Timetabling. Proc. Genet. Evolut. Comput. Conf. 1999 (GECCO'99). Morgan Kaufmann, San Mateo, CA (1999) 635–642
38. Voudouris, C., Tsang, E.P.K.: Guided Local Search and Its Application to the Travelling Salesman Problem. Eur. J. Oper. Res. **113** (1999) 469–499
39. Werra, D.: Graphs, Hypergraphs and Timetabling. Methods Oper. Res. (Germany F.R.) **49** (1985) 201–213
40. Zervoudakis, K., Stamatopoulos, P.: A Generic Object-Oriented Constraint-Based Model for University Course Timetabling. In: Burke, E, Erben W. (Eds.): Practice and Theory of Automated Timetabling III (PATAT 2000, Konstanz, Germany, August, selected papers). Lecture Notes in Computer Science, Vol. 2079. Springer-Verlag, Berlin Heidelberg New York (2001) 28–47

Appendix A. Course Timetabling Problems Specification

Hard constraints:

1. Two courses cannot be scheduled into the same timeslot;
2. A course should be carried out n times a week;
3. Each course has a specific room requirement with type and capacity;
4. There is a specified number of periods for each course timetabling problem.

Soft constraints:

1. One course should be scheduled before or after another;
2. Inclusive/exclusive – a course should/should not be scheduled into a fixed timeslot;
3. Consecutive – a course should/should not be scheduled into a timeslot consecutive to that of another.

Appendix B. Heuristics Used in the System

1. LD – Largest degree first. All the courses not yet scheduled are inserted into an "unscheduled list" in descending order according to the number of conflicts the course has with the other courses. This heuristic tries to schedule the most difficult courses first.

2. LDT – Largest degree first with tournament selection. This heuristic is presented in [40]. It is similar to LD except that a course employing tournament selection is selected from a subset of the "unscheduled list". Here, a probability value of 30% is used to get a subset from the list. This heuristic tries to schedule the most difficult courses first but also give some randomness.
3. HC – Hill climbing. An initial timetable is constructed randomly then is improved by hill climbing.
4. CD – Colour degree. Courses in the "unscheduled list" are ordered by the number of conflicts they have with those courses that are already scheduled in the timetable. Usually, those courses with a large number of such conflicts are harder to schedule than courses with a smaller number of conflicts.
5. SD – Saturation degree.

Courses in the "unscheduled list" are ordered by the number of periods left in the timetable for them to be scheduled validly. This heuristic gives higher priority to courses with fewer periods available.

Appendix C. Initial Features and Their Weights for Cases

f_0: number of hard constraints / number of events,
f_1: number of soft constraints / number of events,
f_2: number of constraints / number of events,
f_3: number of periods / number of events,
f_4: number of rooms / number of events,
f_5: number of not consecutive courses / number of constraints,
f_6: number of consecutive courses / number of constraints,
f_7: number of hard constraints / number of constraints,
f_8: number of soft constraints / number of constraints,
f_9: number of hard constraints / number of periods,
f_{10}: number of soft constraints / number of periods,
normalized weight w_i = factor$_i$ * 1 / sum of weights of all the features,
initial factor$_i$ = 1, 1, 1, 1, 1, 1, 1, 1, 1, 1, 1.

Appendix D. Trained Features and Their Weights

f_0: number of exclusive courses / number of events,
f_1: number of inclusive courses / number of events,
f_2: number of constraints / number of events,
f_3: number of rooms / number of events,
f_4: number of hard constraints / number of periods,
f_5: number of not consecutive courses / number of constraints,
normalized weight w_i = factor$_i$ * 1 / sum of weights of all the features,
factor$_i$ = 45, 10, 10, 15, 30, 6.

Generalizing Bipartite Edge Colouring to Solve Real Instances of the Timetabling Problem

David J. Abraham and Jeffrey H. Kingston

School of Information Technologies,
The University of Sydney,
Sydney 2006,
Australia
jeff@it.usyd.edu.au

Abstract. In this paper we introduce a new algorithm for secondary school timetabling, inspired by the classical bipartite graph edge colouring algorithm for basic class–teacher timetabling. We give practical methods for generating large sets of meetings that can be timetabled to run simultaneously, and for building actual timetables based on these sets. We report promising empirical results for one real-world instance of the problem.

1 Introduction

This paper is concerned with the problem of constructing timetables for secondary schools, in which groups of students meet with teachers in rooms at times chosen so that no student group, teacher, or room attends two or more meetings simultaneously.

One fundamental requirement separates secondary school timetabling from university timetabling: every student is required to be in class during every teaching period. This makes it infeasible for every student to have an individual timetable; instead, the students are placed in groups, and it is these groups that are timetabled, not individual students.

In this paper, a *time* will be a fixed time period (for example, Mondays 9.00–9.40 AM), a *resource* will be either a *student group* taken collectively, such as the Year 7 students, or a *teacher*, or a *room*. A *meeting* is a collection of *time slots*, each capable of holding one time, and a collection of *resource slots*, each capable of holding one resource. Resource slots are always constrained to hold a resource of a particular type, such as an English teacher or Science laboratory room. Both resource slots and time slots may also be *preassigned* to a particular resource or time, meaning that the slot may hold only that resource or time. The timetabling problem then is to assign one value to each slot which satisfies all these constraints and ensures that there are no clashes (pairs of meetings having both a resource and a time in common). There may be other constraints as well, such as limits on the workload of teachers, or the teacher constancy requirement discussed below.

E. Burke and P. De Causmaecker (Eds.): PATAT 2002, LNCS 2740, pp. 288–298, 2003.

In secondary schools known to the authors, the principal technique used for offering students some choice is the *elective*. Suppose that there are 180 students in one year (age cohort), enough to form six separate classes of 30 students each. Late in the previous year the students would be offered a list of subject areas (e.g. French, German, Biology, History, Economics, Business) and required to select exactly one. Depending on their responses, the school management decides how many classes of each type to offer. These classes then run simultaneously the following year.

Electives give rise to meetings containing many resources. Our example elective would contain one student group representing the entire year, six teacher resource slots, and six room resource slots, all constrained to be used at the same set of times. For Mathematics the students are typically grouped by ability, and this requires all the Mathematics classes for a given year to run simultaneously, creating something similar to an elective except that all the classes within it are Mathematics. For the other compulsory subjects the groups of students within a year may often be timetabled independently.

It is not possible to preassign teachers to large elective meetings, since the resulting timetabling problem would be hopelessly over-constrained. Only a few teacher slots (typically in the most senior classes) are preassigned; the rest are assigned as part of the timetabling process, after times have been assigned. Naturally, this teacher assignment phase must assign English teachers to English classes, Economics teachers to Economics classes, etc., so each teacher slot must record the category of teacher it needs. These categories or *teacher types* are not disjoint: some teachers teach several subjects, others are qualified to teach junior subjects but not senior, and so on. Rooms must be assigned too, and they also have categories: Science laboratories, Music studios, ordinary classrooms, etc.

Away from electives the meetings may be much smaller, the natural minimum being a meeting containing three resources: one preassigned student group, one teacher and one room. Manual high school timetabling is usually accomplished by timetabling the large meetings first, then fitting the small ones around them.

Although soft constraints do exist in this problem, concerned with the even spread of classes through the week, not overloading any teacher on any one day, etc., the problem is dominated by the basic hard constraints already described: finding times and qualified teachers which avoid clashes and therefore also keep every student group occupied for every time of the week.

Earlier work on this problem [3,5] has been successful in assigning times to meetings in such a way that, at each time, resources are sufficient to fill all the resource slots of meetings scheduled for that time. This would be a complete solution except for one problem.

The problem is the *teacher constancy requirement*, which states that, when a meeting contains multiple times, any teacher assigned to it must attend for all of those times. We do not want, say, an English teacher slot to be filled by Smith for the first two times, Jones for the next three, and Robinson for the last time. Violations of this requirement, known as *split assignments*, have often been unacceptably frequent when solving the problem using the cited earlier methods. The problem does not arise with student group slots, since they are preassigned, and is usually considered unimportant for room slots, except that when a meeting contains two contiguous times (called a *double period*), it is preferable for the same room to be assigned at both times, so that the class is not disrupted halfway through the double period by a change of room.

If every meeting contained the same number of times, say k, then it would be easy to achieve teacher constancy. Replace the k time slots in each meeting with just one time slot, solve the resulting problem, then duplicate each meeting k-fold. This is equivalent to the method, often used in North American universities, of defining certain patterns of times in advance (e.g. Mondays 9–10 plus Wednesdays 9–10 plus Fridays 9–10) and requiring all meetings to choose one pattern rather than a set of times. Unfortunately, in Australian secondary schools (and elsewhere) the number of times in each meeting depends on the importance of the subject matter. English and Mathematics each require six time slots; other subjects may have 6, 5, 4, 3, 2, or 1 time slot each. The 'time patterns' approach cannot be applied.

It is desirable to assign times to meetings in such a way that pairs of meetings either overlap completely in time or not at all. We say that timetables with this property have good *time coherence* [5]. Good time coherence will minimize the number of pairs of clashing meetings, and should minimize the forces which push teachers into split assignments.

In this paper we present a new algorithm for constructing secondary school timetables. Inspired by the classical edge colouring algorithm for class–teacher timetabling, but designed to handle the general problem, this new algorithm tries to schedule as many meetings as possible into the first time in the week, then as many of the remaining meetings as possible into the second, and so on. This approach seems to be comparable with earlier work in its ability to find suitable times for all time slots, but, unlike earlier work, offers much better prospects for making highly time-coherent timetables, as Section 2 will explain.

We are aware of one piece of closely related prior work, by de Werra [13]. In its general approach, of assigning as many meetings as possible to each time of the week in turn, our algorithm is the same as de Werra's, but our algorithm is more general in being able to handle teacher slots that are not preassigned, and our realization of the approach is very different, being enumerative in character, rather than heuristic.

In addition to the new algorithm, this paper contains an initial empirical study which shows that the new method is promising in practice. As the reader will find, there are several points where different means could be used to achieve the same ends, and we are only at the beginning of the task of exploring these alternatives.

Section 2 introduces the new algorithm, and Section 3 explains how we can test whether a set of meetings can run simultaneously, before assigning resources to the meetings. Sections 4 and 5 explore the two main phases of the algorithm in detail. Section 6 presents our results so far, and Section 7 contains our conclusions and plans for further work. A more detailed exposition of our work appears in [1].

2   Generalizing the Classical Edge Colouring Algorithm

Our new algorithm is inspired by the classical edge colouring algorithm for bipartite graphs, attributed to König [11], and apparently first applied to class–teacher timetabling by Csima [8] (see also [7,9,12]). We begin with a brief recapitulation of that algorithm, then proceed to its generalization.

The edge colouring algorithm applies when each meeting contains one preassigned teacher, one preassigned student group, and any fixed number of time slots. Times are to be assigned to these slots so that no teacher or student group has a clash; this is the only constraint.

The algorithm is based on a *bipartite graph*, which is a graph whose nodes are divided into two sets, which we will call the *left-hand nodes* and the *right-hand nodes*. The edges of a bipartite graph may only join left-hand nodes with right-hand nodes.

Build a bipartite graph by creating one left-hand node for each teacher, one right-hand node for each student group, and one edge for each time slot of each meeting. Each time slot lies in a meeting containing one teacher and one student group, and the corresponding edge connects the nodes corresponding to these two resources. For example, the graph

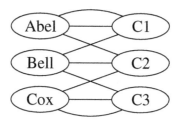

corresponds to three teachers (Abel, Bell and Cox) teaching three student groups (C1, C2 and C3). Abel takes C1 twice, and Cox takes C3 twice.

An *edge colouring* is an assignment of colours, or equivalently integers 1, 2, 3,... to the edges so that no two edges adjacent to any vertex have the same colour. If we interpret these colours as times, then colouring the edges corresponds to finding a timetable, and the rule prohibiting two edges with the same colour from touching any one vertex is equivalent to the timetabling requirement that no resource be required to attend two meetings at the same time.

There is an obvious lower bound on the number of colours needed to solve this problem: the maximum vertex degree. Edge colouring theory proves that this bound can always be achieved, by a polynomial time algorithm based on repeatedly finding maximum matchings.

A *matching* in a graph is a set of edges such that no two edges share an endpoint. A *maximum matching* is a matching with as many edges as possible.

Efficient algorithms for finding maximum matchings exist [2,6]. They are well known and will not be described here.

Here is our example graph with a maximum matching in bold:

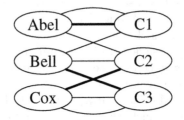

These edges are assigned the first colour, which corresponds with assigning the first available time to the corresponding time slots, then deleted. Because the edges form a matching, no two of them are adjacent, so no teacher or student group can have a clash at this time. A new matching is found and the second colour assigned to its edges, and so on until no edges are left. For minimality it turns out to be necessary to restrict the matching algorithm at each step to the vertices of maximum degree and the edges and vertices adjacent to them.

Edge colouring is not used in practical timetabling because it models too restrictive a version of the problem. In practice, meetings may have many more than two resources, and the resources are not necessarily preassigned. Conversely, some of the time slots may be preassigned. These generalizations make the problem NP-complete [4,9]. Nevertheless, if we interpret the edge colouring algorithm in timetabling terms we obtain an interesting idea for an algorithm for the general problem:

> Timetable as many meetings as possible into the first time of the week, concentrating on those meetings that are hardest to timetable. Delete the assigned time slots, delete any meetings that now have no time slots, and repeat on the second time of the week, then the third, and so on.

This is the algorithm we study in this paper.

To see why this algorithm is likely to deliver the time-coherence promised in Section 1, consider the set of meetings chosen during the first step to occupy the first time of the week. If all of these meetings contain more than one time slot, this exact same set of meetings may be re-used for the second time. In general we cannot expect that all of the meetings chosen will have exactly the same number of time slots, but by encouraging the algorithm to choose sets of meetings with a similar number of time slots, and taking care over what to do with leftover fragments of longer meetings, it should be possible to produce a very time-coherent timetable.

Clearly, the success of this algorithm will partly depend on whether large sets of meetings able to run simultaneously can be found. We have pursued an approach in which all such sets are computed in an initial phase, and this is the subject of Sections 3 and 4. After that, a second phase selects a combination of sets from the first phase that together cover all the meetings. This selection phase is the subject of Section 5.

3   Testing Sets of Meetings for Compatibility

Our first task is to find an efficient test which can tell us whether or not a set of meetings S is *compatible*, that is, whether or not its meetings can run simultaneously.

Meetings contain time slots and resource slots, which may be unconstrained, somewhat constrained, or completely constrained (i.e. preassigned).

Apart from a few preassignments, time slots in school timetabling problems are effectively unconstrained. There is often a soft constraint that the times of a meeting be spread through the week in some desirable pattern, but this does not affect any meeting's ability to run at any particular time.

We restructure the meetings of S to ensure that each meeting either contains exactly one preassigned time slot, or else it contains one or more unconstrained time slots. We do this by repeatedly finding any preassigned time slot s which is not the only time slot in its meeting m, creating a new meeting containing s as its only time slot and copies of all the resource slots of m, and deleting s from m. We then merge meetings whose time slots contain the same preassigned time; there is no need to distinguish meetings that are constrained to run simultaneously. Ignoring resource constraints for the moment, a set of such restructured meetings is compatible if no two parts of what was originally one meeting are involved, and the meetings' time slots do not include preassignments to two different times. (We do not currently have a complete implementation of this part of our test, but the number of preassigned times in our data is so small that it does not matter for present purposes.)

Resource constraints are the main problem. We need to determine whether the supply of resources is sufficient to fill all the resource slots of the set of meetings S. This problem has been solved before [3]; we briefly recapitulate that solution here.

We assume that all resources are available, unless the set of meetings contains a preassigned time (there can be at most one preassigned time after the restructuring above), in which case we leave out resources known to be unavailable at that time.

In practice we always find that each resource slot is constrained independently of the others to be filled by an element of some fixed subset of the available resources: it may require an English teacher, or a Science laboratory, and so on. Preassignment is included in this model: the fixed subset contains just one element.

Build a bipartite graph whose left-hand nodes are all the resource slots in the set of meetings S, and whose right-hand nodes are all the available resources. Connect each slot node to every resource able to fill that slot (these may form an arbitrary subset of the available resources). The meetings are compatible if a matching exists which touches every resource slot, for this matching represents an assignment of resources to all the slots.

For example, suppose we are testing three meetings for compatibility: English, History and Geography. Considering teachers alone, the bipartite graph might look like this:

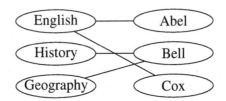

There are enough teachers, there is a qualified teacher for every slot, but there is no matching and these meetings are not compatible.

Although our test for compatibility is reasonably efficient as described, we will be calling it thousands of times, so we have implemented some optimizations. We check that no two slots are preassigned with the same resource, and that for every defined category (e.g. English teacher, Science laboratory) there are at least as many qualified resources as resource slots. Only if the meetings pass these quick tests do we take the time to build the bipartite graph and carry out the full test.

It is also important for efficiency that we can take a known compatible set of meetings for which a matching has been created and stored, as well as the totals needed to implement our quick tests, and efficiently test whether one extra meeting can be added without losing compatibility. The quick tests just add to the totals they keep; the standard bipartite algorithm can build on the existing matching, doing only the relatively small amount of work needed to add in nodes corresponding to the resource slots from the extra meeting. We update the matching as we add the slots, making failed matchings terminate faster and leaving less to undo. Deletion of the extra meeting's nodes after the test, if required, is also efficient.

4   Generating Compatible Sets of Meetings

A compatible set of meetings can be timetabled at any time during the week, except in the rare cases where it contains a preassigned time. Because of this, we decided to try to exhaustively generate all compatible sets of meetings in an initial phase, reasoning that we could select from this collection over and over again to build the timetable.

Our algorithm for generating all compatible sets of meetings is a dynamic programming algorithm based on the matroid recurrence

Every subset of a compatible set of meetings is compatible.

Finding compatible sets of meetings is an NP-hard problem similar to finding independent sets in a graph, and our algorithm is related to algorithms for that problem [2].

We generate all compatible sets of meetings of size (number of meetings) 1, then all compatible sets of meetings of size 2, etc. Our dynamic programming table holds the set of all sets of compatible meetings of size i, structured as a tree, with each subset represented by one path from the root to a leaf:

Sets stored

$\{m_1, m_2, m_3\}$
$\{m_1, m_2, m_5\}$
$\{m_1, m_3, m_5\}$
$\{m_2, m_3, m_5\}$
$\{m_2, m_3, m_6\}$

Tree-structured table

When extending from the set of $\{m_{j_1}, m_{j_2}, \ldots, m_{j_i}\}$ to the larger set of $\{m_{j_1}, m_{j_2}, \ldots, m_{j_i}, m_{j_{i+1}}\}$, by our recurrence we only need test for compatibility those meetings $m_{j_{i+1}}$ for which all i-element subsets of $\{m_{j_1}, m_{j_2}, \ldots, m_{j_i}, m_{j_{i+1}}\}$ are in the table (i.e. are compatible), and the tree structure helps to find the set of all such $m_{j_{i+1}}$ efficiently. Whenever a new set is added, we delete all its proper subsets, ensuring that only maximal compatible sets of meetings are in the table at the end. See [1] for more details.

Despite careful optimization of space and time, we have found that in practice there are too many maximal sets of compatible meetings for us to be able to compute them all, so we have been forced to reduce the number of meetings tested for compatibility by leaving out meetings with three or fewer resource slots. Like manual timetablers, we plan to pack these small meetings around the large ones in a final phase (Section 6).

5   Selecting Compatible Sets of Meetings

At the end of the phase just described, we have a table containing all maximal compatible sets of meetings, excluding the small meetings which we had to leave out. If there are T times in the week, we now need to select T of these sets, ensuring that each meeting m appears in the selected sets as many times as there are time slots in m. Repeated selection of the same set is allowed, and indeed preferred since it leads to time coherence.

This is a set covering problem [2] and many methods are available for solving it. We use a tree search with a greedy heuristic [10] for selecting the set of meetings to try next. We limit the branching factor at deep levels in the tree, and use forward checks to prune futile subtrees.

Our greedy heuristic is to prefer sets of meetings with larger numbers of resource slots. The number of meetings in the set is irrelevant at this stage: what matters is to utilize as many resources as possible.

Two optimizations well known in set covering [2] are implemented. If a meeting m appears in only one set, that set is selected immediately. And if every set containing meeting m_1 also contains m_2, we make sure that any set we select for covering m_2 also covers m_1, since we must cover m_1 eventually and at that time we will definitely be covering m_2 as well, by assumption.

When we select a set, we reduce by one the number of time slots of each of its meetings that remain to be assigned. When this number reaches 0 for some meeting m, we delete m from all remaining compatible sets of meetings, and if this causes any set S to become a proper subset of another, we delete S. Of course, if we backtrack we have to undo these changes. These operations are expensive and data structures to optimize them are required, but they pay off in reducing the amount of data handled at the deeper levels of the tree where most time is spent, and in permitting forward checks, which would otherwise not be based on current information.

Some valuable forward checks are implemented. If there are M meetings remaining to be timetabled and T times remaining to be used, then one of the remaining sets of meetings must contain at least $\lceil M/T \rceil$ meetings – if not, we backtrack immediately. Similarly, if the remaining resource slots of a particular type (e.g. Science laboratory slots) total S times' worth of that type of slot, and there are T times remaining to be used, then one of the remaining sets of meetings must contain at least $\lceil S/T \rceil$ of those slots.

6   Results

We have tested our new algorithm on BGHS98, an instance of the secondary school timetabling problem taken without simplification from a school in Sydney, Australia. BGHS98 contains 40 times, 150 resources (30 student groups, 56 teachers and 64 rooms) and 208 meetings. The number of time slots per meeting varies between 1 and 6, averaging just under 3. On average, we need to schedule 15.4 meetings into each time slot, and the meetings in each time slot must contain 102.6 resources on average. This is considerably less than the 150 resources total, but some parts of the resource load are very tight, notably the student groups (which must attend at every time) and the scarce Science laboratories (whose utilization must be virtually 100%).

Our tests were run on standard hardware. When we omitted meetings with three or fewer resource slots, only 57 of the 208 meetings remained, but these contained 75% of the resource slots. Finding all compatible sets of meetings among the 57 large meetings took only a few minutes,[1] although adding just a few of the omitted meetings caused the program to fail to terminate (not surprisingly, since these small meetings scarcely constrain each other and so

---

[1] An earlier version of this paper stated erroneously that the runs took a few seconds.

combine in exponentially many ways). Finding a cover for the 57 meetings also took just a few minutes. All the parts of our algorithm described in Section 5 contributed to this efficiency, in the sense that leaving any one out produced no useful result in any reasonable time. For example, without backtracking our greedy heuristic with forward checking and optimizations produced a timetable requiring 42 times, two more than the number of times available.

Assignment of the remaining small meetings in a final phase was not as easy as we had hoped it would be. Many methods are of course possible at this stage. Our first attempt, a simple heuristic assignment, was unsuccessful, and a subsequent hill-climber made almost no difference. So we tried meta-matching [3] and this assigned times to all but five meetings. The leftovers were Science meetings unable to find Science laboratories at the limited times that their student group resources were available.

7   Conclusion

We have presented a new algorithm for constructing secondary school timetables, based on the classical edge colouring algorithm for class–teacher timetabling.

Our success in timetabling all but five small meetings in a few minutes on standard hardware is very promising. However, this is work in progress and there is still a lot of work to do.

We need to do more work on the assignment of small meetings. A more elaborate final phase, with a backtracking element for example, might be sufficient. If not, we will need to integrate the smaller meetings into the main assignment phase, perhaps by greedily augmenting the selected compatible sets with compatible small meetings. This would allow the smaller meetings to influence the forward checking and backtracking, and provide a natural way to obtain time coherence among the small meetings.

Then will come the task of evaluating our solutions for time coherence, and tuning our algorithm to enhance it. We may need to encourage the selection of compatible sets of meetings with similar numbers of time slots. Finally, we will need to try out some resource assignment algorithms and verify that improved time coherence really does lead to fewer split assignments, as we expect.

References

1. Abraham, D.J.: The High School Timetable Construction Problem. Honours Thesis, School of Information Technologies, The University of Sydney (2002)
2. Christofides, N.: Graph Theory: an Algorithmic Approach. Academic, New York (1975)
3. Cooper, T.B., Kingston, J.H.: The Solution of Real Instances of the Timetabling Problem. Comput. J. **36** (1993) 645–653
4. Cooper, T.B., Kingston, J.H.: The Complexity of Timetable Construction Problems. In: Proc. 1st Int. Conf. Pract. Theory Timetabling (Napier University, Edinburgh, 1995)

5. Cooper, T.B., Kingston, J.H.: A Program for Constructing High School Timetables. In: Proc. 1st Int. Conf. Pract. Theory Timetabling (Napier University, Edinburgh, 1995) 283–295
6. Cormen, T.H., Leiserson, C.E., Rviest, R.L.: Introduction to Algorithms. MIT Press, Cambridge, MA (1990)
7. Csima, J., Gotlieb, C.C.: Tests on a Computer Method for Constructing School Timetables. Commun. ACM **7** (1964) 160–163
8. Csima, J.: Investigations on a Time-Table Problem. Ph.D. Thesis, School of Graduate Studies, University of Toronto (1965)
9. Even, S., Itai, A., Shamir, A.: On the Complexity of Timetable and Multicommodity Flow Problems. SIAM J. Comput. **5** (1976) 691–703
10. Johnson, D.S.: Approximation Algorithms for Combinatorial Problems. J. Comput. Syst. Sci. **9** (1974) 256–278
11. D. König. Graphok es Alkalmazasuk a Determinansok es a Halmazok Elmeletere. Mathematikai es Termeszettudomanyi **34** (1916) 104–119
12. Schmidt, G., Ströhlein. Timetable Construction – an Annotated Bibliography. Comput. J. **23** (1980) 307–316
13. de Werra, D.: Construction of School Timetables by Flow Methods. INFOR – Can. J. Oper. Res. Inform. Process. **9** (1971) 12–22

Flow Formulations for the Student Scheduling Problem

Eddie Cheng[1], Serge Kruk[1], and Marc Lipman[2]

[1] Department of Mathematics and Statistics, Oakland University,
Rochester, MI 48309-4485, USA
`echeng@oakland.edu,sgkruk@acm.org`
`http://www.oakland.edu/~kruk`
[2] Office of the Dean, School of Arts and Sciences,
Indiana University – Purdue University Fort Wayne,
Fort Wayne, IN 46805, USA
`lipmanm@ipfw.edu`

Abstract. We discuss the student scheduling problem as it generally applies to high schools in North America. We show that the problem is NP-hard. We discuss various multi-commodity flow formulations, with fractional capacities and integral gains, and we show how a number of practical objectives can be accommodated by the models.

1 Student Scheduling

The *Student Scheduling Problem* (we use the terminology found in [3]) is the assignation of students to sections of courses offered at various times during the week. The objective is to fulfil student requests, providing each student with a *conflict-free schedule* (no two assigned sections meeting at the same time), while respecting room capacities and possibly also balancing section sizes (or some other side constraint).

This problem has a very different flavour in a university and a high school. In a typical university, each student has a relatively sparse schedule. Moreover, often, it is the student's responsibility to ensure that the schedule is conflict-free. Of course the large number of students brings its own set of difficulties. On the other hand, in a typical North American high school, student schedules are complete (every hour is accounted for) and student requests for a given course are often binding. One of the authors has been involved for the past 20 years in the development of a commercial package for high school scheduling. We are aware that the practical difficulty of the problem varies tremendously between schools or even between semesters in the same school. Under what may seem like similar conditions, the computing time may range from a few seconds to many hours using the same machine with the same software.

As reported in the survey paper [3], which includes only refereed articles describing algorithms that have been implemented, it seems that flow models have not been used in this particular facet of the timetabling problem. There

E. Burke and P. De Causmaecker (Eds.): PATAT 2002, LNCS 2740, pp. 299–309, 2003.
© Springer-Verlag Berlin Heidelberg 2003

Table 1. Student course selection example

Student	Name	Selections
1001	John Q. Student	101, 126, 134, 156
1002	Susan B. Bright	101, 126, 135, 158

Table 2. Master schedule example

Course	Description	Section	Meetings	Instructor	Room	Size
101	French	01	M8, W8, F8	Cheng	R101	15
		02	M8, W8, F8	Kruk	R201	15
		03	T9, R9, F9	Cheng	L101	15
126	Chinese	01	M8, W8	Lipman	L123	10

have been algorithms based on branch and bound followed by heuristic improvements [15], some greedy approaches moderated by an intelligent ordering of the students [20], simulated annealing [4] and goal programming [19].

In view of the strides of approximation algorithms resulting from polyhedral theory during the past decade, it seems reasonable to revisit the problem and try to reformulate it with an eye towards such approximation algorithms [23]. We are thinking specifically of multi-commodity flow problems and variants [2,10].

Flow models have been used before for different aspects of timetabling: Single-commodity flows have been used by DeWerra [6] and multi-commodity flows were used by Even et al. [7,8] to assign teachers to rooms, times or classes. We are concerned here with a second phase of assigning students to classes already scheduled.

The input to the problem is the student's list of selected courses (as illustration, see Table 1) which we will refer to as the *selection,* and the *master schedule* of course offerings with their multiple sections, each with possibly a number of meeting times, rooms and instructors (see Table 2). The master schedule also contains course description, instructor names and rooms reserved, which may change for different meetings. It also contains the meeting times, which we show here encoded. They may be represented as days and times, or refer to some other table of equivalences, as we illustrate here.

A few comments about constraints are in order. In this example, both students wish to take Chinese (126) along with French (101). The goal of the student scheduling problem will be to satisfy, if possible, both student selections by assigning them to non-conflicting sections, while maintaining the number of students within the maximum size prescribed (10 in the unique Chinese section and 15 in each of the French sections). The first constraint is a hard constraint: assignments must not conflict in time (a student cannot be at two different places at once) and student requests must be satisfied if possible. The size constraint

Table 3. Notation for problem data

Symbol	Interpretation
K	the set of all students
I	the set of all courses
C_k	the set of courses selected by student $k \in K$
S_i	the set of sections of course $i \in I$
T_{ij}	the number of meetings of course $i \in I$, section $j \in S_i$
Z_{ij}	the maximum size of course $i \in I$, section $j \in S_i$

is a softer one: a valid solution minimizes the number of students above the prescribed limit but usually not at the expense of fulfilling selections, though this can vary with schools. The system must usually assign as much of a student selection as possible, dropping some unsatisfiable selection if need be.

2   Combinatorial Formulation of the Decision Problem

To make precise the problem described in the introduction, we now describe a combinatorial formulation. Table 3 establishes the notation.

We will also assume some preprocessing of the data to create a matrix of sections conflicting in time: for all $i \in I$, $j \in S_i$, $\bar{\imath} \in I$, $\bar{\jmath} \in S_{\bar{\imath}}$,

$$M_{ij\bar{\imath}\bar{\jmath}} = \begin{cases} 1, & \text{if course } i, \text{ section } j \text{ conflicts with course } \bar{\imath}, \text{ section } \bar{\jmath}; \\ 0, & \text{otherwise} . \end{cases}$$

This preprocessing is simple and done without loss of generality but could be avoided at the expense of a slightly more complex model. The decision variable is, for all $k \in K$, $i \in I$, $j \in S_i$,

$$y_{ijk} = \begin{cases} 1, & \text{if student } k \text{ is scheduled into course } i, \text{ section } j; \\ 0, & \text{otherwise} . \end{cases}$$

The first constraint indicates that to each course selected by a student corresponds exactly one section of that course. There are $\sum_{k \in K} |C_k|$ such constraints:

$$\forall k \in K, \forall i \in C_k, \quad \sum_{j \in S_i} y_{ijk} = 1 . \tag{1}$$

We then need to enforce that there are no conflicts in the schedule of a student. There are $\sum_{k \in K} \sum_{i \in C_k} |S_i|$ such constraints:

$$\forall k \in K, \forall i \in C_k, \forall j \in S_i, \quad \sum_{\bar{\imath} \in C_k} \sum_{\bar{\jmath} \in S_{\bar{\imath}}} y_{\bar{\imath}\bar{\jmath}k} M_{\bar{\imath}\bar{\jmath}ij} \le 1 . \tag{2}$$

Finally, we indicate that the maximum size per section must not be exceeded. There are $\sum_{i \in I} |S_i|$ such constraints:

$$\forall i \in I, \forall j \in S_i, \quad \sum_{k \in K} y_{ijk} \leq Z_{ij}. \tag{3}$$

Now that we have a precise formulation of the problem, we give a simple proof of its complexity.

Theorem 1. *The student scheduling problem is NP-complete.*

Proof. We will consider the problem of finding a schedule for one student. The proof is by reduction to the *independent set problem* [9]: Given a graph $G = (V, E)$ and an integer n, is there a set of nodes of size at least n, no two adjacent?

From an instance of the independent set problem, we construct a master timetable consisting of n different courses, each has $|V|$ sections (say $s_1, \dots, s_{|V|}$). Section i of all courses have the same meeting times which are constructed to satisfy the following condition: sections s_i and s_j have a meeting in common if and only if vertices v_i of G are adjacent.

We assume one student with a selection comprising all n courses. Then we have a conflict-free schedule for the student if and only if G has an independent set of size at least n since a conflict-free schedule is a set of n sections (one per course), no two sharing a meeting time.

The construction of the meeting times can be done by a search over each node and its neighbours, hence is proportional to $|V||E|$.   □

Notice that this establishes the difficulty of finding a conflict-free schedule for one student, with no room size constraints. The general problem for many students and size constraints is, perforce, not easier.

3   Flow Models of the Relaxed Optimization Problem

The Decision Problem described above is interesting but of little practical value. One problem is that, even for a school of small size, the feasible set is empty, hence the problem, as stated, has no solution. There will almost always be students with unsatisfiable course selections. Therefore, trying to solve the Decision Problem (1)–(3) yields no useful information on the schedules of all students that could see their selections satisfied. In practice the infeasibility of the problem should not detract from the (yet unstated) objective of providing as good a schedule as possible, for some measure of "goodness" for as many students as possible. In this section we present alternative models and show how various objectives can be accommodated.

The models are based on the integral multi-commodity flow problem. Each student represents the source (and sink) of a given commodity and the objective is to maximize the total flow. By Theorem 1, there must be additional constraints since the problem is NP-hard for one student yet the integral flow problem with one source-sink pair is polynomially solvable. The additional difficulty is reflected

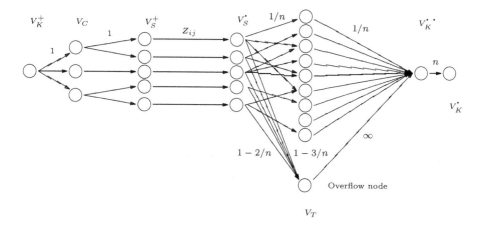

Fig. 1. Multi-commodity fractional flow model for one student

in the following models either as non-integral capacities or as gains on a subset of arcs. For the models to be equivalent to the decision problems, we need to add constraints that enforce integrality on every arc with integer capacity.

3.1   Fractional Capacities

We construct the first network in the following manner (Figure 1 illustrates part of the subnet of a single student with only three course selections). It is a layered network and we indicate by subscripted and superscripted V the names of the node subsets in each layer. The first layer consists of a set of source nodes, one for each student (V_K^+), duplicated in the last layer, as presink (V_K^{--}) and sink nodes (V_K^-). To the source is also associated a supply value, equal to the number of courses selected by a student with a corresponding demand at the student sink. These supply–demand pairs are of different commodities for each student. The objective is for a flow from student source i to go to student sink i and nowhere else. After duplication, there is a total of $3|K|$ of these nodes.

Concentrating now on the sub-network of one student, the first layer ends in a set of selected courses (V_C) with an arc of capacity $c_{ij} = 1$ from the student source. There is such an arc and node if the student has selected the course. There are $\sum_{k \in K} |C_k|$ such nodes.

The second layer consists of the sections of each course duplicated, in the third layer, as source (V_S^+) and sink (V_S^-). The arc from a course to one of its section has capacity $c_{ij} = 1$, while the capacity of a section-source to section-sink is equal to the maximum size of the section ($c_{ij} = Z_{ij}$). Each of the sections of the selected courses of each student are identified so that the flow from all students merge at this layer and diverge afterwards.

The section sinks have arcs to the next layer of timeslots (V_T), indicating the meeting times. If a section occupies t timeslots then the corresponding section sink node has $t + 1$ outgoing arcs. The first t have capacity $c_{ij} = 1/n$ where $n = |V_T|$, and the last has capacity $c_{ij} = 1 - t/n$ and ends at the overflow node. (Only two such arcs are labelled with their correct capacity to avoid clutter).

These timeslot nodes reach the student presink with capacity $c_{ij} = 1/n$, except for the arc coming out of the overflow node with capacity $c_{ij} = \infty$. The presink reaches the sink with capacity $c_{ij} = n$.

The resulting multi-commodity flow problem can be formalized in the following manner. Let $x_{ij}(k)$ be the measure of commodity k flowing on arc (i, j) where $k \in K$, the set of all students. Note that the various commodities are coupled only at the third layer. We therefore have a particularly well-structured multi-commodity flow problem.

Program (4) represents the formulation of a fractional multi-commodity flow problem:

$$\max \sum_{k \in K} \sum_{i \in V_K^+} \sum_{(i,j) \in A} x_{ij}(k) \tag{4a}$$

$$\text{s. to} \sum_{(i,j) \in A} x_{ij}(k) - \sum_{(j,l) \in A} x_{jl}(k) = 0, \quad \forall k \in K, j \in V \setminus \{V_K^+ \cup V_K^-\} \tag{4b}$$

$$\sum_{k \in K} x_{ij}(k) \leq c_{ij}, \qquad\qquad\qquad \forall (i, j) \in A \tag{4c}$$

$$x_{ij}(k) \geq 0 \qquad\qquad\qquad\qquad \forall k \in K, (i, j) \in A \tag{4d}$$

$$x_{ij}(k) \text{ integral}, \qquad\qquad\qquad \forall (i, j) \in \{V_K^+, V_C\}, \tag{4e}$$
$$\forall (i, j) \in \{V_C, V_S^+\},$$
$$\forall (i, j) \in \{V_S^+, V_S^-\},$$
$$\forall (i, j) \in \{V_K^{--}, V_K^-\}.$$

The reader should note that, even though it is possible to multiply all capacity by n and therefore get a simpler flow model, the resulting solution would be meaningless for our purposes. For example, what would be the proper assignment of a student to a section if a flow of value 2 went from a selection node (in V_C) to two different sections of the course (in V_S^+), each with a flow value of 1?

Definition 1. *A flow represents a schedule in the sense that if the flow from some V_K^- to V_K^- passes through a node V_S^+, the student is assigned to the corresponding section.*

Definition 2. *A flow is valid if it is feasible and if it is integral on all arcs with integral capacity.*

We are now is a position to state the useful properties of this model. For the rest of this section, *feasible* means feasibility with respect to program (4). The first result, stated without proof, is obtained from the construction of the model.

Lemma 1. *Given a master schedule, any set of conflict-free student schedules not exceeding section sizes is representable by a valid flow.*

It is possible to add selections to each student schedule (lunch period, home-room periods, "free" time, etc.) in such a manner that every hour of every day is accounted for. If this is done, then a full schedule will use all V_T nodes and we obtain the following easy result which we state without proof.

Lemma 2. *A student schedule is complete (every hour is accounted for) if and only if the arc $(i, j), i \in V_K^{--}, j \in V_K^-$ corresponding to the student is saturated.*

More important for our purposes since we intend to develop algorithms to solve program (4), is the following lemma.

Lemma 3. *Any valid flow represents a conflict-free schedule.*

Proof. Consider a timeslot node $t \in V_T$. To show the schedule is conflict free, we need to show that any flow incoming to t travelled via a single section node. Note that there is only one outgoing arc from t, of capacity $1/|V_T|$. Say there is any flow coming from section-sink node $j \in V_S^-$. We claim this flow has exactly value $1/|V_T|$, hence saturates the only outgoing arc from t. It has that value because any nonzero valid flow has unit value on arc (i, j) and by conservation of flow constraint (4b), must split into flows of value $1/|V_T|$ towards each timeslot with excess going to the overflow node. Therefore no other section can send flow into node t.   □

3.2   From Fractional Capacities to Gains

The first network is correct but can be modified to eliminate the fractional capacities. The first effect of this modification is to allow standard integer programming software to be used to solve the modified model but there are some other benefits as well. Though there are standard transformations we could apply to the first model to use standard integer programming software, they involve many more additional binary variables.

The modification consists in the elimination of the overflow nodes and all incident arcs and the addition of a gain amplifier to the V_S^- node that multiplies the flow into the node by an integer equal to the number of arcs out of the nodes. Since these arcs are incident to the timeslots nodes, the effect is the same as with the overflow nodes, namely, picking-up all the timeslots corresponding to a given section. The capacity of these last arcs is 1. Figure 2 illustrates for one student.

This variation has fewer arcs and all capacities are integral and we will show that both networks are equivalent, so that the results obtained with the previous network are still valid. We give the complete formulation of the modified problem in (5) where $N(j)$ means the downstream neighbours of node j,

$$N(j) := \{i \in V \mid (j, i) \in A\},$$

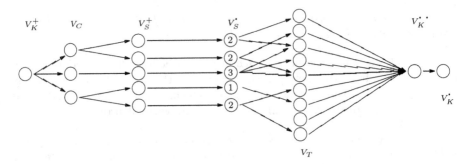

Fig. 2. Gain variation model for one student. The node annotations represent the gains. Arcs have unit capacity everywhere except at the common layer (V_S^+, V_S^-) where it is $c_{ij} = Z_{ij}$ and at rightmost layer (V_K^{--}, V_K^-), where it is $c_{ij} = n$

and the gains are given by

$$T_{ij} := \begin{cases} |N(j)| - 1 & j \in V_S^+, \\ 1 & \text{elsewhere}. \end{cases}$$

These gain values correspond to the fraction of the unit flow that was previously directed towards the overflow node in program (4):

$$\max \quad \sum_{k \in K} \sum_{i \in V_K^+} \sum_{(i,j) \in A} x_{ij}(k) \tag{5a}$$

$$\text{s. to} \quad \sum_{(i,j) \in A} T_{ij} x_{ij}(k) - \sum_{(j,l) \in A} x_{jl}(k) = 0, \quad \forall k \in K, j \in V \setminus \{V_K^+ \cup V_K^-\} \tag{5b}$$

$$\sum_{k \in K} x_{ij}(k) \le c_{ij}, \qquad \forall (i,j) \in A \tag{5c}$$

$$x_{ij}(k) \ge 0 \text{ and integral} \qquad \forall k \in K, (i,j) \in A. \tag{5d}$$

Lemma 4. *There is a one-to-one correspondence between a valid flow in program (4) (implicitly corresponding to the fractional capacitated flow model of Figure 1) and a valid flow in program (5) (implicitly corresponding to the gain model of Figure 2).*

Proof. Consider a node $j \in V_S^-$ with k neighbours in V_T. An incoming valid flow in the fractional model will have either value zero or unit value. Consider only the unit incoming flow. The k neighbours will each receive $1/|V_T|$ flow value with the excess $(1 - k/|V_T|)$ going to the overflow node. The corresponding valid flow in the gain model has the same unit flow incoming to the node and the multiplier is k, ensuring that each of the neighbours gets a unit flow. □

Network with gains have been studied for a while [13,16,18,11,12], and some specialized solution techniques are known. One important property of this particular network is that there are no cycles, hence no endogenous flows.

Since we have shown that both networks correctly model the student scheduling problem, we have the following theorem.

Theorem 2. *Let D be a directed network with both integral and non-integral capacities (or, equivalently, integral gains). Let s and t be two vertices in D. Let k be an integer. Then the question "does there exist a flow of integral value on all arcs of integral capacity, from s to t of value at least k" is NP-complete.*

3.3 Various Objectives

In both variations above, (4) and (5), we have written the objective function as

$$\max \quad \sum_{k \in K} \sum_{i \in V_K^+} \sum_{(i,j) \in A} x_{ij}(k) \, .$$

This choice maximizes the number of courses assigned, a reasonable choice but not the only one. A slight modification yields

$$\max \quad \sum_{k \in K} \sum_{j \in V_K^*} \sum_{(i,j) \in A} x_{ij}(k) \, ,$$

which maximizes the number of timeslots assigned, hence the "occupation" level of the students.

Other possibilities are to minimize the number of students with incomplete schedules. This is also a reasonable objective since a student with an incomplete schedule, however incomplete, will likely need to change his selections and therefore whatever was assigned will need to be de-assigned. This goal can be modelled with the following nonlinear objective function in the first network:

$$\max \quad \sum_{k \in K} \sum_{j \in V_K^*} \sum_{(i,j) \in A} f(x_{ij}(k)) \, ,$$

where

$$f(x) := (0 - x)\left(\frac{1}{n} - x\right)\left(\frac{2}{n} - x\right) \cdots \left(\frac{n-1}{n} - x\right) \, ,$$

so that $f(x) = 0$ unless the flow in the sink layer is at capacity. This relies on the fact that the a maximum flow will be a multiple of $\frac{1}{n}$.

Another approach is to move the size constraints in the objective function. From the second network, eliminate constraint (5c) and let the objective be

$$\min \sum_{k \in K} \sum_{j \in V_S^*} \left(\sum_{(i,j) \in A} x_{ij}(k) - Z_{ij} \right)^+ - \sum_{k \in K} \sum_{i \in V_K^+} \sum_{(i,j) \in A} x_{ij}(k) \, ,$$

where the nonlinearity appears in the function

$$(x)^+ := \begin{cases} x & \text{if } x \geq 0, \\ 0 & \text{otherwise}. \end{cases}$$

This formulation has the advantage of creating a separable multi-commodity flow problem, which is much more amenable to solutions via standard nonlinear convex programming techniques [1,5,14,17,21,22].

Other variations could be developed to account for the various goals of distinct schools. For example, the distinction between a required course and an elective could be handled via a weight factor on each student selection.

4   Conclusion

Our intention in this paper has been to develop mathematically useful models for a difficult, important problem. In this context, difficult means NP-complete for even the simplest case. Important means that the problem occurs in its full complexity across North America thousands of times each year. Mathematically useful means that the flow models can be adapted to satisfy different optimization criteria (types of partial solutions) for a variety of applications, and further, that these problem formulations are amenable to attack by approximation methods. Since approximation algorithms for multi-commodity flow problems are well-studied, and our network models are layered and coupled at only one layer, a sensible approach is to specialize existing algorithms to these models, taking advantage of the topology of the underlying networks. Experimentation has begun and will be reported in future work.

Acknowledgements. We wish to thank the three anonymous referees for their helpful suggestions.

References

1. Bertsekas, D.P., Polymenakos, L.C., Tseng, P.: An ϵ-relaxation Method for Separable Convex Cost Network Flow Problems. SIAM J. Optim. **7** (1997) 853–870
2. Brunetta, L., Conforti, M., Fischetti, M.: A Polyhedral Approach to an Integer Multicommodity Flow Problem. Discr. Appl. Math. **101** (2000) 13–36
3. Carter, M.W., Laporte, G.: Recent Developments in Practical Course Timetabling. In: Burke, E, Carter, M. (Eds.): Practice and Theory of Automated Timetabling II (PATAT 1997, Toronto, Canada, August, selected papers). Lecture Notes in Computer Science, Vol. 1408. Springer-Verlag, Berlin Heidelberg New York (1998) 3–19
4. Davis, L., Ritter, L.: Schedule Optimization with Probabilistic Search. In: Proc. 3rd IEEE Conf. on Artif. Intell. Appl. IEEE, (1987) 231–236
5. De Leone, R., Meyer, R.R., Zakarian, A.: A Partitioned ϵ-relaxation Algorithm for Separable Convex Network Flow Problems. Comput. Optim. Appl. **12** (1999) 107–126

6. de Werra, D.: Construction of School Timetables by Flow Methods. INFOR – Can. J. Oper. Res. and Inf. Process. **9** (1971) 12–22
7. Even, S., Itai, A., Shamir, A.: On the Complexity of Timetable and Multicommodity Flow Problems. In: Proc. 16th Annu. Symp. Found. Comput. Sci. (Berkeley, CA, 1975). IEEE Computer Society, Long Beach, CA (1975) 184–193
8. Even, S., Itai, A., Shamir, A.: On the Complexity of Timetable and Multicommodity Flow Problems. SIAM J. Comput. **5** (1976) 691–703
9. Garey, M.R., Johnson, D.S.: Computers and Intractability. Freeman, San Francisco, CA (1979)
10. Goemans, M.X., Williamson, D.P.: Improved Approximation Algorithms for Maximum Cut and Satisfiability Problems Using Semidefinite Programming. J. ACM **42** (1995) 1115–1145
11. Grinold, R.C.: Calculating Maximal Flows in a Network with Positive Gains. Oper. Res. **21** (1973) 528–541
12. Jensen, P.A., Bhaumik, G.: A Flow Augmentation Approach to the Network with Gains Minimum Cost Flow Problem. Manage. Sci. **23** (1976/77) 631–643
13. Jewell, W.S.: Optimal Flow through Networks with Gains. Oper. Res. **10** (1962) 476–499
14. Klincewicz, J.G.: A Newton Method for Convex Separable Network Flow Problems. Networks **13** (1983) 427–442
15. Laporte, G., Desroches, S.: The Problem of Assigning Students to Course Sections in a Large Engineering School. Comput. Oper. Res. **13** (1986) 387–394
16. Malek-Zavarei, M., Aggarwal, J.K.: Optimal Flow in Networks with Gains and Costs. Networks. **1** (1971/72) 355–365
17. Marín Gracia, Á.: Optimization of Nonlinear, Convex and Separable, Networks. In: Proc. 1st Int. Seminar Oper. Res. Basque Provinces (Zarauz, 1986). Univ. Pais Vasco, Bilbao (1986) 173–203
18. Maurras, J.F.: Optimization of the Flow through Networks with Gains. Math. Program. **3** (1972) 135–144
19. Miyaji, I., Ohno, K., Mine, H.: Solution Method for Partitioning Students into Groups. Eur. J. Oper. Res. **33** (1981) 82–90
20. Sabin, G.C.W., Winter, G.K: The Impact of Automated Timetabling on Universities – a Case Study. J. Oper. Res. Soc. **37** (1986) 689–693
21. Sun, J., Kuo, H.: Applying a Newton Method to Strictly Convex Separable Network Quadratic Programs. SIAM J. Optim. **8** (1998) 728–745 (electronic)
22. Tseng, P., Bertsekas, D.P.: An ϵ-relaxation Method for Separable Convex Cost Generalized Network Flow Problems. Math. Program. Ser. A **88** (2000) 85–104
23. Vazirani, V.V.: Approximation Algorithms. Springer-Verlag, Berlin Heidelberg New York (2001)

University Course Timetabling with Soft Constraints

Hana Rudová[1] and Keith Murray[2]

[1] Faculty of Informatics, Masaryk University,
Botanická 68a, Brno 602 00, Czech Republic
hanka@fi.muni.cz
[2] Space Management and Academic Scheduling, Purdue University,
400 Centennial Mall Drive, West Lafayette, IN 47907-2016, USA
kmurray@purdue.edu

Abstract. An extension of constraint logic programming that allows for weighted partial satisfaction of soft constraints is described and applied to the development of an automated timetabling system for Purdue University. The soft-constraint solver implemented in the proposed solution approach allows constraint propagation for hard constraints together with preference propagation for soft constraints. A new repair search algorithm is proposed to improve upon initially generated (partial) assignments of the problem variables. The model and search methods applied to the solution of the large lecture room component are presented and discussed along with the computational results.

1 Introduction

This paper describes the design approach and solution techniques being developed for an automated timetabling system at Purdue University. The initial problem considered here is the design of an intelligent system to assist with construction of the large lecture component of the university's master class schedule. The design anticipates expanding the scope of the problem to accommodate a demand-driven approach to timetabling all classes at the University. In *demand-driven timetabling*, student course selections are utilized to construct a timetable that attempts to maximize the number of satisfied course requests. In the initial problem we consider the course demands of almost 29 000 students enrolled in approximately 750 classes, each taught several times a week.

A solution to the timetabling problem is being developed using *constraint logic programming* (CLP) [28,18]. CLP is a respected technology for solving hard problems which include many complicated (non-linear) constraints [15]. Its main advantages over other frameworks are the declarative nature of problem descriptions via logical constraints, and a constraint propagation technique for reducing the search space.

Timetabling problems [8,27] are often over-constrained, which is the case with our problem since it is not possible to satisfy all requests of students for enrolment to specific courses. Preferential requirements for time and room assignment

E. Burke and P. De Causmaecker (Eds.): PATAT 2002, LNCS 2740, pp. 310–328, 2003.

may also lead to the problem being over-constrained. *Soft constraints* [10,5] can be applied to define these requirements declaratively rather than encapsulating many of them into the control part of the problem solution. In our problem solution, we have applied a weighted CSP [10] approach which considers weights/costs for each constraint and minimizes the weighted sum of unsatisfied constraints.

Our work includes development of a new solver for soft constraints. The solver was implemented as an extension of the CLP(FD) library [6] of SICStus Prolog. This approach is of particular importance for the construction of demand-driven schedules where complete satisfaction of all constraints is not feasible. Soft constraints have also been applied to accommodate the preferences of instructors with respect to time and room assignments for their classes. We will also describe a new search algorithm based on backtracking with constraint propagation. This search procedure allows the return of a partial solution even when the problem is over-constrained.

The following section of this paper presents a description of our timetabling problem. Section 3 explains the soft-constraint solver that was implemented. The new search algorithm developed for this problem is detailed in Section 4. This is followed by a description of how the problem has been modelled, including the representation of soft and hard constraints. In addition, the methods applied to the search for a feasible solution are discussed. Computational results are discussed in Section 6. Comparisons with other approaches to solving demand-driven timetabling problems and applying soft constraints to the problem solution are presented in Section 7. The final section reviews the results of our work and looks to future extensions of the problem solution and soft-constraint solver improvements.

2   Problem Description

At Purdue University, the timetabling process currently consists of constructing a master class schedule prior to student registration. The timetable for large lecture classes is constructed by a central scheduling office in order to balance the requirements of many departments offering large classes that serve students from across the university. Smaller classes, usually focused on students in a single discipline, are timetabled by "schedule deputies" in the individual departments. This process has been tailored to the political realities of a decentralized university, where a faculty can be quite put off by the idea of having a central office tell them when to teach, or even by providing such an office with much information about when they are available to teach.

A natural decomposition of the university timetabling problem has therefore resulted, consisting of a central large lecture timetabling problem and 74 disciplinary problems. Construction of the timetable for large lectures is the primary focus of this paper. This problem consists of approximately 750 classes having a high density of interaction that must fit within 41 lecture rooms with capacities up to 474 students. Course demands of almost 29 000 students out of a total enrolment of 38 000 must also be considered. The departmental problems range from only a few classes up to almost 700, with an average size of slightly more than 100 organized classes. The largest departmental problems are simplified by having many sections of the same course that are offered at multiple times.

The timetable maps classes (students, instructors) to meeting locations and times. A major objective in developing an automated system is to minimize the number of potential student course conflicts which occur during this process. This requirement substantially influences the automated timetable generation process since there are many specific course requirements in most programmes of study offered by the University.

To minimize the potential for time conflicts, Purdue has historically subscribed to a set of standard meeting patterns. With few exceptions, 1 hour x 3 day per week classes meet on Monday, Wednesday, and Friday at the half hour. 1.5 hour x 2 day per week classes meet on Tuesday and Thursday during set time blocks. 2 or 3 hours x 1 day per week classes must also fit within specific blocks, etc. Generally, all meetings of a class should be taught in the same location. Conforming to this set of standard meeting patterns will be seen to have a strong influence on both the problem definition and the solution process, since the meeting patterns defined for each class introduce hard restrictions on the acceptability of any generated solution.

Another important constraint on the problem solution is instructor availability. Balancing instructor time preferences was found to be a critical factor in developing acceptable solutions since an earlier attempt [20] to automate timetabling at Purdue was unsatisfactory largely because solutions heavily favoured instructors who imposed the most constraints.

Room availability is also a major constraint for Purdue. In addition to room capacity, it was necessary to consider specific equipment needs and the suitability of the room's location. Historically, a limited number of classrooms has been used to force a wide distribution of class times. Increased enrolments, however, have left the university with little excess room capacity.

Another aspect of the timetabling problem that must be considered here is the need to perform an initial student sectioning. Most of the classes in the large-lecture problem (about 75%) correspond to single-section courses. Here we have exact information about all students who wish to attend a specific class. The remaining courses are divided into multiple sections. In this case, it is necessary to divide the students enrolled to each course into sections that will constitute the classes. Without this initial sectioning it is not easy to measure the desirability or undesirability of having classes overlap in the timetable. Our current approach sections students in lexicographic order before joint enrolments between classes are computed. This gives us the worst-case possibility. The university currently processes a precise student schedule after the master class schedule is created, however, which should introduce some improvements. Possible directions for improving this solution are discussed in the final section.

3   Solver for Soft Constraints

Constraint propagation algorithms for the soft constraints are implemented as a part of the soft-constraint solver [24]. This constraint solver is built on top of the CLP(*FD*) solver of SICStus Prolog [6] and implemented with the help of attributed variables. An advantage of this implementation is the ability to include both hard constraints from the CLP(*FD*) library and soft constraints from a new *soft-constraint solver*.

3.1   Preference Variables and Preference Propagation

The soft-constraint solver handles preferences for each value in the domain of the variable which will be called the *preference variable*. Each preference corresponds to a natural number indicating the degree to which any soft constraint dependent on the domain value is violated. An increase of the preference during the computation with soft constraints is called *preference propagation*. Note the difference from constraint propagation, which removes values from the domain of the domain variable during the computation of hard constraints. Removal of domain values may also occur with preference variables. This corresponds to violation of a hard constraint.

We may also set the degree of acceptable violation for any preference variable. If the preference associated with a value in the domain of the preference variable should exceed this limit, it is removed from the domain. This possibility is of particular interest for time (or classroom) variables since all classes (each represented by a preference variable) should be relatively equal in importance.

Zero preference means complete satisfaction of the constraint for the corresponding value in the domain of the variable. Any higher preference expresses a degree of violation that would result from the assignment of this value to the variable. All values which are not present in the domain of the preference variable have the infinite preference **sup**. Preferences for each value in the domain of the variable may be initialized with a natural number. This allows us to handle initial preferences of values in the domain of the variable.

Example 1. The unary soft constraint pref(PA, [7-5, 8-0, 10-0],...) creates the preference variable PA with initial domain containing values 7, 8 and 10 and preferences 5, 0 and 0, resp. It means that the value 7 is discouraged w.r.t. other values. Preferences for the remaining values are assumed as infinite, indicating complete unsatisfaction.

3.2   Binary Soft Constraints

Two binary soft constraints have been implemented in the problem solution:

```
soft_different( PA, PB, Cost )\\
soft_disjunctive( PStart1, Duration1, PStart2, Duration2, Cost ).
```

The soft_different constraint expresses that the two preference variables PA, PB should have different values. The constant Cost gives us the cost for violation of this constraint. The soft_disjunctive constraint asks for the non-overlapping of the two tasks specified by the preference variables PStart1, PStart2 and the constant durations Duration1, Duration2. Again the Cost is the weight of this constraint.

Algorithms for both constraints are based on the partial forward checking algorithm and inconsistency counts [10] which are stored in the preferences for each value of the preference variable. Let us take a look at the soft_different constraint. Once the first preference variable is instantiated to some value X, the inconsistency count for the second variable and the value X should be increased by Cost, i.e. the preference propagation is processed for this variable and value. The soft_disjunctive would process preference propagation for all values in the interval X...(X+Duration-1).

3.3   Cost Function

For each preference variable, the soft-constraint solver maintains an additional domain variable (*cost variable*) having the current best preference of the preference variable as its lower bound. The initial upper bound is set to infinity. Any preference propagation results in an increase of the current best preference, with the lower bound of the cost variable being increased accordingly.

Preferences for each value of the preference variables are used to store initial preferences and any changes in the inconsistency counts for these values. This information is reflected in the cost variables. The sum of all cost variables gives us the total cost of the solution (*cost function*). The cost function can be then applied during labelling and optimization.

For efficiency, only the bound consistency is processed for all cost variables. This means that only changes to the lower and upper bounds are maintained. Any change in the current best preference is stored in the lower bound and, during optimization, the upper bound may be used to prune the search space.

4   Limited Assignment Number Search Algorithm

The aim of our problem solution is to be able to search for the complete assignment of the preference variables giving the best possible satisfaction of all soft constraints. Since the evaluation of the assignment is given by the sum of the corresponding cost variables, it may seem possible to apply a classical branch and bound algorithm. Unfortunately, it is not easy to find such a complete assignment. Mistakes in the assignment of some variable(s) may lead to a time consuming exploration of the search space with no complete solution. A complete assignment may not even exist due to conflicts among the hard constraints. Because of these disadvantages, we have proposed a new non-systematic iterative search algorithm based on chronological backtracking – *limited assignment number (LAN) search* algorithm [29]. It attempts to find some initial *partial assignment* of the variables and subsequently *repair* it such that all, or at least most, of the variables are assigned a value.

For each variable, the LAN search algorithm maintains a count of how many times a value has been assigned. A *limit* is set on this count. If the limit is exceeded, the variable is left unassigned and the search continues with the other variables. Labelling of unassigned variables is not processed even during backtracking. As a result of this search, a partial assignment of variables is obtained together with the set of the remaining *unassigned variables*.

Limiting the number of attempts to assign a value to each variable ensures the finiteness of this incomplete search. The current limit is set to the maximal domain size d of any labelled variable. As each of n variables can be tried d times, one iteration of the LAN search is of *linear complexity* $\mathcal{O}(dn)$.

Results of the LAN search process are used in subsequent iterations of the search. The following variable and value ordering heuristics are developed based on the previous iteration:

- values of successfully assigned variables are used as initial assignments in the next iteration – once a suitable value for a variable has been found, it remains a promising assignment;
- unsuccessfully attempted values for any variable left unassigned are demoted in the ordering so that they will be tried last in the subsequent iteration – a suitable value is more likely to remain among those that have not been tried;
- any variable left unassigned is labelled first in the subsequent iteration – it may be difficult to assign a value to the variable, therefore it should be given preference in labelling.

In the first iteration of the algorithm (*initial search*) we have a choice of using either problem-specific heuristics or standard heuristics, such as first-fail [28], for variable ordering. The most promising values are used for value ordering. In successive iterations (*repair searches*), heuristics based on the previous iteration are primarily used. Any ties are broken in favour of the initial search heuristics.

The user may also manually modify the results after each iteration to influence the behaviour of the heuristics, relax constraints to eliminate contradictory requirements, or change the problem definition. The options available for continuing the search are as follows:

1. processing the automated search as proposed above;
2. defining other values to be tried first or last based on user input, and process the repair search directed by the updated value ordering heuristics;
3. relaxing some hard constraints based on user input, and processing the initial search or the repair search;
4. addition or deletion of variables or constraints based on user input, and applying the repair search to reuse results of the former solution.

The first possibility is aimed at automated generation of a better assignment. Approach 2 allows the user to direct the search into parts of the search space where a complete assignment of variables might more easily be found. Step 3 can be useful if the user discovers a conflict among some hard constraints based on a partially generated output. The repair search can reuse most of the results from the former search and permute only some parts of the last partial assignment. Let us note that there is often an advantage to a user-directed search in timetabling problems, since the user may be able to detect inconsistencies or propose a suitable assignment based on a partially generated timetable. The last possibility introduces a new direction for the development of the algorithm aimed at incorporating changes to the problem input. This will be studied in detail as a part of our future work.

5   Problem Solving

We would like to describe a model for the timetabling problem which consists of variables for the time and room assignments of each class and of both hard and soft constraints, applying an approach described in the previous section. We also explore the control portion of the solution, which consists of the application of the proposed initial and repair searches.

5.1   Time and Classroom Variables

The domain of the time variables is represented by the natural numbers $0 \ldots 104$, corresponding to 5 days of 21 half-hours. The domain of the classroom variables is represented by the natural numbers $1 \ldots$ Number_Of_Classrooms.

Each class consists of between one and five meetings per week (typically two or three). All meetings have the same duration and are typically taught at the same time of day. Valid combinations of the number of meetings and the duration are called *meeting patterns*. Each meeting pattern (e.g. 1 hour x 3 meetings) has a defined set of days on which the meetings may be scheduled (e.g. Monday, Wednesday, Friday for 1 hour x 3 meetings). Interestingly, the start time of the first meeting of a class differs from the start times of the following meetings by a constant factor for most combinations (see Table 1). Excepting the MF (Monday and Friday) combination for two meetings per week and the four meeting patterns (includes less than 1% of classes), one time variable is sufficient to contain the complete information about the start time of classes. This is a preference variable indicating the start time of the first meeting (T1). It will be referred to as the *time preference variable*. The starting times of all remaining meetings (T2, ..., Tn) are domain variables only, and may be related to the time preference variable by the simple constraint

$$\text{Ti} \# = \text{T1} + \text{Constant} * (i - 1). \tag{1}$$

Preferences associated with each value in the domain of the time preference variable allow us to express the degree to which any time assignment for a class is preferred or discouraged. The remaining domain variables may be referenced

Table 1. Maximal sets of the possible combinations of days for class with given number of meetings per week (e.g., TTh means that course can have its meetings on Tuesday and Thursday)

Number of meetings	Possible combination of days
1	M or T or W or Th or F
2	MW or TTh or WF or MF
3	MWF
4	TWThF or MWThF or MTThF or MTWF or MTWTh
5	MTWThF

in the hard constraints (e.g. `serialized`), but they do not require the more expensive processing by the soft-constraint solver.

Since all class meetings should be taught in the same room, we suffice with only one common classroom variable for all meetings (called the *classroom preference variable*). As a preference variable, it associates a preference with each classroom expressing how desirable or undesirable it is for a given class.

5.2 Hard Constraints

Let us summarize the requirements which are implemented in the system using hard constraints:

1. meeting pattern specification;
2. prohibited or required times for classes;
3. class requires room with sufficient seating capacity;
4. class requires or prohibits some building(s) or room(s);
5. class requires or prohibits classroom of a specified generic type (computer, computer projection, audio recording, document camera, ...);
6. classes taught by the same instructor do not overlap;
7. sections of the same course do not overlap;
8. additional constraints over selected sets of classes: classes must be taught at the same times, on the same days, in the same classrooms, ...

Meeting pattern constraints relate the domain variables for all class meetings as was described in (1). In addition, the domain of the time preference variable is reduced such that all invalid values are removed.

Example 2. A 1.5 hour x 2 meetings class is represented by the two variables T1, T2. The first of these is the time preference variable with the initial domain $(0 \ldots 104)$ reduced to the values $21, 24, \ldots, 39$ because the TTh combination is valid only. The second domain variable is related with T1 by the constraint T2 #= T1 + (21*2)*1. The constant separating start times here is 21 periods x 2 days.

A 2 hours x 2 meetings class has MW, TTh, and WF as valid meeting day combinations. It is represented by the two variables T1, T2 related by the same constraint as before. The reduced domain of T1 corresponds to the values 0,4,8,12,16, 21,25,29,33,37, 42,46,50,54,58.

Requirements 2–5 are implemented by domain reduction in the corresponding domains of the time and classroom preference variables. Requirements 6 and 7 are included with help of the constraint `serialized` which constrains tasks, each with a start time and duration, so that no tasks ever overlap. Built-in constraints of CLP(*FD*) library of SICStus Prolog are used to implement various requirements over selected sets of classes as mentioned in item 8.

Additional hard constraints must be posted to ensure that each class is assigned to just one suitable classroom. This requirement could be implemented via the `disjoint2` constraint, which ensures non-overlapping of a set of rectangles. In our case, the rectangle is defined by the start time variable (`Time`) and the duration (`Duration`) of each meeting, and by the classroom variable (`Classroom`) for the corresponding class:

```
disjoint2([ rectangle(Time, Duration, Classroom, 1) | _ ]).
```

The number 1 represents the requirement of *one* classroom for each meeting.

A different type of propagation among the time variables is achieved via the `cumulative` constraint. It ensures that a resource can run several tasks in parallel, provided that the discrete resource capacity is not exceeded. If there are N tasks, each starting at a certain time (`StartI`), having a certain duration (`DurationI`) and consuming a certain amount of resource (`ResourceI`), then the sum of resource usage of all the tasks must not exceed resource limit (`ResourceLimit`) at any time:

```
cumulative([Start1,...,StartN], [Duration1,...,DurationN],
           [Resource1,...,ResourceN], ResourceLimit).
```

The `cumulative` constraint helps to assign a classroom of sufficient size to each meeting while allowing smaller classes to be assigned to larger classrooms.

Example 3. Let us imagine a small example with 2 rooms for 40 students, 3 rooms for 20 students, and 1 room for 10 students. The set of `cumulative` constraints follows:

```
cumulative(Time_meetings_with_size_40, Dur_40, ListOf1, 2),
cumulative(Time_meetings_with_size_20_40, Dur_20_40, ListOf1, 5),
cumulative(Time_all_meetings, Dur_all, ListOf1, 6).
```

The first constraint ensures that the largest classes are accommodated into the largest rooms, the second constraint allows medium-sized classes to be placed into rooms for 20 students, and also into rooms for 40 students if they are not already asked for by the first constraint. The third constraint allows movement of small classes between all rooms, subject to the condition that they are not occupied by any larger classes at the same time.

More precisely, we can post one cumulative(Starts, Durations, ListOf1, Limit) constraint for each possible size of classroom denoted by Size. The constant ListOf1 denotes a list of 1 representing a unit resource requirement (one classroom) by each course. Durations represents the durations of classes with the start time Starts. Variables Starts and Limit should satisfy the following properties:

Starts = {Start | meeting(Start,Duration,Capacity) ∧ Capacity ≥ Size}
Limit = card{Id | classroom(Id,Capacity) ∧ Capacity ≥ Size}.

Actually, it is sufficient to post this constraint only for some specific sizes of classrooms. Classrooms of similar size are grouped together to achieve better efficiency.

Another possibility for taking cumulative constraints into account consists of splitting classrooms into *groups* by size. Each class would be included in the group of corresponding size only. Such division can be useful if we do not want to put smaller classes into larger classrooms of other group at any time (e.g., smaller classes must be in the classrooms with a capacity smaller than 400 students).

5.3 Soft Constraints

Three types of soft constraints are currently handled by the system which will be discussed in this section:

1. unary constraints on time variables – faculty time preferences;
2. unary constraints on classroom variables – faculty preferences on the classroom selection for classes;
3. binary constraints for each joint enrolment between two classes.

Instructors may specify preferences for the days, hours, or parts of days they wish to encourage or discourage. This specification is transformed into a list of integer preferences corresponding to the possible start time of each class. We have seen that the initial selection of start times for each class is determined by its meeting pattern. The domain size of this set of start times can differ greatly among meeting patterns (it ranges from 5 to 50 possible values). This causes the relative effect of any given preference to vary greatly among the meeting patterns. To compensate for this effect, the number of preference points associated with instructor time preferences differ based on the meeting pattern.

Each class is associated with a time preference variable with preferences initialized either as specified by the instructor or to a set of default preferences. These default preferences are very important – their exclusion would lead to the construction of timetables which discriminate against classes for which no preferences have been provided. Many such classes would be placed in undesirable times, which no human timetabler would want to do.

Instructors may also specify positive or negative preferences towards the room selection for each class. It is possible to prefer or discourage particular classrooms, buildings, or properties of the room (e.g. "I prefer classrooms with a computer" or "I discourage classrooms without windows"). Each value in the domain of the classroom preference variable has either the specified preference or the neutral preference specification.

Any two classes potentially have a number of students who are enrolled in both at the same time. We seek to control their degree of overlap in the timetable by a generalization of the soft disjunctive constraint (see `soft_disjunctive` in Section 3.2). Such binary soft constraints include two time preference variables for corresponding classes, with the cost given by the number of students enrolled in both. Since each class may have several meetings, such a generalized disjunction needs to propagate preferences to all the values of the uninstantiated preference time variable which could be affected by the overlap of any of the meetings.

5.4   Cost Functions

There are two types of cost functions in our problem formulation. The first is related to the assignment of time preference variables. The second is dependent on the classroom preference variables.

Preferences associated with the time preference variables are influenced by the times faculty wish to teach or not teach, and by the soft constraints on joint enrolments. Time preferences are initialized based on faculty input. Constraints on joint enrolments propagate (increase) preferences during computation of the cost function. The sum of the cost variables (see Section 3.3) for the time preference variables gives this cost function, i.e. the solution cost w.r.t. time assignment. As a consequence, we need to balance the number of student joint enrolments from the third constraint with the number of preference points assigned by the first constraint, (e.g., the sum of preference points associated with an instructor for one class corresponds to an overlapping of 20 students). The relationship between preference points and joint enrolments is specified as part of the input data.

The sum of the cost variables related with the classroom preference variables represents the second cost function. Its value is dependent on the satisfaction of faculty preferences on classroom selection.

Both cost functions are independent of each other and introduce two different criteria in our problem. Minimizing student conflicts and accommodating the time preferences of classes were judged to be a more critical aspect of the problem than meeting preferences for classrooms.

5.5   Labelling

Labelling consists of two parts. Time preference variables are processed first, followed by room preference variables. This ordering corresponds to the relative importance of the cost functions defined over particular sets of variables (see Section 5.4).

One iteration of the LAN search is applied to find a partial assignment of time variables. A branch and bound algorithm is then used to find an optimal solution over classroom variables. If no solution for classroom variables is found within the time limit, one iteration of the LAN search over the classroom variables is processed. Once a partial assignment is generated by this process, the search for a complete solution may continue by repeating these steps. The user may also provide input to influence the behaviour of the search (see Section 4).

Initial and repair searches using the LAN algorithm over time variables are processed as described in Section 4. The initial search is always processed for classroom variables however (i.e. the limit on the number of assignments is set but new heuristics are not developed). Information about an unassigned classroom variable reflects back upon the corresponding time preference variable. Any classroom variable non-assignment is the result of no classroom being available for the corresponding time. To reflect this fact, the value assigned to a time variable with a corresponding unassigned classroom variable is discouraged in the subsequent search.

Different initial heuristics were used for time and classroom variable labelling but the main idea remains always the same: a variation of first-fail was applied for variable ordering and we have chosen the most preferred values.

Our first-fail heuristics to determine the ordering of classes for time assignment selects the most highly constrained variables with respect to both hard and soft constraints. First, we select among the variables having the smallest domain. Ties are broken based on the greatest number of soft constraints related to this variable. If not selected early, the domain of such a variable may become too small to select a sufficiently preferred value, or it may even become empty and cause a backtracking. Early propagation of soft constraints is also encouraged, so as not to discover mistakes too late. A specific class time assignment was selected among the most preferred values, i.e. we have chosen an optimistic approach for the value selection.

Just the first-fail approach was used to choose a class to be placed into a classroom. The most preferred classroom was selected and ties were broken by the selection of the smallest available classroom so as not to waste available resources.

6   Computational Results

Our data set from fall semester 2001 includes 747 classes to be placed into 41 classrooms. The classes included represent 81 328 course requirements for 28 994 students. A complete data set in the form of Prolog facts can be downloaded from http://www.fi.muni.cz/~hanka/purdue_data. In the future, we intend to add data from other semesters.

The results presented here were computed by SICStus Prolog 3.9.1 on a PC with AMD Athlon/850 MHz processor and with 128 MB of memory.

Figure 1 illustrates the number of classes with either time or classroom variables left assigned during subsequent iterations. It can be seen that the automated repair search was able to substantially improve on the initial solution. Only one class remained unassigned after eight iterations. Assignment of this

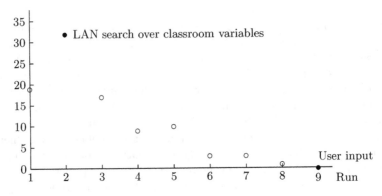

Fig. 1. Change in unassigned classes during the subsequent iterations

Table 2. Results of particular iterations

Run	1	2	3	4	5	6	7	8	9
Satisfied time (%)	83.8	81.8	79.9	82.2	80.7	80.7	80.6	80.7	80.7
Unsatisfied time (%)	4.2	4.6	4.4	4.5	4.5	4.5	4.2	4.2	4.5
Student conflicts (%)	1.5	1.6	2.0	1.9	2.1	2.2	2.3	2.3	2.3
Preferred classrooms (%)	74.2	45.3	52.1	47.0	52.7	56.3	49.7	52.3	52.3

class was successfully completed with the help of user input. The increase in the number of unassigned classes during the second iteration occurs as a result of the branch and bound search over classroom variables being replaced by the initial iteration of the LAN search.

Table 2 shows the computational results for the initial search and the subsequent automated repair searches. *Satisfied time* gives the percentage of how many encouraged times for classes were selected. *Unsatisfied time* refers to the percentage of discouraged times that were selected for classes. *Student conflicts* estimates the percentage of unsatisfied requirements for courses by students. *Preferred classrooms* measures the percentage of classes for which encouraged classrooms were selected. The current data set does not include any preferential requirements which discouraged specific classrooms.

We can see that most of preferential requirements of instructors were satisfied during assignment of time variables. The unsatisfied time percentage mostly illustrates that a number of classes must be taught at unpopular times due to the limited number of rooms available. Let us also note that these results include default preferences for classes with no preferred times.

One iteration took about 2–3 minutes for time labelling. One step of the branch and bound search for classroom variables took 1–2 seconds. The time labelling takes longer due to the preference propagation and more complex constraints (e.g. cumulative) posted on time variables.

Originally we started to solve the problem using a built-in backtracking search algorithm with a variety of variable and value ordering heuristics. This attempt did not lead to any solution after 10 hours of computations however. Too many failed computations were repeated exploring parts of the search tree with no solution.

One of the more important lessons learned here is that the disjoint2 and cumulative constraints must be used together in a redundant manner to find an acceptable solution (for description of both constraints see Section 5.2). The cumulative constraints were able to introduce additional constraint propagation for the time variables by informing them of the available room resources. Neither cumulative nor disjoint2 constraints alone were able to find an acceptable solution. Results using only the cumulative constraints left about 20 unassigned classes. Using only the disjoint2 constraint resulted in 50 unassigned classes.

7 Related Work

Currently the timetable for Purdue University is constructed by a manual process. An earlier approach examined for automating construction of the Purdue University timetable modelled the room–time assignment problem as a multiple-choice quadratic vertex packing and utilized a tabu search algorithm [20]. This approach was further developed into a prototype system used to create a schedule for large lecture classes in spring 1994, but was never adopted by university schedulers due to inadequacies in the way it handled instructor time preferences and student conflicts.

7.1 Soft Constraints

Various approaches to soft constraints have been introduced and studied [23]. These include basic frameworks over particular types of preferences (e.g. weighted [10], fuzzy CSPs [9]) and also meta-frameworks (e.g. partial [10], semiring-based CSPs [5]). Solving algorithms for soft constraints include extensions of the branch and bound algorithm [10,19] and local search methods, e.g. tabu search [12]. However, there are still few tools [4,13] available for soft constraint solving. To these authors' knowledge, no library can be used together with an existing CLP(FD) solver [6,17]. This paper presents the main ideas behind our proposal for a new soft-constraint solver built on top of the SICStus Prolog CLP(FD) solver [6]. We have further extended this work. A detailed description can be found in [24].

Soft constraints are often applied to solve timetabling problems with the help of constraint satisfaction. Unfortunately, they are mostly applied via the standard constraint satisfaction method, which offers no special support for more effective resolution of the soft constraints. Golz et al. [14] apply the typical solution – the given unary soft constraints, with priorities, are integrated into the solution search through value and variable ordering heuristics. An optimization constraint was applied for solving medium-sized problems [16]. Support for soft constraints is included in Abdennadher et al. [1]. They describe the solution of a department-sized problem, including soft constraint solver, for a cost-based approach implemented with Constraint Handling Rules [11].

7.2 Constraint Programming and Demand-Driven Timetabling

There have not been many attempts [25,3] to apply constraint programming to the solution of demand-driven timetabling problems where it is not possible to satisfy all requests of students. Such over-constrained problems require enhancements to the classical constraint satisfaction approach. Once these are developed, we can apply all of the advantages of constraint programming, including a declarative description of the problem together with strong propagation techniques.

We have previously constructed a demand-driven schedule for the Faculty of Informatics at Masaryk University [25] having 270 classes and about 1250 students. Conflicts of students between classes were controlled using a similar cost function as in our current approach. CLP allowed implementation of a variety of constraints available in ILOG Scheduler [21]. Soft constraints were implemented with the help of special variable and value ordering heuristics defined by the preferences of particular variables in the constraints. The timetable constructed was able to satisfy 94% of the demands of students and more than 90% of the preferential requirements of teachers. Unfortunately, it was not easy to extend this implementation to larger problems due to the bound consistency algorithms in ILOG Scheduler. Since these algorithms only propagate changes over the bounds of the domain variables, both constraint and preference propagations were much weaker than they are now.

A solution of the section assignment sub-problem is included in Banks et al. [3] via iterative addition of constraints into a CSP representation. Inconsistent constraints are not included in the final CSP representation, which allows solution of an over-constrained problem. This implementation, including its own constraint satisfaction solver, was verified using random timetabling problems based on problems from high schools in Edmonton, Alberta, Canada. The largest data set included requirements of 2000 students and 200 courses. They were able to satisfy 98% of student demand on more than half of the experiments. The solution presented is influenced by a special set of times that must be assigned to each course. It conforms well to high schools, but is rather different from the situation in university course timetabling. University class meeting patterns are not as strict, which results in a problem with a variety of additional requirements and preferences. Their inclusion into the problem solution can be more easily accomplished with the help of the embedded CLP system rather than implementing new procedures.

7.3   Other Approaches for Demand-Driven Timetabling

The comprehensive university timetabling system described by Carter [7] is characterized by problem decomposition with respect to both type and size of final sub-problems. They were able to solve the problem for 20 000 students and 3000 course sections. The system was used for 15 years at the University of Waterloo.

Aubin and Ferland [2] propose an iterative heuristic method to solve the problem which alternately assigns times and students to course sections until no further improvements to the solution can be found. The system was tested on data from a High School in Montreal including demands of 3300 students and 1000 courses.

Robert and Hertz [22] decompose the problem into a series of easier sub-problems corresponding to time, section, and classroom assignments and solve them via tabu search methods. The method presented is able to generate an initial solution which can be incrementally improved after problem redefinition (negotiation on initial constraints with teachers and students). The initial solution for about 500 students and 340 course sections satisfied about 80% of the preferential requirements of teachers. Forty-five students were involved in overlapping situations.

The local search heuristic procedure of Sampson et al. [26] solves a problem having a smaller solution space with 89 course sections and 230 students. They were able to meet 94% of the student scheduling requirements at the Graduate School of Business Administration at the University of Virginia.

8   Conclusion

We have proposed and implemented a solution to a large-scale university timetabling problem and have constructed a demand-driven schedule which is able to reflect diverse requirements of students during course enrolment. Our solution is able to satisfy the course requests of 98% of students. About 80% of preferential requirements on time variables were also met with only a small number of classes taught at discouraged times (about 4%). The automated search was able to find suitable times and classrooms for almost all classes. One remaining class was assigned with the help of the built-in support for user input.

Our proposal included a new solver for soft constraints which is of particular interest for timetabling problems where the costs in the problem are directly related to the present values for time and room assignments of classes. We have proposed a new search algorithm which allows us to find a solution to the problem (even when it is over-constrained). We have also discussed a special set of `cumulative` constraints which, together with the `disjoint2` constraint, processes stronger constraint propagation.

Our future research will include an extension of the problem solution together with improvements to the soft-constraint solver and search algorithm that have been described. Also, our approach must be validated using data sets from other semesters.

We are working on improvements to the search algorithm originally proposed for Purdue timetabling. First results, including experiments on random problems, can be found in [29].

Our current work also concerns pattern-oriented heuristics. They are aimed at improving the solution with the help of a pattern matching mechanism based on the sets of meeting patterns in the problem.

A new approach for making initial student section assignments for courses with multiple sections is currently under development. The proposal and implementation is based on Carter's [7] homogeneous sectioning, which tends to result in fewer classes having joint enrolments with others. This simplifies the task of finding non-conflicting assignments and appears to be an accurate representation when a final sectioning process will take place after construction of the timetable.

Purdue University currently relies on a completely manual process for constructing its timetable. A detailed comparison of results between the approach described in this paper and the manual process for the full large lecture problem is one of the next steps in our work. However, timetables are under continual revision by the timetablers who try to detect any possible problems in the generated solution. This allows us to extend the system and adjust it to be accepted by the community.

The primary focus of our future work will be support for making changes to the generated timetable. Once timetables are published they require many changes based on additional input. These changes should be incorporated into the problem solution with minimal impact on any previously generated solution.

We feel that the solution methods used for the large lecture problem should be directly applicable to construction of the 74 academic unit timetables. Some solution refinements may be necessary to simplify time assignments for introductory courses with large numbers of sections. Additional system architecture work will also be necessary to allow unit timetablers to use local preference files, and to work co-operatively if there is a high degree of inter-relationship between classes offered by the units. The administration of the university feels this work is very promising and has funded continuing work based on this approach.

Acknowledgements. This work is partially supported by the Grant Agency of the Czech Republic under the contract 201/01/0942 and by Purdue University.

We would like to thank our students Vlastimil Křápek, Aleš Prokopec and Kamil Veřmiřovský, who are assisting with the solution of this problem, and Purdue staff who have helped in many ways.

Our thanks also go to the Supercomputer Centre, Brno where experiments with the search algorithm were conducted.

References

1. Abdennadher, S., Marte, M.: University Course Timetabling Using Constraint Handling Rules. J. Appl. Artif. Intell. **14** (2000) 311–326
2. Aubin, J., Ferland, J.A.: A Large Scale Timetabling Problem. Comput. Oper. Res. **16** (1989) 67–77
3. Banks, D., van Beek, P., Meisels, A.: A Heuristic Incremental Modeling Approach to Course timetabling. In: Proc. Artif. Intell. '98 (Canada) (1998)
4. Bistarelli, S., Frühwirth, T., Martel, M., Rossi, F.: Soft Constraint Propagation and Solving in Constraint Handling Rules. In: Proc. ACM Symp. Appl. Comput. (2002.)
5. Bistarelli, S., Montanari, U., Rossi, F.: Semiring-Based Constraint Solving and Optimization. J. ACM **44** (1997) 201–236
6. Carlsson, M., Ottosson, G., Carlson, B.: An Open-Ended Finite Domain Constraint Solver. In: Programming Languages: Implementations, Logics and Programming. Lecture Notes in Computer Science, Vol. 1292. Springer-Verlag, Berlin Heidelberg New York (1997)
7. Carter, M.W.: A Comprehensive Course Timetabling and Student Scheduling System at the University of Waterloo. In: Burke, E, Erben W. (Eds.): Practice and Theory of Automated Timetabling III (PATAT 2000, Konstanz, Germany, August, selected papers). Lecture Notes in Computer Science, Vol. 2079. Springer-Verlag, Berlin Heidelberg New York (2001) 64–82
8. Carter, M.W., Laporte, G.: Recent Developments in Practical Course Timetabling. In: Burke, E, Carter, M. (Eds.): Practice and Theory of Automated Timetabling II (PATAT 1997, Toronto, Canada, August, selected papers). Lecture Notes in Computer Science, Vol. 1408. Springer-Verlag, Berlin Heidelberg New York (1998) 3–19
9. Dubois, D., Fargier, H., Prade, H.: Possibility Theory in Constraint Satisfaction Problems: Handling Priority, Preference and Uncertainty. Appl. Intell. **6** (1996) 287–309
10. Freuder, E.C., Wallace, R.J.: Partial Constraint Satisfaction. Artif. Intell. **58** (1992) 21–70
11. Frühwirth, T.: Constraint Handling Rules. In: Podelski, A. (Ed.): Constraint Programming: Basics and Trends (Châtillon-sur-Seine Spring School, France, May 1994). Lecture Notes in Computer Science, Vol. 910. Springer-Verlag, Berlin Heidelberg New York (1995)
12. Galinier, P., Hao, J.-K.: Tabu Search for Maximal Constraint Satisfaction Problems. In: Smolka, G. (Ed.): Proc. 3rd Int. Conf. on Principles and Practice of Constraint Programming. Lecture Notes in Computer Science, Vol. 1330. Springer-Verlag, Berlin Heidelberg New York (1997) 196–208
13. Georget, Y., Codognet, P.: Compiling Semiring-Based Constraints with clp(FD,S). In: Maher, M., Puget, J.-F. (Eds.): Principles and Practice of Constraint Programming (CP'98). Lecture Notes in Computer Science, Vol. 1520. Springer-Verlag, Berlin Heidelberg New York (1998) 205–219
14. Goltz, H.-J., Küchler, G., Matzke, D.: Constraint-Based Timetabling for Universities. In: Proc. INAP'98, 11th Int. Conf. on Applications of Prolog (1998) 75–80
15. Guéret, C., Jussien, N., Boizumault, P., Prins, C.: Building University Timetables Using Constraint Logic Programming. In: Burke, E, Ross, P. (Eds.): Practice and Theory of Automated Timetabling I (PATAT 1995, Edinburgh, Aug/Sept, selected papers). Lecture Notes in Computer Science, Vol. 1153. Springer-Verlag, Berlin Heidelberg New York (1996) 130–145

16. Henz, M., Würtz, J.: Using Oz for College Timetabling. In: Burke, E, Ross, P. (Eds.): Practice and Theory of Automated Timetabling I (PATAT 1995, Edinburgh, Aug/Sept, selected papers). Lecture Notes in Computer Science, Vol. 1153. Springer-Verlag, Berlin Heidelberg New York (1996) 162–177

17. IC-Park. ECL^iPS^e *Constraint Library Manual, Release 5.5,* 2002. http://www.icparc.ic.ac.uk/eclipse.

18. Jaffar, J., Maher, M.J.: Constraint Logic Programming: a Survey. J. Logic Program. **19** (1994) 503–581

19. Meseguer, P., Larrosa, J., Sánchez, M.: Lower Bounds for Non-binary Constraint Optimization Problems. In: Walsh, T. (Ed.): Principles and Practice of Constraint Programming (CP'01). Lecture Notes in Computer Science, Vol. 2239. Springer-Verlag, Berlin Heidelberg New York (2001) 317–331

20. Mooney, E.: Tabu Search Heuristics for Resource Scheduling. Ph.D. Thesis, Purdue University (1991)

21. Le Pape, C.: Implementation of Resource Constraints in ILOG SCHEDULE: a Library for the Development of Constraint-Based Scheduling Systems. Intell. Syst. Engng **3** (1994) 55–66

22. Robert, V., Hertz, A.: How to Decompose Constrained Course Scheduling Problems into Easier Assignment Type Subproblems. In: Burke, E, Ross, P. (Eds.): Practice and Theory of Automated Timetabling I (PATAT 1995, Edinburgh, Aug/Sept, selected papers). Lecture Notes in Computer Science, Vol. 1153. Springer-Verlag, Berlin Heidelberg New York (1996) 364–373

23. Rudová, H.: Constraint Satisfaction with Preferences. Ph.D. Thesis, Faculty of Informatics, Masaryk University (2001). http://www.fi.muni.cz/~hanka/phd.html

24. Rudová, H.: Soft CLP(*FD*). In: Haller, S., Russell, I. (Eds.): Proc. 16th Int. Florida Artif. Intell. Symp. (FLAIRS'03). AAAI Press (2003) (Accepted for publication)

25. Rudová, H., Matyska, L.: Constraint-Based Timetabling with Student Schedules. In: Burke, E, Erben W. (Eds.): Practice and Theory of Automated Timetabling III (PATAT 2000, Konstanz, Germany, August, selected papers). Lecture Notes in Computer Science, Vol. 2079. Springer-Verlag, Berlin Heidelberg New York (2001) 109–123

26. Sampson, S.E., Freeland, J.R., Weiss, E.N.: Class Scheduling to Maximize Participant Satisfaction. Interfaces **25** (1995) 30–41

27. Schaerf, A.: A Survey of Automated Timetabling. Technical Report CS-R9567. CWI, Amsterdam (1995)

28. Van Hentenryck. P.: Constraint Satisfaction in Logic Programming. MIT Press, Cambridge, MA (1989)

29. Veřmiřovský, K., Rudová, H.: Limited Assignment Number Search Algorithm. In: Student Res. Forum of the 29th Annu. Conf. on Curr. Trends in Theory and Practice of Informatics (SOFSEM'2002) (2002). See http://www.fi.muni.cz/~hanka/publications

A Comparison of the Performance of Different Metaheuristics on the Timetabling Problem

Olivia Rossi-Doria[1], Michael Sampels[2], Mauro Birattari[3], Marco Chiarandini[3], Marco Dorigo[2], Luca M. Gambardella[4], Joshua Knowles[2], Max Manfrin[2], Monaldo Mastrolilli[4], Ben Paechter[1], Luis Paquete[3], and Thomas Stützle[3]

[1] School of Computing, Napier University,
10 Colinton Road, Edinburgh, EH10 5DT, Scotland
{o.rossi-doria|b.paechter}@napier.ac.uk
[2] IRIDIA, Université Libre de Bruxelles, CP 194/6,
Av. Franklin D. Roosevelt 50, 1050 Bruxelles, Belgium
{msampels,mdorigo,jknowles,mmanfrin}@ulb.ac.be
[3] Intellektik, Technische Universität Darmstadt,
Alexanderstr. 10, 64283 Darmstadt, Germany
{mbiro,machud,lpaquete,tom}@intellektik.informatik.tu-darmstadt.de
[4] IDSIA, Galleria 2, 6928 Manno, Switzerland
{luca,monaldo}@idsia.ch

Abstract. The main goal of this paper is to attempt an unbiased comparison of the performance of straightforward implementations of five different metaheuristics on a university course timetabling problem. In particular, the metaheuristics under consideration are Evolutionary Algorithms, Ant Colony Optimization, Iterated Local Search, Simulated Annealing, and Tabu Search. To attempt fairness, the implementations of all the algorithms use a common solution representation, and a common neighbourhood structure or local search. The results show that no metaheuristic is best on all the timetabling instances considered. Moreover, even when instances are very similar, from the point of view of the instance generator, it is not possible to predict the best metaheuristic, even if some trends appear when focusing on particular instance classes. These results underline the difficulty of finding the best metaheuristics even for very restricted classes of timetabling problem.

1 Introduction

This work is part of the Metaheuristic Network[1], a European Commission project undertaken jointly by five European institutions, whose aim is to empirically compare and analyse the performance of various metaheuristics on different combinatorial optimization problems including timetabling.

Course timetabling problems arise periodically at every educational institution, such as schools and universities. A general problem consists in assigning a set of events (classes, lectures, tutorials, etc.) into a limited number of timeslots,

---

[1] http://www.metaheuristics.net

E. Burke and P. De Causmaecker (Eds.): PATAT 2002, LNCS 2740, pp. 329–351, 2003.

so that a set of constraints are satisfied. Constraints are usually classified as hard or soft. Hard constraints are constraints that must not be violated under any circumstances, e.g. students cannot attend two classes at the same time. Soft constraints are constraints that should preferably be satisfied, but can be accepted with a penalty associated to their violation, e.g. students should not attend three classes in a row. The general course timetabling problem is NP-hard. A considerable amount of research has dealt with the problem, and comprehensive reviews can be found in [6,21].

We consider here a reduction of a typical university timetabling problem. We aim at an unbiased comparison of the performance of straightforward implementations of five different metaheuristics on this problem. In order to attempt a fair and meaningful analysis of the results of the comparison we have restricted all the algorithms to the use of a common direct solution representation and search landscape. Moreover, all use the same library, programming language and compiler, and experiments are run on the same hardware.

The stress here is on the comparison of the different methods under similar conditions. More freedom in the use of more efficient representations and more heuristic information may give different results.

The rest of the paper is organized as follows. In Section 2 a description of the particular timetabling problem considered is given and the classes of instances used for the experiments are presented. In Section 3 we describe the common representation and search landscape used for all the implementations of metaheuristics. In Section 4 we give a description of the general features of each metaheuristic under consideration and details of our implementations. Finally, in Section 5 we outline the results and conclusions of our study.

2   The University Course Timetabling Problem

The timetabling problem considered here is a reduction of a typical university course timetabling problem. It has been introduced by Ben Paechter to reflect aspects of Napier University's real timetabling problem.

2.1   Problem Description

The problem consists of a set of events or classes E to be scheduled in 45 timeslots (5 days of 9 hours each), a set of rooms R in which events can take place, a set of students S who attend the events, and a set of features F satisfied by rooms and required by events. Each student attends a number of events and each room has a size. A feasible timetable is one in which all events have been assigned a timeslot and a room so that the following hard constraints are satisfied:

- no student attends more than one event at the same time;
- the room is big enough for all the attending students and satisfies all the features required by the event;
- only one event is in each room at any timeslot.

Table 1. Parameters used to produce the different instance classes

Class	small	medium	large
Num_events	100	400	400
Num_rooms	5	10	10
Num_features	5	5	10
Approx_features_per_room	3	3	5
Percent_feature_use	70	80	90
Num_students	80	200	400
Max_events_per_student	20	20	20
Max_students_per_event	20	50	100

In addition, a candidate timetable is penalized equally for each occurrence of the following soft-constraint violations:

- a student has a class in the last slot of a day;
- a student has more than two classes in a row;
- a student has a single class on a day.

Note that the soft constraints have been chosen to be representative of three different classes: the first one can be checked with no knowledge of the rest of the timetable; the second one can be checked while building a solution, taking into account the events assigned to nearby timeslots; and finally the last one can be checked only when the timetable is complete, and all events have been assigned a timeslot.

The objective of the problem is to minimize the number of soft constraint violations in a feasible solution. All infeasible solutions are considered worthless.

2.2 Problem Instances

A generator is used to produce problem instances with different characteristics for different values of given parameters. All instances produced have a perfect solution, i.e. a solutions with no constraint violations, hard or soft. The generator takes eight command line parameters which specify various characteristics of the instance, and a random seed. Using the same seed will produce the same problem instance – a different seed will produce a different instance with the same characteristics.

Three classes of instances of different size have been selected for comparison purposes, respectively called small, medium and large. They are generated with the sets of parameters reported in Table 1.

Each class of problem has been determined experimentally to be given a specified time limit. The time limits for the problem classes are respectively 90, 900 and 9000 seconds for the small, medium and large class. These timings refer to durations on a specific piece of hardware (see Section 5).

3   Common Search Landscape

All metaheuristics developed here for the Metaheuristics project employ the same direct solution representation and search landscape, as described in the following (see also [20]). In particular, we used a common local search in an evolutionary algorithm, an ant colony optimization algorithm, and an iterated local search. A simulated annealing, and a tabu search were restricted to the same neighbourhood structure.

3.1   The Solution Representation

We chose a direct solution representation to keep things as simple as possible. A solution consists of an ordered list of length $|E|$ where the positions correspond to the events (position i corresponds to event i for $i = 1, \ldots, |E|$). An integer number between 1 and 45 (representing a timeslot) in position i indicates the timeslot to which event i is assigned.

The room assignments are not part of the explicit representation; instead we use a matching algorithm to generate them. For every timeslot there is a list of events taking place in it, and a preprocessed list of possible rooms to which these events can be assigned according to size and features. The matching algorithm gives a maximum cardinality matching between these two sets using a deterministic network flow algorithm. If there are still unplaced events left, it takes them in label order and puts each one into the room of correct type and size which is occupied by the fewest events. If two or more rooms are tied, it takes the one with the smallest label. This procedure ensures that each event–timeslot assignment corresponds uniquely to one timetable, i.e. a complete assignment of timeslots and rooms to all the events.

3.2   The Neighbourhood Structure and Local Search

The solution representation described above allows us to define a neighbourhood using simple moves involving only timeslots and events. The room assignments are taken care of by the matching algorithm.

The neighbourhood is the union of two smaller neighbourhoods, N_1 defined by an operator that moves a single event to a different timeslot, and N_2 defined by a second operator that swaps the timeslots of two events.

The Local Search is a stochastic first improvement local search based on the described neighbourhood. It goes through the list of all the events in a random order, and tries all the possible moves in the neighbourhood for every event involved in constraint violations, until improvement is found. It solves hard-constraint violations first, and then, if feasibility is reached, it looks at soft-constraint violations as well. Delta evaluation of solutions is extensively used to allow a faster search through the neighbouring timetables. A more detailed description of the local search is outlined in the following:

1. Ev-count ← 0;
   Generate a circular randomly-ordered list of the events;
   Initialize a pointer to the left of the first event in the list;
2. Move the pointer to the next event;
   Ev-count ← Ev-count + 1;
   if (Ev-count = $|E|$) {
      Ev-count ← 0;
      goto 3.; }
   a) **if** (current event NOT involved in hard-constraint violation (hcv))
      { **goto** 2.; }
   b) **if** (∄ an untried move for this event) { **goto** 2.; }
   c) Calculate next move (first in N_1, then N_2) and[2] generate resulting potential timetable;
   d) Apply the matching algorithm to the timeslots affected by the move and delta-evaluate the result;
   e) **if** (move reduces hcvs) {
         Make the move;
         Ev-count ← 0;
         goto to 2.;}
   f) **else goto** 2.(b);
3. **if** (∃ any hcv remaining) END LOCAL SEARCH;
4. Move the pointer to the next event;
   Ev-count ← Ev-count + 1;
   if (Ev-count = $|E|$) END LOCAL SEARCH;
   a) **if** (current event NOT involved in soft-constraint violation (scv))
      { **goto** 4.; }
   b) **if** (∄ an untried move for this event) { **goto** 4.; }
   c) Calculate next move (first in N_1, then N_2)** and generate resulting potential timetable;
   d) Apply the matching algorithm to the timeslots affected by the move and delta-evaluate the result;
   e) **if** (move reduces scvs without introducing a hcv) {
         Make the move;
         Ev-count ← 0;
         goto 4.; }
   f) **else goto** 4.(b);

Since the described local search can take a considerable amount of CPU time, it could be more effective within the context of some of the metaheuristics to use this time in a different way. We therefore introduced in the local search a parameter for the maximum number of steps allowed, which was left free for the different metaheuristic implementations.

---

[2] That is, for the event being considered, potential moves are calculated in strict order. First, we try to move the event to the next timeslot, then the next, then the next etc. If this search through N_1 fails then we move through the N_2 neighbourhood, by trying to swap the event with the next one in the list, then the next one, and so on.

4   Metaheuristics

In the following we briefly describe the basic principles of each metaheuristic under consideration and give details of the implementations for the timetabling problem described in Section 2 which are used for the comparison.

4.1   Evolutionary Algorithm

Evolutionary algorithms (EAs) are based on a computational model of the mechanisms of natural evolution [3]. EAs operate on a population of potential solutions and comprise three major stages: selection, reproduction and replacement. In the selection stage the fittest individuals have a higher chance than those less fit of being chosen as parents for the next generation, as in natural selection. Reproduction is performed by means of recombination and mutation operators applied to the selected parents: recombination combines parts of each of two parents to create a new individual, while mutation makes usually small alterations in a copy of a single individual. Finally, individuals of the original population are replaced by the new created ones, usually trying to keep the best individuals and deleting the worst ones. The exploitation of good solutions is ensured by the selection stage, while the exploration of new zones of the search space is carried out in the reproduction stage, based on the fact that the replacement policy allows the acceptance of new solutions that do not necessarily improve existing ones.

EAs have been successfully used to solve a number of combinatorial optimization problems, including timetabling. State-of-the-art algorithms often use problem-specific information to enhance their performance, such as heuristic mutation [19] or some heuristically guided constructive technique [17].

Here, for the benefit of the comparison and understanding of the role of each component of the algorithm, we propose a basic implementation that uses only the problem-specific heuristic information coming from the local search. It is characterized by a steady-state evolution process, i.e. at each generation only one couple of parent individuals is selected for reproduction. A generational genetic algorithm, where the entire population is replaced at each generation, was also implemented, but the steady-state scheme gave better results. Tournament selection is used: that is, a number of individuals are chosen randomly from the current population and the best one in terms of fitness function is selected as parent. The fitness function $f(s)$ for a solution s is given by the weighted sum of the number of hard-constraint violations hcv and soft-constraint violations scv

$$f(s) := \# \ hcv(s) * C + \# \ scv(s) \,,$$

where C is a constant larger than the maximum possible number of soft-constraint violations. The crossover used is a uniform crossover on the solution representation, where for each event a timeslot's assignment is inherited either from the first or from the second parent with equal probability. The timeslot assignment corresponds uniquely to a complete timetable after applying the matching

---

Algorithm 1 Evolutionary algorithm

---
input: A problem instance I
for $i = 1$ **to** n **do**
    {generate a random population of solutions}
    $s_i \leftarrow$ random initial solution
    $s_i \leftarrow$ solution s_i after local search
    sort population by fitness
end for
while time limit not reached **do**
    Select two parents from population by tournament selection
    $s \leftarrow$ child solution after crossover with probability α
    $s \leftarrow$ child solution after mutation with probability β
    $s \leftarrow$ child solution after applying local search
    $s_n \leftarrow$ child solution s replaces worst member of the population
    sort population by fitness
    $s_{best} \leftarrow$ best solution in the population s_1
end while
output: An optimized solution s_{best} for I

---

algorithm. Mutation is just a random move in the neighbourhood defined by the local search extended with three-cycle permutations of the timeslots of three distinct events, which corresponds to the complete neighbourhood defined in [20]. The offspring replaces the worst member of the population at each generation. The algorithm is outlined in Algorithm 1.

The algorithm is a memetic algorithm using the local search described in Section 3. The local search is run with maximum number of steps 200, 1000 and 2000 respectively for the small, medium and large instances. The problem is to find a balance between a reasonable number of steps for the local search and a sufficient number of generations for the evolutionary algorithm to evolve while the local search is not abruptly cut too often and can effectively help to reach local optima.

The initial population is built assigning randomly, for each individual, a timeslot to each event according to a uniform distribution, and applying the matching algorithm. Local search is then applied to each member of the initial population. The population size n is 10, the tournament size is 5, crossover rate is $\alpha = 0.8$ and mutation rate is $\beta = 0.5$.

4.2   Ant Colony Optimization

Ant colony optimization (ACO) is a metaheuristic proposed by Dorigo et al. [10]. The inspiration of ACO is the foraging behaviour of real ants. The basic ingredient of ACO is the use of a probabilistic solution construction mechanism based on stigmergy. ACO has been applied successfully to numerous combinatorial optimization problems including the quadratic assignment problem, satisfiability problems, scheduling problems etc. The algorithm presented here is the first implementation of an ACO approach for a timetabling problem. It follows the ACS

Algorithm 2 Ant Colony System

$\tau(e,t) \leftarrow \tau_0 \ \forall \ (e,t) \in E \times T$

input: A problem instance I

calculate $c(e,e') \ \forall \ (e,e') \in E^2$

calculate $d(e), f(e), s(e) \ \forall \ e \in E$

sort E according to $\prec$, resulting in $e_1 \prec e_2 \prec \cdots \prec e_n$

$j \leftarrow 0$

while time limit not reached **do**

   $j \leftarrow j + 1$

   for $a = 1$ **to** m **do**

      {construction process of ant a}

      $A_0 \leftarrow \emptyset$

      for $i = 1$ **to** n **do**

         choose timeslot t randomly according to probability distribution P for event e_i

         perform local pheromone update for $\tau(e_i, t)$

         $A_i \leftarrow A_{i-1} \cup (e_i, t)$

      end for

      $s \leftarrow$ solution after applying matching algorithm to A_n

      $s \leftarrow$ solution after applying local search for $h(j)$ steps to s

      $s_{best} \leftarrow$ best of s and C_{best}

   end for

   global pheromone update for $\tau(e,t) \ \forall \ (e,t) \in E \times T$ using C_{best}

end while

output: An optimized candidate solution s_{best} for I

---

branch of the ACO metaheuristic, which is described in detail in [4] and which showed good results for the travelling salesman problem [9].

The basic principle of an ACS for tackling the timetabling problem is outlined in Algorithm 2. At each iteration of the algorithm, each of m ants constructs, event by event, a complete assignment of the events to the timeslots. To make a single assignment of an event to a timeslot, an ant takes the next event from a pre-ordered list, and probabilistically chooses a timeslot for it, guided by two types of information: (1) heuristic information, which is an evaluation of the constraint violations caused by making the assignment, given the assignments already made, and (2) stigmergic information in the form of a "pheromone" level, which is an estimate of the utility of making the assignment, as judged by previous iterations of the algorithm. The stigmergic information is represented by a matrix of "pheromone" values $\tau : E \times T \to \mathbf{R}_{\geq 0}$, where E is the set of events and T is the set of timeslots. These values are initialized to a parameter τ_0, and then updated by local and global rules; generally, an event–timeslot pair which has been part of good solutions in the past will have a high pheromone value, and consequently it will have a higher chance of being chosen again in the future. At the end of the iterative construction, an event–timeslot assignment is converted into a candidate solution (timetable) using the matching algorithm. This candidate solution is further improved by the local search routine. After all m

ants have generated their candidate solution, a global update on the pheromone values is performed using the best solution found since the beginning. The whole construction phase is repeated, until the time limit is reached.

The single parts of Algorithm 2 are now described in more detail. The following data are precalculated for events $e, e' \in E$:

$$c(e, e') := \# \text{ students attending both } e \text{ and } e',$$
$$d(e) := |\{e' \in E \setminus \{e\} \mid c(e, e') \neq 0\}|,$$
$$f(e) := \# \text{ features required by } e,$$
$$a(e) := \# \text{ students attending } e.$$

We define a total order[3] $\prec$ on the events by

$$e \prec e' \Leftrightarrow d(e) > d(e') \vee$$
$$d(e) = d(e') \wedge f(e) < f(e') \vee$$
$$d(e) = d(e') \wedge f(e) = f(e') \wedge a(e) > a(e') \vee$$
$$d(e) = d(e') \wedge f(e) = f(e') \wedge a(e) = a(e') \wedge l(e) < l(e').$$

Here, $l : E \to \mathbf{N}$ is an injective function that is only used to handle ties. We define $E_i := \{e_1, \ldots, e_i\}$ for the totally ordered events denoted as $e_1 \prec e_2 \prec \cdots \prec e_n$.

For the construction of an event–timeslot assignment each ant assigns sequentially timeslots to the events, which are processed according to the order $\prec$. This means that it constructs assignments $A_i : E_i \to T$ for $i = 0, \ldots, n$.

We start with the empty assignment $A_0 = \emptyset$. After A_{i-1} has been constructed, the assignment A_i is constructed as $A_i = A_{i-1} \cup \{(e_i, t)\}$ where t is chosen randomly out of T with the following probabilities:

$$P(t = t' \mid A_{i-1}, \tau) = \frac{\tau(e_i, t')^\alpha \cdot \eta(e_i, t')^\beta \cdot \pi(e_i, t')^\gamma}{\sum_{u \in T} \tau(e_i, u) \cdot \eta(e_i, u)^\beta \cdot \pi(e_i, u)^\gamma} .$$

The parameters β and γ control the weight of the heuristic information corresponding to hard- and soft-constraint violations, respectively. These heuristic functions η and π are defined as follows:

$$\eta(e_i, t') := \frac{1}{1 + \sum_{e \in A_{i-1}^{-1}(t')} c(e_i, e)}$$

is used to give higher weight to those timeslots that produce fewer student clashes. In order to give higher weight to those timeslots that produce fewer soft-constraint violations, we use

$$\pi(e_i, t') := \frac{1}{1 + L + S + R_{before} + R_{around} + R_{after}}$$

---

[3] We are aware of the fact that it could make more sense to order the events with respect to the features the other way around, but actually in this case it does not make a significant difference in terms of results.

with

$$L := \begin{cases} a(e_i) & \text{if } t' \text{ is the last timeslot of the day} \\ 0 & \text{otherwise,} \end{cases}$$

$S := \#$ students attending event e_i, but no other events belonging to the same day as t' in A_{i-1},

$R_{before} := \#$ students attending event e_i and also events in the two timeslots before t' on the same day,

$R_{around} := \#$ students attending event e_i and also events in both timeslots before and after t' on the same day,

$R_{after} := \#$ students attending event e_i and also events in the two timeslots after t' on the same day.

After each construction step, a local update rule on the pheromone matrix is applied for the entry corresponding to the current event e_i and the chosen timeslot t_{chosen}:

$$\tau(e_i, t_{chosen}) \leftarrow (1 - \psi) \cdot \tau(e_i, t_{chosen}) + \psi \cdot \tau_0 .$$

Parameter $\psi \in [0, 1]$ is the pheromone decay parameter, which controls the diversification of the construction process. The higher its value the smaller the probability to choose the same event–timeslot pair in forthcoming steps.

After assignment A_n has been completed, the matching algorithm for the assignment of rooms is executed, in order to generate a candidate solution s. The local search routine is applied to s for a number of steps $h(j)$ depending on the current iteration number $j \in \mathbf{N}$.

The global update rule for the pheromone matrix τ is performed after each iteration as follows. Let A_{best} be the assignment of the best candidate solution s_{best} found since the beginning. For each event–timeslot pair (e, t) we update:

$$\tau(e, t) \leftarrow \begin{cases} (1 - \rho) \cdot \tau(e, t) + \rho \cdot \frac{Q}{1 + q(s_{best})} & \text{if } A_{best}(e) = t \\ (1 - \rho) \cdot \tau(e, t) & \text{otherwise,} \end{cases}$$

where Q is a parameter controlling the amount of pheromone laid down by the update rule, and the function q measures the quality of a solution s as the sum of hard-constraint violations hcv and soft-constraint violations scv:

$$q(s) := \# \; hcv(s) + \# \; scv(s) .$$

The parameters for the algorithm described above were chosen after several experiments on the given test problem instances and are reported in Table 2.

4.3 Iterated Local Search

ILS [15] is based on the simple yet powerful idea of improving a local search procedure by providing new starting solutions obtained from perturbations of

Table 2. Parameters for the ACO algorithm

	small	medium	large
m	15	15	10
τ_0	0.5	10	10
ρ	0.1	0.1	0.1
α	1	1	1
β	3	3	3
γ	2	2	2
ψ	0.1	0.1	0.1
$h(j)$	$\begin{cases} 5\,000 \ j = 1 \\ 2\,000 \ j \geq 2 \end{cases}$	$\begin{cases} 50\,000 \ j \leq 10 \\ 10\,000 \ j \geq 11 \end{cases}$	$\begin{cases} 150\,000 \ j \leq 20 \\ 100\,000 \ j \geq 21 \end{cases}$
Q	10^5	10^{10}	10^{10}

a current solution, often leading to far better results than when using random restart [12,15,16,18,22]. To apply ILS, four components have to be specified. These are a GenerateInitialSolution procedure that generates an initial solution s_0, a Perturbation procedure, that modifies the current solution s leading to some intermediate solution s', a LocalSearch procedure that returns an improved solution s'', and a procedure AcceptanceCriterion that decides to which solution the next perturbation is applied. A scheme for ILS is given below.

Algorithm 3 Iterated local search

$s_0 = $ GenerateInitialSolution()
$s = $ LocalSearch(s_0)
while termination condition not met **do**
    $s' = $ Perturbation(s, *history*)
    $s'' = $ LocalSearch(s')
    $s = $ AcceptanceCriterion(s, s'', *history*)
end while

In our implementation for the university course timetabling problem the LocalSearch procedure was the common local search described in Section 3. GenerateInitialSolution generates initial random solutions according to a uniform distribution, so that no problem-specific information is used. We implemented the following three types of perturbation moves:

P1: choose a different timeslot for a randomly chosen event;
P2: swaps the timeslots of two randomly chosen events;
P3: choose randomly between the two previous types of moves and a three-exchange move of timeslots of three randomly chosen events.

All random choices were taken according to a uniform distribution. Each different move is applied k times, where k is chosen of the set $\{1, 5, 10, 25, 50, 100\}$.

Hence, it determines the strength of the perturbation. The `Perturbation` is applied to the solution returned by the `AcceptanceCriterion`. We considered three different methods for accepting solutions in `AcceptanceCriterion`:

Random Walk. This method always accepts the new solution s'' returned by `LocalSearch`.

Accept if Better. The new solution s'' is accepted if it is better than s. This leads to a first improvement descent in the space of the local optima.

Simulated Annealing. The new solution s'' is always accepted if it is better than the current one. Otherwise s'' is accepted with a probability based on the evaluation function $f(s)$, but infeasible new solutions are never accepted when the current one is feasible. $f(s)$ is the number of hard-constraint violations if both s and s'' are infeasible, or the number of soft-constraint violations if they are both feasible. Two methods for calculating this probability were applied:

$$\text{SA1: } P_1(s, s'') = e^{-\frac{(f(s)-f(s''))}{T}}$$
$$\text{SA2: } P_2(s, s'') = e^{-\frac{(f(s)-f(s''))}{T \cdot f(s_{best})}}$$

where T is a parameter called *temperature* and s_{best} is the best solution found so far. The value of T is kept fixed during the run, and it is chosen from $\{0.01, 0.1, 1\}$ for SA1 and $\{0.05, 0.025, 0.01\}$ for SA2.

Finally, we even considered applying ILS without `LocalSearch`. In this implementation, the `Perturbation` switched between moves P1 and P2. The values of k tested were chosen from the set $\{25, 50, 100, 200\}$. This implementation is also known as Reduced Variable Neighbourhood Search.

We ran all combinations of parameters for the `small` and `medium` instances in an automated parameter tuning procedure, the racing algorithm proposed by Birattari et al. [2]. This method, based on the Friedman statistical test, empirically evaluates a set of candidate configurations discarding bad ones as soon as statistically sufficient evidence is gathered against them. The instances used in the race were generated with the problem instance generator described in Section 2.2. The best resulting configurations of parameters for each instance class are summarized as follows:

Small instances:
    Type of `Perturbation`: P1
    $k = 1$
    `AcceptanceCriterion`: SA2 with $T = 0.025$

Medium instances:
    Type of perturbation: P1
    $k = 5$
    `AcceptanceCriterion`: SA1 with $T = 0.1$

For the `large` instances the automated tuning would have required a very large amount of time given the actual computational environment available. Consequently, the same parameter setting found for the `medium` instances is used also for the `large` ones.

4.4   Simulated Annealing

Simulated Annealing (SA) is a local search inspired by the process of annealing in physics [7,14]. It is widely used to solve combinatorial optimization problems, especially to avoid getting trapped in local optima when using simpler local search methods [1]. This is done as follows: an improving move is always accepted while a worsening one is accepted according to a probability which depends on the amount of deterioration in the evaluation function value, such that the worse a move is, the less likely it is to accept it. Formally a move is accepted according to the following probability distribution, dependent on a virtual temperature T, known as the Metropolis distribution:

$$p_{accept}(T, s, s') = \begin{cases} 1 & \text{if } f(s') \leq f(s) \\ e^{-\frac{(f(s') - f(s))}{T}} & \text{otherwise}, \end{cases}$$

where s is the current solution, s' is the neighbour solution and $f(s)$ is the evaluation function. The temperature parameter T, which controls the acceptance probability, is allowed to vary over the course of the search process.

We tackle the course timetabling problem in two distinct phases. In a first phase only hard constraints are considered and reduced. When a feasible solution is reached, which means no hard constraints are violated any more, a second phase starts and tries to minimize the number of soft-constraint violations. Going back from a feasible to an infeasible solution is not allowed. In the first phase the evaluation function f is given by the number of hard-constraint violations, hcv, while in the second phase by the number of soft-constraint violations, scv.

Algorithm 4 outlines the global procedure with the two phases of SA, where T_h is the temperature in the infeasible region, and T_s is the temperature in the feasible region.

We implemented versions of SA that differ in the following components: neighbourhood exploration strategy, initial temperature, cooling schedule and temperature length. The different variants and different parameters for these components have been object of an automated tuning for finding the best configuration, using the same racing algorithm as described in Section 4.3. The considered choices for the four components are the following:

Neighbourhood exploration. We considered two strategies for generating a neighbouring solution:

1. Strategy 1 considers moves from neighbourhoods N_1 and N_2 in the same order as in the local search outlined in Section 3.2. Only events involved in constraint violations are considered. Yet, different from the local search procedure, after trying all possible moves for each event of the timetable, the algorithm does not end but continues with a different order of events.
2. Strategy 2 abandons the local search framework and uses a completely random move selection strategy. At each step the proposed move is generated randomly from the union of N_1 and N_2.

Algorithm 4 Simulated annealing
___
input: A problem instance I
$s \leftarrow$ random initial solution
{Hard Constraints phase}
$T_h \leftarrow T_{h0}$;
while time limit not reached and $hcv > 0$ **do**
    Update temperature;
    $s' \leftarrow$ Generate a neighbouring solution of s
    if $f(s') < f(s)$ **then**
        $s \leftarrow s'$;
    else
        $s \leftarrow s'$ with probability $p(T, s, s') = \mathrm{e}^{-\frac{(f(s')-f(s))}{T}}$
    end if
    $s_{best} \leftarrow$ best between s and s_{best}
end while
{Soft Constraints phase}
$T_s \leftarrow T_{s0}$
while time limit not reached and $scv > 0$ **do**
    Update temperature
    $s' \leftarrow$ Generate a neighbouring solution of s
    if $hcv = 0$ in s' **then**
        if $f(s') < f(s)$ **then**
            $s \leftarrow s'$
        else
            $s \leftarrow s'$ with probability $p(T, s, s') = \mathrm{e}^{-\frac{(f(s')-f(s))}{T}}$
        end if
        $s_{best} \leftarrow$ best between s and s_{best}
    end if
end while
output: An optimized solution s_{best} for I
___

Initial temperature. Two possibilities were considered:

1. Use the temperature that provides a probability of $1/e$ for accepting a move that worsens by 2% the evaluation function value of a randomly generated solution s_r. Formally, choose T such that

$$p = \frac{1}{e} = \mathrm{e}^{-\left(\frac{0.02 \cdot f(s_r)}{T}\right)}$$

   i.e. $T = 0.02 \cdot f(s_r)$

2. Sample the neighbourhood of a randomly generated initial solution, compute the average value of the variation in the evaluation function produced by the sampled neighbours, and multiply this value by a given factor to obtain the initial temperature. We fix the size of the sample to 100 neighbours.

Because the first method produces initial temperatures which do not scale well with the hardness of the instances, the latter one is preferred. Two different multiplier factors (`TempFactHcv` and `TempFactScv`) have to be considered in the tuning of the algorithm, one for the initial value of T_h and the other one for the initial value of T_s.

Cooling schedule. We use a non-monotonic temperature schedule realized by the interaction of two strategies: a standard geometric cooling and a temperature re-heating. The standard geometric cooling computes the temperature T_{n+1} in iteration $n+1$ by multiplying the temperature T_n in iteration n with a constant factor α (cooling rate):

$$T_{n+1} = \alpha \times T_n, \qquad 0 < \alpha < 1.$$

This schedule is expected to be competitive with the adaptive cooling as proposed for the Graph Partitioning Problem by Johnson et al. [12] and in many successful implementations for timetabling where parameters were obtained by experimentations. A sort of adaptation to the behaviour of the search process, however, is included in our implementation by re-heating the temperature when the search seems to be stagnating. Indeed, according to a rejection ratio given by the number of moves rejected on the number of moves tested, the temperature is increased to a value equal to the initial one when the ratio exceeds a given limit. This inspection is done every fixed number of iterations, in our case three times the temperature length. Cooling rate and rejection ratio are thus other parameters which need tuning; effectively these are four parameters α_h, α_s, `RejLimHcv`, `RejLimScv` w.r.t. the phase undertaken. For the sake of simplicity we fix $\alpha_h = \alpha_s = \alpha$.

Temperature length. The number of iterations at each temperature is kept proportional to the size of the neighbourhood, as suggested by Johnson et al. [13], who remarked that this seems to be necessary in order to obtain high-quality solutions. The rate of the neighbourhood size (`NeighRate`) is another parameter to optimize. We keep it the same for the two phases.

Suggested by experimental observations, we also tested a version of SA in which the temperature is kept constant over the whole search process. The idea is that the SA acceptance criterion is useful at the end of the search for getting away from local optima, allowing the acceptance of worsening moves. Therefore maintaining a certain probability of accepting worsening moves during the whole search, and above all during the end of the process, could produce competitive results with less effort for properly tuning the parameters since in this case only the constant temperature needs to be considered. We tested this version using the first neighbourhood exploration strategy.

In all the cases the stopping criterion is the time limit imposed by the experimental set-up.

Table 3 summarizes the components and the values of the different parameters that were tuned with the racing algorithm for the SA. The 70 different configurations tested in the race were generated from all the possible combinations of these values.

The result of the race is that the best implementation, out of the three described, for both the small and medium instances, is the SA with complete

Table 3. SA parameters for the three versions considered. `NeighRate` is the proportion of the neighbourhood examined at each temperature, α is the cooling rate, `TempFactHcv` and `TempFactScv` are the multiplier factors for the initial temperature, respectively for hard and soft constraints, `AccLimHcv` and `AccLimScv` are the acceptance ratio limits respectively for the hard- and soft-constraint loops

			SA in LS		
`NeighRate`	α	`TempFactHcv`	`TempFactScv`	`RejLimHcv`	`RejLimScv`
0.1	0.8	0.1	0.3	0.98	0.97
0.2	0.9	0.2	0.62		
	0.97		0.7		

		SA in LS with fixed temperature			
`NeighRate`	α	`TempFactHcv`	`TempFactScv`	`AccLimHcv`	`AccLimScv`
–	–	0.1	0.05	–	–
		0.2	0.1		
			0.15		
			0.2		
			0.3		

		SA random			
`NeighRate`	α	`TempFactHcv`	`TempFactScv`	`AccLimHcv`	`AccLimScv`
0.1	0.8	*0.1*	*0.3*	*0.98*	*0.97*
0.2	**0.9**	**0.2**	0.62		
	0.95				

random move selection. In Table 3 the values for the winning configurations are indicated by italic face for the `small` instances and by bold face font for the `medium` instances. As in the ILS case, the configuration proposed for the `large` instances is the same as used for the `medium` instances.

Figure 1 shows the behaviour over the search process for the temperature, the acceptance ratio, soft-constraint violations and hard-constraint violations of the best configuration found in a run on a `medium` instance. The behaviour on `small` instances is similar.

4.5   Tabu Search

Tabu search (TS) is a local search metaheuristic which relies on specialized memory structures to avoid entrapment in local minima and achieve an effective balance of intensification and diversification. TS has proved remarkably powerful in finding high-quality solutions to computationally difficult combinatorial optimization problems drawn from a wide variety of applications [1,11]. More precisely, TS allows the search to explore solutions that do not decrease the objective function value, but only in those cases where these solutions are not forbidden. This is usually obtained by keeping track of the last solutions in terms of the move used to transform one solution to the next. When a move is per-

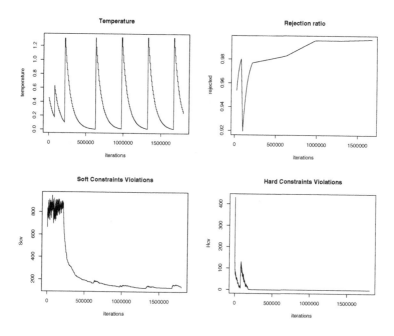

Fig. 1. Behaviour of SA components over the search process in a run of the best configuration found on a medium instance. We remark that we check the rejection rate only every three times the temperature length

formed the reverse move is considered *tabu* for the next l iterations, where l is the tabu list length. A solution is forbidden if it is obtained by applying a tabu move to the current solution.

The implementation of TS proposed for the university course timetabling problem is described in the following. According to the neighbourhood operators described in Section 3.2, a move is defined by moving one event or by swapping two events. We forbid a move if at least one of the events involved has been moved less than l steps before. The tabu status length l is set to the number of events divided by a suitable constant k (we set $k = 100$). With the aim of decreasing the probability of generating cycles and reducing the size of the neighbourhood for a faster exploration, we consider a variable neighbourhood set: every move is a neighbour with probability 0.1. Moreover, in order to explore the search space in a more efficient way, tabu search is usually augmented with some aspiration criteria. The latter are used to accept a move even if it has been marked tabu. We perform a tabu move if it improves the best known solution.

The TS algorithm is outlined in Algorithm 5, where L denotes the tabu list. In summary, it considers a variable set of neighbours and performs the best move that improves the best known solution, otherwise it performs the best non-tabu move chosen among those belonging to the current variable neighbourhood set.

Algorithm 5 Tabu search

---

input: A problem instance I
$s \leftarrow$ random initial solution
$L \leftarrow \emptyset$
while time limit not reached **do**
    for i=0 to 10% of the neighbours **do**
        $s_i \leftarrow s$ after i-th move
        compute fitness $f(s_i)$
    end for
    if $\exists s_j | f(s_j) < f(s)$ and $f(s_j) \leq f(s_i) \forall i$ **then**
        $s \leftarrow s_i$
        $L \leftarrow L \cup E_i$ where E_i is the set of events moved to get solution s_i
    else
        $s \leftarrow$ best non-tabu moves between all s_i
        $L \leftarrow L \cup E_b$ where E_b is the set of events moved by the best non-tabu move
        $s_{best} \leftarrow$ best solution so far
    end if
end while
output: An optimized solution s_{best} for I

---

5   Evaluation

We tested the five algorithms on a PC with an AMD Athlon 1100 Mhz on five small instances with running time 90 seconds, in 500 independent trials per metaheuristic per instance; five medium instances with running time 900 seconds, in 50 independent trials per metaheuristic per instance; and two large instances with running time 9000 seconds, in 20 independent trials per metaheuristic per instance.

The complete results of the experiments, the test instances and all the algorithms can be found at http://iridia.ulb.ac.be/~msampels/ttmn.data.

Results for one instance of each class small and medium are summarized in Figures 2 and 3. The results of all trials on a single instance are ordered by the quality of the solution (number of soft-constraint violations) and the rank of the solution in all solutions. An invalid solution (with hard-constraint violations) is considered to be worse than any valid solution. Thus it is ordered behind them. The solutions are grouped by the metaheuristic used. In the boxplots a box shows the range between the 25% and the 75% quantile of the data. The median is indicated by a bar. The whiskers extend to the most extreme data point which is no more than 1.5 times the interquantile range from the box. Outliers are indicated as circles.

Figures 4 and 5 show the results for the two large instances with additional diagrams to report the distribution of the valid solutions and the percentage of invalid solutions that were found in the 20 independent trials of each implemented metaheuristics.

In the following we give a few highlights and comments on the results on each class of instances.

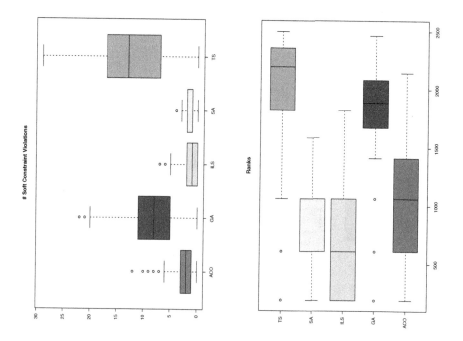

Fig. 2. Results for the `small03` instance

On the `small` instances all the algorithms reach feasibility in every run. ILS generally performs best, followed closely by SA and ACO. GA is definitely worse, but TS shows the worst overall performance.

SA is best on `medium` instances, even if it does not reach feasibility in some runs. ILS is still very good and more reliable in terms of feasibility. GA and TS give similar bad results, and ACO shows the worst performance.

For the first `large` instance `large01` most metaheuristics do not even find feasibility. TS reaches feasibility for about 8% of the trials, ILS for a bit more, and, when it does, results for soft constraints are definitely better than the TS ones. ILS is again best for the `large02` instance, where it finds feasibility for about 97% of the trials against only 10% of ACO and GA. SA never reaches feasibility, while TS gives always feasible solutions but with worse results than ILS and ACO in terms of soft constraints.

The results presented here have to be read bearing in mind the context to which they belong. Strong restrictions have been made on the implementations of the metaheuristics, as the use of a single representation and a single search landscape, and a minimal use of problem-specific heuristics. The use of a different representation, indirect and/or constructive, a different neighbourhood structure and local search, or more freedom in the use of additional heuristic information, might give different results.

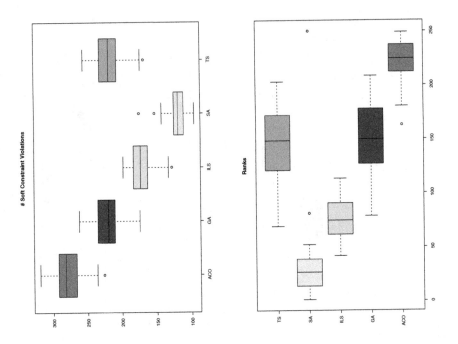

Fig. 3. Results for the medium01 instance

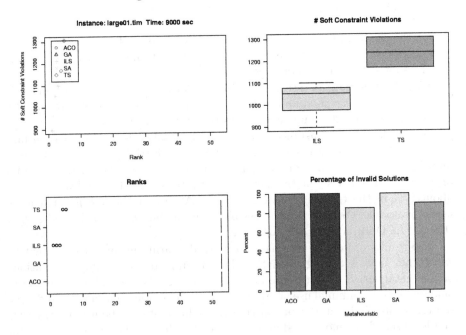

Fig. 4. Results for the large01 instance

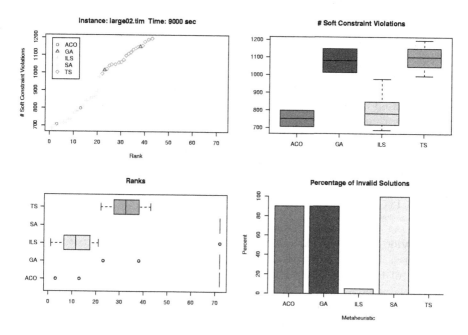

Fig. 5. Results for the `large02` instance

6   Conclusion

Based on the full set of results presented in the previous section, we can make the following general conclusions, also confirmed by the analysis made in [8]:

1. problem instance difficulty varies (sometimes significantly) between problem instances, across categories, and to a lesser extent, within a category (i.e. where all parameters to the generator except the random seed were the same), in terms of the observed aggregated performance of the metaheuristics. This is what we had expected, and reflects real data, where some specific problem (e.g. a particular choice of subjects by a particular student) can make timetabling much more difficult in a particular year;
2. the absolute performance of a single metaheuristic varies (sometimes significantly) between instances, within and, to a lesser extent, across categories;
3. the relative performance of any two metaheuristics varies (sometimes significantly) between instances, within and, to a lesser extent, across categories;
4. the performance of a metaheuristic with respect to satisfying hard constraints and satisfying soft constraints may be very different.

These conclusions lead us to believe that it will be very difficult to design a metaheuristic that can tackle general instances, even from the restricted class of problems provided by our generator. However, our results suggest that a hybrid algorithm consisting of at least two phases, one taking care of feasibility, the other taking care of minimizing the number of soft-constraint violations, is a promising research direction.

Additionally, we confirmed that knowing certain aspects of an instance does not guarantee that we will know about the structure of the search space, nor does it suggest a priori that we will know which metaheuristic will be best. This suggests the importance, in future, of trying to measure the search space characteristics directly; the aim here would be to try and match algorithms/parameter settings based on measurements of these characteristics, so that some of the a priori uncertainty of performance is removed.

This is ongoing work. In order to progress further in understanding these problems we have organized an international competition [23] based on our generator. We hope that this will shed more light on these difficult problems. We have also begun to analyse search space characteristics and relate these to metaheuristic performance.

Acknowledgements. This work was supported by the *Metaheuristics Network*, a Research Training Network funded by the Improving Human Potential programme of the CEC, grant HPRN-CT-1999-00106. The information provided is the sole responsibility of the authors and does not reflect the Community's opinion. The Community is not responsible for any use that might be made of data appearing in this publication.

References

1. Aarts, E.H.L., Lenstra J.K. (Eds.): Local Search in Combinatorial Optimization. Wiley, Chichester (1997)
2. Birattari, M., Stützle, T., Paquete, L., Varrentrapp, K.: A Racing Algorithm for Configuring Metaheuristics. Technical Report, Intellektik, Technische Universität Darmstadt, Germany (2002)
3. Baeck, T., Fogel, D., Michalewicz Z.: Evolutionary Computation 1: Basic Algorithms and Operators. Institute of Physics Publishing, Bristol (2000)
4. Bonabeau, E., Dorigo, M., Theraulaz, G.: From Natural to Artificial Swarm Intelligence. Oxford University Press, Oxford (1999)
5. Burke, E, Carter, M. (Eds.): Practice and Theory of Automated Timetabling II (PATAT 1997, Toronto, Canada, August, selected papers). Lecture Notes in Computer Science, Vol. 1408. Springer-Verlag, Berlin Heidelberg New York (1998)
6. Carter, M.W., Laporte, G.: Recent Developments in Practical Course Timetabling. In: Burke, E, Carter, M. (Eds.): Practice and Theory of Automated Timetabling II (PATAT 1997, Toronto, Canada, August, selected papers). Lecture Notes in Computer Science, Vol. 1408. Springer-Verlag, Berlin Heidelberg New York (1998) 3–19
7. Cerný, V.: A Thermodynamical Approach to the Traveling Salesman Problem. J. Optim. Theory Appl. **45** (1985) 41–51
8. Chiarandini, M., Stützle, T.: Experimental Evaluation of Course Timetabling Algorithms. Technical Report, FG Intellektik, TU Darmstadt (2002)
9. Dorigo, M., Gambardella, L.M.: Ant Colony System: A Cooperative Learning Approach to the Traveling Salesman Problem. IEEE Trans. Evolut. Comput. **1** (1997) 53–66

10. Dorigo, M., Maniezzo, V., Colorni, A.: The Ant System: Optimization by a Colony of Cooperating Agents. IEEE Trans. Syst. Man Cybern. **26** (1996) 29–41
11. Glover, F., Laguna, M.: Tabu Search. Kluwer, Boston, MA (1998)
12. Johnson, D.S., McGeoch, L.A.: The Traveling Salesman Problem: A Case Study in Local Optimization. In: Aarts, E.H.L., Lenstra, J.K. (Eds): Local Search in Combinatorial Optimization. Wiley, New York (1997) 215–310
13. Johnson, D.S., Aragon, C.R., McGeoch, L.A., Schevon, C.: Optimization by Simulated Annealing: an Experimental Evaluation I: Graph Partitioning. Oper. Res. **37** (1989) 865–892
14. Kirkpatrick, S., Gelatt, C.D., Vecchi, M.P.: Optimization by Simulated Annealing. Science **220** (1983) 671–680
15. Lourenço, H.R., Martin, O., Stützle, T.: Iterated Local Search. In: Glover, F., Kochenberger, G. (Eds): Handbook of Metaheuristics. Int. Series in Oper. Res. Management, Vol. 57. Kluwer, Dordrecht (2002) 321–353
16. Martin, O., Otto, S.W.: Partitioning of Unstructured Meshes for Load Balancing. Concurrency: Pract. Exper. **7** (1995) 303–314
17. Paechter, B., Rankin, R.C., Cumming, A., Fogarty, T.C.: Timetabling the Classes of an Entire University with an Evolutionary Algorithm. Parallel Problem Solving from Nature (PPSN). Lecture Notes in Computer Science, Vol. 1498. Springer-Verlag, Berlin Heidelberg New York (1998) 865–874
18. Paquete, L., Stützle, T.: Experimental Investigation of Iterated Local Search for Coloring Graphs. In: Cagnoni, S., Gottlieb, J., Hart, E., Middendorf, M., Raidl, G.(Eds): Applications of Evolutionary Computing (Proc. EvoWorkshops 2002). Lecture Notes in Computer Science, Vol. 2279. Springer-Verlag, Berlin Heidelberg New York (2002) 122–131
19. Ross, P., Corne, D., Fang, H.: Improving Evolutionary Timetabling with Delta Evaluation and Directed Mutation. In: Schwefel, H.P., Davidor, Y., Manner, R. (Eds.): Parallel Problem Solving from Nature (PPSN) III. Lecture Notes in Computer Science, Vol. 866. Springer-Verlag, Berlin Heidelberg New York (1994) 560–565
20. Rossi-Doria, O., Blum, C., Knowles, J., Sampels, M., Socha, K., Paechter, B.: A Local Search for the Timetabling Problem. In: Proc. 4th Int. Conf. Pract. Theory Automated Timetabling (PATAT 2002), Gent, Belgium, 124–127
21. Schaerf, A.: A Survey of Automated Timetabling. Artif. Intell. Rev. **13** (1999) 87–127
22. Stützle, T.: Local search Algorithms for Combinatorial Problems – Analysis, Improvements, and New Applications. Ph.D. Thesis, TU Darmstadt, Germany (1998)
23. http://www.idsia.ch/Files/ttcomp2002

Other Timetabling
Presentations

Other Timetabling Presentations

Other conference presentations are listed below together with the addresses of the authors.

Title: Theory and Practice of the Shift Design Problem
Author: W. Slany
Address: Institut für Informationssysteme, Technische Universitat Wien, Favoritenstrasse 9–11, A-1040 Wien, Austria

Title: Interactive Timetabling: Concepts, Techniques and Practical Results
Authors: T. Muller, R. Bartak
Address: Charles University, Department of Theoretical Computer Science, Malostranske namesti 2 25, Praha 1, Czech Republic

Title: Timetabling Using a Steady State Generic Algorithm
Authors: E. Ozcan, A. Alkan
Address: Yeditepe University, Department of Computer Engineering, 81120 Kayisdagi/Istanbul, Turkey

Title: Timetabling for Further Education Institutions Using Constraint Logic Programming
Authors: D. Matzke
Address: FhG-FIRST, Fraunhofer Institute for Computer Architecture and Software Technology, Kekulestrasse 7, D-12489, Berlin, Germany

Title: Functional Model of the Timetable Problem
Authors: V. G. Abramov, V. Franchak
Address: Department of Computer Science, Lomonosov Moscow State University. GSP-2, Leninskie, Russia

Title: A GA Evolving Instructions for a Timetable Builder
Authors: C. Blum, S. Correia, M. Dorigo, B. Paetcher, O Rossi-Doria, M. Snoek
Addresses: C. Blum, M Dorigo: IRIDIA, Université Libre de Bruxelles, Belgium; S. Correia: Chronopost International, Paris, France; B. Paechter: School of Computing, Napier University, Edinburgh, Scotland; M. Snoek: Department of Computer Science, University of Twente, Twente, The Netherlands

Title: A Local Search for the Timetabling Problem
Authors: O. Rossi-Doria, C. Blum, J. Knowles, M. Sampels, Ksocha, B. Paechter

Addresses: O. Rossi-Doria, B. Paechter: School of Computing, Napier University, Edinburgh, Scotland. C. Blum, J. Knowles, M. Sampels, K. Socha: IRIDIA, Universite Libre de Bruxelles., CP 194/6, Av. Franklin D. Rossevelt 50, 1050 Bruxelles, Belgium

Title: Is Genetic Programming a Sensible Research Direction for Timetabling?
Authors: E. K. Burke, S. Gustafson, G. Kendall
Address: School of Computer Science and IT, The University of Nottingham, Jubilee Campus, Nottingham, UK

Title: Addressing the Availability-Based Laboratory/Tutorial Timetabling Problem with Heuristics and Metaheuristcs
Authors: D. Corne and J. Kingston
Addresses: D. Corne: School of Computer Science, Cybernetics and Electronic Engineering, University of Reading, Reading, UK; J. Kingston: Department of Computer Science, University of Sydney, Australia

Title: An Average Case Approximation Bound for Course Scheduling by Greedy Bipartite Matching
Authors: G. Lewandowsky, P. Ohja, J. Rizzo, A. Walker
Addresses: G. Lewandowsky, R. Rizzo, A. Walker: Xavier University, Mathematics and Computer Science Department, Cincinnati OH 45207-4441, USA; P. Ohja: Hope College, Computer Science Department, Holland, MI 49423, USA

Title: Timetabling with No 8 Fencing Wire
Authors: J. Baumfield, B Graves, B. Pawson
Address: 21 Juliana Place, Palmerston North, New Zealand

Title: A Timetabling System for the German "Gymnasium"
Author: M. Lohnertz
Address: Institut für Informatik, University of Bonn, Germany

Title: School Timetabling for Compact Student and Teacher Schedules
Authors: T. Birbas, S. Daskaliki, E. Housos
Address: Department of Electrical and Computer Engineering, Patras, Greece

Title: Solving Real/Class Teacher Timetabling Problems using Neural Networks
Authors: M. P. Carrasco, M. V. Pato
Address: Centro de Investigacao Operacional, Faculdade de Ciencias, University of Lisbon, Portugal

Title: Subproblem-centric Algorithms for the Nurse Scheduling Problem
Authors: A. Ikegami, A. Niwa

Address: Seikei University, Tokyo 180-8633, Japan

Title: Efficient Generation of Cyclic Schedules
Author: R. Hope
Address: Department of Computer Science, Hedmark University College, Rena, Norway

Title: Agent Technology for Timetabling
Authors: P. De Causmaecker, P. Demeester, Y. Lu and G. Vanden Berghe
Address: KaHo Sint-Lieven, Information Technology, Gebr. Desmetstraat 1, 9000 Gent, Belgium

Title: Possible Models for Timetabling at Tertiary Institutions
Authors: T. Nepal, M. I. Ally
Address: Department of Computer Studies, ML Sultan Technikon, PO Box 1334, Durban, South Africa

Title: An Implicit Enumeration Based Heuristics for the Course Timetabling Problem
Author: S. Ketabi
Address: University of Isfahan, Iran

Title: Using Web Standards for Timetabling
Authors: P. De Causmaecker, P. Demeester, Y. Lu and G. Vanden Berghe
Address: KaHo Sint-Lieven, Information Technology, Gebr Desmetstraat 1, 9000 Gent, Belgium

Title: A Survey and Case Study of Practical Examination Timetabling Problems
Authors: P. Cowling, G. Kendall and N. M. Hussin
Addresses: P. Cowling: MOSAIC, University of Bradford, Bradford, UK; G. Kendall and N. M. Hussin, ASAP, School of Computer Science and IT, The University of Nottingham, Jubilee Campus, Nottingham, UK

Title: A Review of Existing Interfaces of Automated Examination and Lecture Scheduling Systems
Authors: B. McCollum, S. Ahmadia, E. K. Burke, R. Barone, P. Cheng, P. Cowling
Addresses: B. McCollum: School of Computer Science, Queen's University, Belfast, Northern Ireland; S. Ahmadia, E. K. Burke: ASAP, School of Computer Science and IT, The University of Nottingham, Nottingham, UK; R. Barone, P. Cheng: School of Psychology, University of Nottingham, Nottingham, UK; P. Cowling: MOSAIC, Department of Computing, University of Bradford, Bradford, UK

Title: Integrating Human Abilities and Automated Systems for Timetabling: A Competition using STARK and HuSSH Representations at the PATAT 2002 Conference
Authors: S. Ahmadi, R. Barrone, E. K. Burke, P. Cheng, P. Cowling, B. McCollum
Addresses: S. Ahmadi, E. K. Burke: ASAP, School of Computer Science and IT, The University of Nottingham, Nottingham, UK; R. Barrone, P. Cheng: CREDIT research group, School of Psychology, University of Nottingham, Nottingham, UK; P. Cowling, MOSAIC, Department of Computing, University of Bradford, Bradford, UK; B. McCollum: School of Computer Science, The Queen's University of Belfast, Belfast, BT7 1NN, UK

Title: Whose Fault Is It Anyway?
Author: R. C. Rankin
Address: School of Computing, Napier University, Edinburgh, UK

Title: A Design Pattern: "Test Conditions" Which Could Be Used in Timetable Construction Software
Authors: R. Gonzales Rubio and Y. Syam
Address: Departement de genie electrique et de genie informatique, Université de Sherbrooke

Title: Timetabling at the University of Sheffield, UK – an Incremental Approach to Timetable Development
Author: S. Geller
Address: 285, Glossop Road, Sheffield, UK

Title: Educational Timetabling – Experience, Practice and Improvements
Authors: B. R. Doughty and D. Whigham
Addresses: B. R. Doughty: Division of Accounting and Finance; D. Whigham: Division of Economics and Enterprise; both from Glasgow Caledonian University, Cowcaddens Road, Glasgow, UK

Title: Creating a New University Timetable Containing Mixed Structure Types with (New) Software
Author: T de W Jooste
Address: Potchefstroom University for Christian Higher Education, Computer, Mathematical and Statistical Sciences, Potchefstroom, 2520 South Africa

Title: A Generalised Class-Teacher Model for Some Timetabling Problems
Authors: A. Asratian, D. de Werra
Addresses: A. Asratian: Department of Mathematics, Lulea University of Technology, S-971 87, Lulea, Sweden; D. de Werra: Department of Mathematics, École Polytechnique Fédérale de Lausanne, CH 1015 Lausanne, Switzerland

Title: A Generate-and-Test Heuristic Inspired by Ant Colony Optimization for the Traveling Tournament Problem
Authors: H. Crauwels, D. Van Oudheusden
Addresses: H. Crauwels: Hogeschool voor Wetenschap & Kunst, De Nayer Instituut, B-2860 Sint-Katelijne-Waver, Belgium; D. Van Oudheusden: K.U. Leuven, Centre for Industrial Management, Celestijnenlaan 300A, B-3001 Heverlee, Belgium

Title: Generating Fair and Attractive Football Timetables
Authors: T. Bartsch, A. Drexl, S. Kroger
Addresses: T. Bartsch: SAP Portals Europe GmbH, Neurottstrase 16, 69190 Waldorf, Germany; A. Drexl: Institut für Betriebswirtschaftslehre, Christian-Albrechts-Universitat zu Kiel. S. Glas: Hattenbergstr. 10, 55122 Mainz, Germany; S. Kroger: Hattenbergst, 10, 55122 Mainz, Germany

Title: Solving Sports Scheduling Problems Using Network Structure
Authors: A. Suzuka, Y. Saruwatari, A. Yoshise
Addresses: A. Suzuka: Graduate School of Systems and Information Engineering, University of Tsukuba, Ibaraki, Japan; Y. Saruwatari, A. Yoshise: Institute of Policy and Planning Sciences, University of Tsukuba, Ibaraki, Japan

Title: A Broker Algorithm for Timetabling Problems
Author: S. L. M. Lin
Address: IC-Parc, William Penney Laboratory, Imperial College, London, UK

Title: An Evolutionary Approach for the Examination Timetabling Problems
Author: K. Sheibani
Address: Tadbir Institute for Operations Research, Tehran, Iran

Title: Examination Timetabling with Ants
Authors: K. A. Dowsland, N. Pugh, J. Thompson
Addresses: K. A. Dowsland: Gower Optimal Algorithms Ltd; Jonathan Thompson: Department of Chemistry, The University of Nottingham, Nottingham, UK; N. Pugh: The University of Wales

Title: Recolour, Shake and Kick: A Recipe for the Examination Timetabling Problem
Author: L. di Gaspero
Address: Dipartimento di Matematica e Informatica, Universita di Udine, via dell Scienze 206, I-33100. Udine, Italy

Title: A Case Based Heuristic Selection Investigation of Hill Climbing, Simulated Annealing and Tabu Search for Exam Timetabling Problems
Authors: E. K. Burke, A. J. Eckersley, B. McCollum, S. Petrovic and R. Qu
Addresses: E. K. Burke, A. J. Eckersley, S. Petrovic, R. Qu: ASAP, School of

Computer Science and IT, The University of Nottingham, Nottingham, UK; B. McCollum: Queens University of Belfast, Belfast, Northern Ireland

Title: Empirical Analysis of Tabu Search for the Lexicographic Optimisation of the Examination Timetabling Problem
Authors: L. Paquete, T. Stützle
Address: Intellektik, Technische Universität Darmstadt, Alexanderstr. 10, 64283 Darmstadt, Germany

Title: Decision Support Without Magic or Mind Reading for Assigning Magistrates to Sessions of the Amsterdam Criminal Court
Author: J. A. M. Schreuder
Address: University of Twente – TWRC-A205, Postbus 217, 7500 AE Enschede, Netherlands

Title: The Cost of Flexibility in Vehicle Routing and Scheduling
Authors: W. Dullaert, B. Johannessen, O. Braysy and T. Dahl
Addresses: W. Dullaert: Ufsia-Ruca Faculty of Applied Economics, University of Antwerp, Prinsstraat 13, 2000 Antwerp, Belgium; B. Johannessen, O. Braysy, T. Dahl: SINTEF Applied Mathematics, Department of Optimisation, PO Box 124 Blindern, N-0314 Oslo, Norway

Author Index

Lecture Notes in Computer Science

For information about Vols. 1–2674
please contact your bookseller or Springer-Verlag